Post Harvest Handling and Value Addition of Temperate Fruits and Nuts

The Authors

Dr Desh Beer Singh presently posted as Director, ICAR-CITH. Srinagar. J & K. He significantly contributed in development of high yielding varieties of temperate and tropical horticultural crops and developed various technologies on minimal processing of fruits and vegetables for quality maintenance, processing technologies for making novel value added products of temperate and arid fruits and their further licensing and commercialization, registration of newly identified/developed varieties. Further he has obtained advanced training from USDA, USA in the field of conservation of plant genetic resources and from RAS, Moscow (Russia) in the field of post harvest management of horticultural produce. He has been awarded by the SAARC Agriculture Centre (SAC) for developing the Farm Technology in SAARC countries on Round the Year Pineapple production in Andaman and Nicobar Islands and its processing for making various products. He has published more than 110 research papers and contributed 14 bulletins, 15 book chapters, 06 books besides 57 presentations in national and international conferences.

Dr Om Chand Sharma after completing B.Sc. (Horticulture), M.Sc. (Horticulture-Fruit Breeding and Genetic Resources) and Ph.D (Horticulture-Fruit Breeding and Genetic Resources) from Dr Y S Parmar University of Horticulture and Forestry, Nauni-Solan. Dr Sharma Presently working as Principal Scientist (Horticulture). He has been awarded with many awards by various societies and agencies. Published about 80 research papers in journals of national and international repute, 3 review papers, 1 book, 11 book chapters, 8 popular articles and 34 bulletins/pamphlets for the benefit of researchers and farmers. Many technologies have been generated and varieties were identified for commercial production and adoption.

Dr Javid Iqbal Mir is presently working as Scientist-ARS Biotechnology at ICAR-Central Institute of Temperate Horticulture, Srinagar, J & K. He had his graduation in Agricultural Sciences from SKUAST-K, Master's degree from CCSHAU Hisar and Ph.D in biotechnology from University of Kashmir, Srinagar. He published more than 70 research papers, eight books, fifteen extension bulletins and more than 45 book chapters. Dr Mir developed and released nine varieties in temperate horticultural crops. He is the recipient of many awards like Outstanding Achievement Award, Best Researcher Award, Outstanding Participant, Best Project Award *etc.*

Post Harvest Handling and Value Addition of Temperate Fruits and Nuts

Desh Beer Singh
Director

O. C. Sharma
Principal Scientist (Horticulture)

J. I. Mir
Senior Scientist

ICAR-Central Institute of Temperate Horticulture, Srinagar
Old Air Field, Rangreth, Srinagar – 191 132, J&K

2020

Daya Publishing House®

A Division of

Astral International Pvt. Ltd.

New Delhi – 110 002

ISBN 9789390371464 (Int. Edition)

Published by : **Daya Publishing House®**
A Division of
Astral International Pvt. Ltd.
– ISO 9001:2015 Certified Company –
4736/23, Ansari Road, Darya Ganj,
New Delhi-110 002
Ph. 011-43549197, 23278134
E-mail: info@astralint.com
Website: www.astralint.com

Preface

India is second largest producer of fruit in the world due to its rich biodiversity and varied agro climatic conditions spread from true tropical to extreme cold temperate region. Unlike other agricultural crops, the horticulture crops especially fruits are highly perishable in nature having limited shelf life. The post harvest management of fruits has assumed a great significance in view of high losses of perishable commodities at post harvest stage. These crops undergo a rapid transformation between the harvest and consumption which results spoilage and reduces market value. The spoilage has been estimated to be nearly 30-40 per cent in most of the produce which account for more than 25,000 Crores of rupees every year. Such losses occur at several stages such as harvesting, grading, packing, transport, storage and marketing. This is not only a loss to the growers but a net loss of huge human nutrition and wastage of inputs involved. Proper post harvest management practices can reduce these losses to great extent. Post harvest technology is therefore a very important component in the total production and marketing system of horticultural crops. Work on post harvest technology cannot be considered in isolation and must be linked with other activities such as soil, water and nutrient management, suitable varieties and genetic improvement, insect pest and disease management, pollination, maturity index, processing attributes of other quality characteristics of the commodity, which greatly influence post production condition of perishable fruits. For increasing the income by getting more returns from fruits we need to have better post harvest handling for enhancing market value with higher consumer acceptability.

The publication entitled post harvest handling and value addition of temperate fruits and nut covers comprehensive information generated in area of post harvest management of fruits. The publication emphasizes both basic information as well as applied aspect of post harvest handling, packaging, storage and processing system and value addition of horticultural crops. The content of this book covers very important aspects required for scientific management of fruits during post harvest the area which needs more attention for achieving better values and fetching better prices.

The book will prove to be very useful to the post harvest technologists, horticulturists, food scientists, entrepreneurs, teachers, under graduate and post graduate students.

Desh Beer Singh
O.C. Sharma
J.I. Mir

Contents

Chapter 1

Almond

Introduction

The sweet almond (*Amygdalus communis*) originated in the Mediterranean basin and has been cultivated from ancient times. Their exact ancestry is unknown, but almonds are thought to have originated in China and Central Asia. Explorers ate almonds while traveling the "Silk Road" between Asia and the Mediterranean. One race of the almond, which produces bitter nuts, is grown in some part of Europe, largely for the manufacture of flavoring extracts. The soft-shelled cultivars are commercially more important than the hard-shelled ones. The principal, commercial producers are United States, Italy, Iran, Tunisia and Turkey. The almond tree was brought to California from Spain in the mid-1700's by the Franciscan Padres. The moist, cool weather of the coastal missions, however, did not provide optimum growing conditions. It wasn't until the following century that trees were successfully planted inland. By the 1870's, research and cross-breeding had developed several of today's prominent almond varieties. By the turn of the 20th century, the almond industry was firmly established in the Sacramento and San Joaquin areas of California's great Central Valley. Throughout history, almonds have maintained religious, ethnic and social significance.

Botany

The almond plant is classified in the genus *Prunus,* which also includes plums, cherries, peaches, and apricots. There are around 430 species of *Prunus* spread throughout the northern temperate regions of the globe. Almond is a small deciduous tree, *Prunus amygdalus* (syn. *Prunus dulcis,* or *Amygdalus communis*) belonging to the subfamily Prunoideae of the family Rosaceae. It is also the name of the ellipsoidal nut-like kernal of this tree. There are two main types of almonds. One variety (*Prunus amygdalus* var. *dulcis*) produces sweet almonds, which are edible, and may be eaten raw or roasted or pressed for the almond oil. The other variety (*Prunus amygdalus* var. *amara*) produces bitter almonds, which are used for almond oil, but the raw also may yield dangerous amounts of prussic acid (hydrogen cyanide) (from the enzyme emulsin acting on a soluble glucoside, amygdalin).The almond contributes values to the ecosystem and to human beings. Ecologically, the almond flowers have a mutually beneficial relationship with bees in terms of pollination. For people, almonds supply almond oil, almond syrup, and a highly nutritious food, that is used in a variety of dishes. Almonds have been described as "a nutritional powerhouse," offering a rich source of protein, riboflavin, vitamin E, copper, niacin, calcium, fiber, folic acid, magnesium, and potassium, among other nutrients. There are also various claimed medicinal benefits of almonds.The almond tree is a deciduous tree which can grow as high as 10 meters. The leaves of the almond tree are broad and serrated, and the sprouting flowers seen in early spring are either white or pale pink. By the time autumn comes (some 7-8 months after flowering), the almond is mature, ripe, and ready to be harvested.

Almond trees become productive and begin bearing fruit after five years. The fruit mature in the autumn, 7–8 months after flowering. The sweet fleshy outer covering of other members of *Prunus,* such as the plum and cherry, is replaced by a leathery coat called the hull, which contains inside a hard shell the edible kernel, commonly called a nut in culinary terms. However, in botanical terms, an almond is not a true nut. In botanical parlance, the reticulated hard stony shell is called an endocarp, and the fruit, or exocarp, is a drupe, having a downy outer coat.

It is a small tree, growing to 4–9 meters tall. The leaves are lanceolate, 6–12 centimeters long, and serrated at the edges. The flowers are white or pale pink, 3–5 cm diameter with five petals, produced before the leaves in early spring.

Sweet and Bitter Almonds

There are two forms of the plant, one (often with white flowers) producing sweet almonds, and the other (often with pink flowers) producing bitter almonds. The kernel of the former contains a fixed oil and emulsion. As late as the early twentieth century, the oil was used internally in medicine, with the stipulation that it must not be adulterated with that of the bitter almond. It remains fairly popular in alternative medicine, particularly as a carrier oil in aromatherapy, but has fallen out of prescription among doctors.The bitter almond is rather broader and shorter than the sweet almond, and contains about 50 percent of the fixed oil that also occurs in sweet almonds. It also contains the enzyme emulsin which, in the presence of water, acts on a soluble glucoside, amygdalin, yielding glucose, cyanide, and the essential oil of bitter almonds or benzaldehyde. Bitter almonds may yield from six to eight percent of hydrogen cyanide (prussic acid). Extract of bitter almond was once used medicinally but even in small doses effects are severe and in larger doses can be deadly. The prussic acid (hydrogen cyanide) must be removed before consumption.

Use of Almonds

Many times, almonds are toasted before eating; other times they are simply eaten raw. However, there are a wide variety of culinary uses for raw almonds, as well. It is most commonly sprinkled on ice cream sundaes or other desserts such as marzipan, French macaroons, and even more. Almonds can even be processed into "almond milk", which is ideal for vegans and those that are lactose-intolerant, and the drupe can be used to produce "almond syrup". Additionally, the oil of sweet almonds may even be used as a substitute for olive oil.

To complement its wide variety of uses, raw almonds also have a great deal of nutritional value, too. A handful of almonds (28 grams/one recommended serving) are enough to provide the human body with high levels of Vitamin E, magnesium, and fiber, apart from being a great source of monounsaturated fat, protein, potassium, and even calcium.

Almond Oil

"Oleum Amygdalae," the fixed oil from the almond kernel, is prepared from either variety of almond. It is a glyceryl oleate, with a slight odor and a nutty taste. It is almost insoluble in alcohol but readily soluble in chloroform or ether. It may be used as a substitute for olive oil.

The sweet almond oil is obtained from the dried kernel of the plant. This oil has been traditionally used by massage therapists to lubricate the skin during a massage session, being considered by many to be an effective emollient.

Almond Syrup

Historically, almond syrup was an emulsion of sweet and bitter almonds usually made with barley syrup (orgeat syrup) or in a syrup of orange-flower water and sugar.

Grocer's Encyclopedia notes that "Ten parts of sweet almonds are generally employed to three parts of bitter almonds," however due to the cyanide found in bitter almonds, modern syrups generally consist of only sweet almonds.

Nutritional and Culinary Uses

Raw Almonds

Almonds are a nutritional powerhouse." Almonds are a rich source of protein, Vitamin B_2 (riboflavin), vitamin E, copper, niacin, calcium, fiber, folic acid, magnesium, and potassium and a good source of iron and zinc. The almond also is a source of vitamin B_1, are rich in monounsaturated fat, one of the two "good" fats responsible for lowering LDL cholesterol.

Carbohydrates

Almonds are relatively low in carbs. 28 grams has six grams of carbs, most of which are fiber. They are also low in sugar, with only 1.2 grams per ounce. Almonds have a low glycemic index, so they should not cause major spikes in blood sugar levels. This makes them suitable for diabetics.

Fiber

Almonds are high in soluble and insoluble fiber, which makes up about 12.5 per cent of their weight. 28 gms of almonds contains 3.5 grams of fiber. Fiber helps moderate blood sugar by slowing absorption of carbohydrates. It has many other benefits as well, such as improves digestive health.

Bottom Line

Almonds are low in carbohydrates, but high in fiber. They should not have major effects on blood sugar levels.

Proteins

Almonds are relatively high in protein, with 6 grams of protein 28 gms.

Protein is essential for growth and various body functions.

However, almonds are lacking in the essential amino acids lysine, methionine and threonine.

Therefore, they are not considered a complete protein source.

That being said, almonds are a good source of the amino acid arginine. Arginine can help heal wounds, treat infections and maintain a healthy heart.

Bottom Line

Almonds are relatively high in protein but are not considered a complete protein source. Nevertheless, they are a good source of the amino acid arginine.

Nutritional Composition

Table 1.1: Nutritional Composition of Almond (Value per 100g)

Nutrient	*Unit*	*Amount*
Proximates		
Water	g	4.41
Energy	kcal	579
Protein	g	21.15
Total lipid (fat)	g	49.93
Carbohydrate, by difference	g	21.55
Fiber, total dietary	g	12.5
Sugars, total	g	4.35
Minerals		
Calcium, Ca	mg	269
Iron, Fe	mg	3.71
Magnesium, Mg	mg	270
Phosphorus, P	mg	481
Potassium, K	mg	733
Sodium, Na	mg	1

Nutrient	Unit	Amount
Zinc, Zn	mg	3.12
Vitamins		
Vitamin C, total ascorbic acid	mg	0.0
Thiamin	mg	0.205
Riboflavin	mg	1.138
Niacin	mg	3.618
Vitamin B-6	mg	0.137
Folate, DFE	µg	44
Vitamin B-12	µg	0.00
Vitamin A, RAE	µg	0
Vitamin A, IU	IU	2
Vitamin E (alpha-tocopherol)	mg	25.63
Vitamin D (D2 + D3)	µg	0.0
Vitamin D	IU	0
Vitamin K (phylloquinone)	µg	0.0
Lipids		
Fatty acids, total saturated	g	3.802
Fatty acids, total monounsaturated	g	31.551
Fatty acids, total polyunsaturated	g	12.329
Fatty acids, total trans	g	0.015
Cholesterol	mg	0
Others		
Caffeine	mg	0

Fats

By weight, almonds are roughly 50 per cent fat.

About 90 per cent of almond fat consists of mono- and polyunsaturated fatty acids, which have been linked to a reduced risk of heart disease.

The main fatty acids in almonds are oleic acid and the omega-6 fatty acid linoleic acid.

However, about 10–15 per cent of the fat in almonds isn't absorbed by the body.

Bottom Line

Almonds are high in monounsaturated and polyunsaturated fats, which have been linked to reduced risk of heart disease. About 10-15 per cent of the fat in almonds isn't absorbed by the body

Vitamins and Minerals

Almonds contain high amounts of several important vitamins and minerals, including:

- **Vitamin E:** This antioxidant is essential for immune function and red blood cell formation. Almonds are one of the world's best sources of vitamin E.
- **Vitamin B2:** Also known as riboflavin, vitamin B2 is important for growth and red blood cell production.
- **Magnesium:** This mineral is essential for all living organisms. Magnesium is important for blood pressure, bone health and many other important processes.
- **Manganese:** This nutrient is needed for the immune system, reproduction, digestion and bone growth.
- **Phosphorus:** This mineral is very important for teeth and bone health.
- **Iron:** A metallic element that serves many important functions. It is an essential component of hemoglobin, a protein in red blood cells that carries oxygen throughout the body.

Almonds also contain decent amounts of calcium, potassium and zinc, as well as vitamins B1, B3 and B9.

Bottom Line

Almonds contain many important nutrients, including vitamin E, vitamin B2 (riboflavin), magnesium, manganese, phosphorus and iron.

Plant Compounds

Almonds contain beneficial plant compounds, including many powerful antioxidants:

- **Resveratrol:** Also found in red wine, resveratrol may have benefits for heart health, diabetes and cancer.
- **Catechin:** Catechin is one of over 20 types of flavonoid antioxidants found in almond skin.
- **Epicatechin:** Also found in cocoa and green tea, epicatechin may have beneficial effects on heart and brain health.
- **Kaempferol:** This antioxidant is linked to a reduced risk of chronic diseases and cancer.
- **Quercetin:** A beneficial plant compound with anti-inflammatory and anti-bacterial effects.

The combination of vitamin E and flavonoid antioxidants in almonds may be particularly effective, having synergistic effects.

Many of the antioxidants in almonds are concentrated in the skin. Therefore, you should eat the whole nut to maximize the health benefits.

Bottom Line

Almonds contain many beneficial plant compounds, including powerful antioxidants like resveratrol, catechin, epicatechin, kaempferol and quercetin.

Health Benefits of Almonds have been linked with several health benefits. They have been shown to aid in weight loss, promote heart health, help protect against type 2 diabetes and improve digestive health.

Weight Loss

Almonds have several characteristics that make them a weight loss friendly food.

About 10–15 per cent of the fat in them isn't absorbed by the body, so fewer calories are actually consumed.

Additionally, almonds are relatively high in both protein and fiber. These nutrients are known to promote a feeling of fullness and help people reduce calorie intake.

There is also some evidence that eating almonds, as well as other nuts, may slightly boost metabolism.

One study found that a low-calorie diet resulted in 62 per cent more weight loss when supplemented with three ounces of almonds per day.

Another study followed 100 overweight women and found that almonds helped them lose more weight and belly fat, compared to women who were not eating nuts.

Bottom Line

Almonds are very filling, since they are high in protein and fiber. They may also boost metabolism. Several studies show that almonds can promote weight loss and help reduce belly fat.

Heart Health

Several studies link nut consumption to a lower risk of heart disease.

Eating almonds on a regular basis has been shown to lower levels of LDL (the "bad") cholesterol.

Consuming almonds also appears to be beneficial for people who are already taking cholesterol-lowering drugs. Almonds have been shown to help increase the effects of these drugs.

These nuts also protect the LDL particles from oxidative damage, which is linked to heart disease. One study found that eating almonds every day reduced oxidized LDL cholesterol by 14 per cent.

Bottom Line

Eating almonds has been shown to help lower LDL (the "bad") cholesterol and protect the LDL particles from oxidative damage. Studies show that people who eat nuts have a lower risk of heart disease.

Diabetes

Almonds are an excellent food for people with type 2 diabetes. They are low in carbs, but high in fiber, protein and healthy fats.

They are also loaded with magnesium, a mineral that many diabetics are deficient in magnesium plays an important role in insulin sensitivity and blood sugar control.

Several studies show that nuts, like almonds, can lead to significant reductions in blood sugar levels in people with type 2 diabetes.

Additionally, studies show that almonds may reduce the inflammation and oxidative stress caused by diabetes. This indicates a reduced risk of diabetes-related complications, such as heart disease.

Bottom Line

Almonds are loaded with magnesium and other nutrients that are beneficial for diabetics. Eating almonds on a regular basis may also help moderate blood sugar levels and protect against type 2 diabetes.

Digestive Health

The bacteria that reside in the digestive system are incredibly important for overall health, and have been linked to a reduced risk of many of diseases.

One study shows that eating almonds can improve the balance of beneficial bacteria in the digestive system.

Bottom Line

Whole almonds are beneficial for digestion. They can also improve the balance of beneficial bacteria in your digestive system.

Adverse Effects Almonds are well tolerated by most people.

However, some people may be allergic to them, and bitter almonds are poisonous.

They may also contribute to kidney stone formation.

There are two types of almonds:

- ☆ **Sweet almonds** are edible and widely grown for human consumption.
- ☆ **Bitter, or wild, almonds** are inedible. They contain a substance called amygdalin, which is toxic. Eating even a few bitter almonds can be fatal.

Almonds that are improperly stored may also develop mold that produces aflatoxin, a dangerous chemical.

Kidney Stones

Kidney stones can develop from concentrated minerals in the urine, and 75 per cent of them are mainly made of calcium oxalate.

Almonds are high in oxalates, and these oxalates tend to be well absorbed by humans. For this reason, people with a tendency to form kidney stones should not consume too many almonds.

Mineral Deficiencies

Almonds contain phytic acid, a substance that can inhibit the absorption of calcium, iron and zinc. However, this effect lasts only for the same meal.

People that consume foods rich in phytic acid with every meal may become deficient in essential minerals.

Bottom Line

Almonds may cause allergic reactions in some individuals, and bitter almonds are poisonous. They may also contribute to kidney stone formation, and too much phytic acid may cause mineral deficiencies.

Medicinal Uses

Claimed medicinal benefits of almonds include improved complexion, improved movement of food through the colon, and the prevention of cancer. Recent research associates inclusion of almonds in the diet with elevating the blood levels of high density lipoproteins and of lowering the levels of low density lipoproteins.

In Ayurveda, the Indian System of Medicine, almond is considered a nutritive for the brain and nervous system. It is said to induce high intellectual level and longevity. Almond oil is called *Roghan Badam* in both Ayurveda and *Unani Tibb* (the Greco-Persian System of Medicine). It is extracted by cold process and is considered a nutritive aphrodisiac both for massage and internal consumption. Recent studies on ayurvedic herbs have reported that the constituents of almond have anti-inflammatory, immunity boosting, and anti-hepatotoxicity effects.

Edgar Cayce, an American psychic who some regard as a father of American holistic medicine, highly favored the almond. In his readings, Cayce often recommended that almonds be included in the diet.

Varieties/Cultivars

There are number of varieties available in almond. But there are four major varieties of export quality *viz.*, Non Pareil, California Paper Shell, IXL and Merced. In addition to the exotic cultivars, recently released indigenous cultivars like Shalimar, Makhdoom and Waris have also shown very promising results under Kashmir conditions. The salient features of important cultivars are:

Table 1.2: Commercial Varieties Grown in India

Name of Variety	*Salient Features*	*Recommended Areas*
Shalimar	Trees are medium and spreading, nuts are long in size and bold with tapering at curved pointed apex. Nut colour is creamy brown to slightly whitish, soft shelled and mid season maturity.	Jammu and Kashmir, Uttarakhand
Mukhdoom	Trees are large and spreading, mid blooming, nuts are broad at shoulder with bold and slightly curved at apex. Nut colour is brown, soft shelled and mid to late season maturity.	Jammu and Kashmir, Uttarakhand

Name of Variety	*Salient Features*	*Recommended Areas*
Waris	Trees are upright in growth and medium in vigour, mid blooming. Nuts are medium in size, bold and bulged at shoulder with sharply pointed apex. Nut colour is brown to creamy whitish, soft shelled and mid to late season maturity.	Jammu and Kashmir, Uttarakhand
Non-Pareil	Trees are moderately vigourous, upright to spreading, mid blooming, nuts are medium, bold and light brown in colour, thin shelled and early season maturity.	Jammu and Kashmir, Himachal Pradesh, Uttarakhand
Drake	Trees are low in vigour, spreading, mid blooming. Nuts are small to medium in size, bold and roundish with pointed apex and light creamy whitish brown in colour. Semi soft shelled and mid season maturity.	Jammu and Kashmir, Himachal Pradesh, Uttarakhand
Ne-Plus Ultra	Trees are vigourous and spreading, mid blooming, nuts are medium to large flattened, bold and light brown in colour, paper shelled and mid season maturity.	Jammu and Kashmir, Himachal Pradesh, Uttarakhand
Merced	Trees are medium in vigour, upright, mid blooming, nuts are medium, bold, slightly flattened and light brown in colour. Paper shelled and mid to late season maturity.	Jammu and Kashmir, Uttarakhand
Pranyaj	Trees are moderate in vigour and spreading, mid blooming, nuts are medium, brown and flattened to bulge, kernel is medium to large. Papery shelled and mid season maturity.	Jammu and Kashmir, Uttarakhand
Primorskij	Trees are spreading and moderately vigorous, mid to late blooming, nuts are medium to large, bold, slightly flattened and brown in colour, kernel medium to large. Soft paper shelled and late season maturity.	Jammu and Kashmir, Uttarakhand
California Paper Shell	Trees are upright and medium in vigour, mid blooming, nuts are long, light brown in colour, Paper shelled and mid to late season maturity.	Jammu and Kashmir, Himachal Pradesh, Uttarakhand
IXL	Trees are spreading and intermediate in vigour, mid bloomer, nuts are medium, bold, brown in colour, kernel is medium, soft shelled and mid to late season maturity.	Jammu and Kashmir, Himachal Pradesh, Uttarakhand

Maturity Indices

In almond, the date of "hull-split" (HS) signals the beginning of fruit maturity. A goal of the 8-year study was to examine the correlation between the length of the period between FB and HS in almond and temperature accumulation after the start of FB, and thus, base the model on the collection and analysis of this type of data. Data analyses indicated that the length of the period from FB to HS was negatively correlated with the accumulation of Growing Degree Days between FB and 90 DAFB (GDD90). Calculations of GDD90 were based on the single sine method. Results of the 8-year study indicate that temperatures in the first 90 DAFB are a primary factor influencing the time of nut maturity in almond cultivars in California. This is in contrast to previous studies with peach, nectarine and plum, which indicated a strong relationship between fruit development and Growing Degree Hours in the first 30 days after full bloom (GDH30). Data on 28 California almond cultivars were evaluated in the study, but the predictive model is currently applicable to these 12 important cultivars: 'Nonpareil', 'Sonora', 'Price', 'Ruby', 'Wood Colony', 'Padre', 'Butte', 'Aldrich', 'Winters', 'Monterey', 'Mission' and 'Carmel'.

Harvesting and Handling

Almonds are harvested manually or mechanically. Before mechanization, hand-held mallets were used to strike the major limbs to remove nuts. Canvas sheets, either moved manually or dragged from tree to tree by tractor power are used to catch the falling nuts. Most commercial almond orchards presently employ interia limb shakers to remove nuts, windrowers to concentrate them, and pickup machines to gather them. The shake-catch systems are preferred to avoid contact of buts with toxic molds that produce aflatoxins.

Start almond nut harvesting when 95 percent of the drupes on the tree split. The first step in harvesting almond nuts is to gather the drupes that have already split and fallen. After that, spread a tarp beneath the tree. Start picking almond nuts from the branches you can reach on the tree.

A successful almond harvest begins with a clean, dry orchard floor.

Almonds should be harvested as soon as possible after they have matured to avoid quality loss and to minimize exposure to navel orangeworm and the subsequent potential for aflatoxin contamination.

After the crop is mechanically shaken to the ground, the nuts are raked into windrows and allowed to dry naturally, ideally to a hull moisture content that does not exceed 12 per cent, or a kernel moisture content that does not exceed 6 per cent, which can take anywhere from a few days up to two weeks.

To avoid insect damage, stockpiles should be monitored and fumigated if necessary.

Storage

In-shell almonds can be stored for about 7-8 months at room temperature below 70 per cent relative humidity (RH), if the nuts are dried properly. For prolonged storage, in-shell almonds should be held at 10°C or below. Almonds can remain in excellent marketable condition for as long as 2 years if stored at 0°C with about 75 per cent RH. Although shelling generally increases the perishability of all nuts, shelled almonds are comparatively less perishable than most other nuts. Air storage at 0°C with 60-75 per cent RH is satisfactory to hold almonds for 15-16 months.

It was reported that modified atmosphere (vacuum, CO_2 or N_2) at 0°C was superior to air for maintenance of flavor during prolonged storage. However, at 21.1°C, vacuum-packed almonds were superior in color to those stored in CO_2, N_2 or air after 16 months of storage. Good quality was maintained for 8 months in vacuum CO_2 or N_2 at 26.7°C, whereas almonds stored in air at this temperature turned black in 8 months. Controlled-atmosphere (CA) storage was found to be equal to or better than normal-atmosphere storage for almonds. Stored almonds may be infested with Paramyelcis transitella Walker, which initiates its infestation in the orchard. During storage, various storage insects may infest the nuts. Insect control is accomplished by spraying with malathion, fumigating with methyl bromide or phosphein or treating with primiphosmethyl.

Processing and Value Addition

While the sweet almond is most often eaten on its own, raw or toasted, it is used in some dishes. It, along with other nuts, is often sprinkled over desserts, particularly sundaes and other ice cream based dishes. It also is used in making baklava and nougat, as well as marzipan (traditional European candy) and macaroons. There is also almond butter, a spread similar to peanut butter, popular with peanut allergy sufferers and for its less-salty taste.

The young, developing fruit of the almond tree can also be eaten as a whole ("green almonds"), when it is still green and fleshy on the outside, and the inner shell has not yet hardened. The fruit is somewhat sour, and is available only from mid-April to mid-June; pickling or brining extends the fruit's shelf life.

The sweet almond itself contains practically no carbohydrates and may therefore be made into flour for cakes and biscuits for low carbohydrate diets or for patients suffering from diabetes mellitus or any other form of glycosuria. A standard serving of almond flour, 1 cup, contains 20 grams of carbohydrates, of which 10 gram is dietary fiber, for a net of 10 grams of carbohydrate per cup. This makes almond flour very desirable for use in cake and bread recipes by people on carbohydrate-restricted diets.

Almonds can be processed into a milk substitute simply called almond milk; the nut's soft texture, mild flavor, and light coloring (when skinned) make for an efficient analog to dairy, and a soy-free choice, for lactose intolerant people, vegans, and so on. Raw, blanched, and lightly toasted almonds all work well for different production techniques, some of which are very similar to that of soymilk and some of which actually use no heat, resulting in "raw milk."

The Marcona variety of almond, which is shorter, rounder, sweeter, and more delicate in texture than other varieties, originated in Spain and is becoming popular in North America and other parts of the world. Marcona almonds are traditionally served after being lightly fried in oil, and are also used by Spanish chefs to prepare a dessert called *turrón*.

In China, almonds are used in a popular dessert when they are mixed with milk and then served hot. In Indian cuisine, almonds are the base ingredient for pasanda-style curries.

The majority of almonds are consumed as ingredients in almond manufactured goods, and granola bars, almonds in cereals, ice cream,wheat-free crackers made of nuts, confectionery and baked goods have been introduced. while remaining consumption takes place in the form of snacks, in home baking and food service outlets.

Almond milk is a beverage made from ground almonds, often used as a substitute for dairy milk. It contains neither cholesterol nor lactose. it is suitable for vegans and vegetarians who abstain from dairy products. Commercial almond milk products often come in plain, vanilla or chocolate flavors. Almond milk can also be made at home by grinding almonds with water in a blender and the vanilla flavoring and sweeteners are often added.

Use the not-yet-ripe kernels of green almonds which harvest early, and the liquid kernels used to add a delicate taste on high-end dishes.

Almond butter may be crunchy or smooth, and is generally "stir" (susceptible to oil separation) or "no-stir" (emulsified). Almond butter may be either raw or roasted, describing the almonds themselves prior to grinding(also can use peanut butter machine). It is recommended that almond butter be refrigerated once opened to prevent spoilage and oil separation.

Almond oil used as an emollient, applied regularly, it helps keep skin well protected from dryness. Additionally, the oil is used in cooking in Iran, and Turkey. It is also used as "carrier or base oil" in traditional medicines in aromatherapy, in pharmaceutical, and cosmetic industries.

Almonds are either sold as in-shell or as processed, which can include shelling, dry roasting, blanching, slicing, chopping and conversion into flour, paste (marzipan) or flavorings. The majority of almonds are shelled during processing, and the excess casings (the hull and shell) are used in the livestock industry for both feed and bedding material.

REFERENCES

A.A. Kader and J.F. Thompson. 2002. Post-harvest handling systems: tree nuts. A.A. Kader (Ed.), Post-harvest Technology of Horticultural Crops (3rd ed), Division of Agriculture and Natural Resources, University of California, Oakland, USA, pp. 399-406.

Abdallah, M.H. Ahumada and T.M. Gradziel. 1998. Oil content and fatty acid composition of almond kernels from different genotypes and California production regions. *Journal of the American Society for Horticultural Science*, 123, pp. 1029-1033.

Bender, D. A., and A. E. Bender. 2005. *A Dictionary of Food and Nutrition* New York: Oxford University Press. ISBN 0198609612.

Bolling, G. Dolnikowski, J. Blumberg and C-Y.O. Chen. 2010. Polyphenol content and antioxidant activity of California almonds depend on cultivar and harvest year. *Food Chemistry*, 122: 819-825.

Browicrz, K. and D. Zohary, 1996. The genus *Amygdalus* L. (*Rosaceae*): Species relationships, distribution and evolution under domestication. Genet. Resou. Crop Evol, 43: 229–247. Google Scholar.

Cantor, D., J. Fleischer, J. Green, and D. L. Israel. 2006. "The fruit of the matter." *Mental Floss* 5(4): 12.

D.P. Richardson, A. Astrup, A. Cocaul, P. Ellis. 2009. The nutritional and health benefits of almonds: a healthy food choice. *Food Science and Technology Bulletin: Functional Foods*, 6 (4): 41-50.

Davis, P. A., and C. K. Iwahashi. 2001. Whole almonds and almond fractions reduce aberrant crypt foci in a rat model of colon carcinogenesis. *Cancer Letters* 165(1): 27-33. Retrieved August 26, 2007.

Denisov, V.P., 1988. Almond genetic resources in the USSR and their use in production and breeding. Acta. Hort. 244: 2990306.Google Scholar.

Diamond, J. 1999. *Guns, Germs, and Steel: The Fates of Human Societies.* New York: Norton. ISBN 0393317552.

F. Bartolozzi, M.L. Warburton, S. Arulseakr and T.M. Gradzeil. 1998. Genetic characterization and relatedness among California almond cultivars and breeding lines detected by randomly amplified polymorphic DNA (RAPD) analysis. *Journal of the American Society for Horticultural Science,* 123: 381-387.

G.D. Nanos, I. Kazantzis, P. Kefalas, C. Petrakis, G.G. Stavroulakis. 2002. Irrigation and harvest time affect almond kernel quality and composition. *Scientia Horticulturae,* 96: 249-256.

G.V. Civille, K. Lapsley, G. Huang, S. Yada, J. Seltsam. 2010. Development of an almond lexicon to assess the sensory properties of almond varieties. *Journal of Sensory Studies,* 25: 146-162.

Herbst, S. T. 2001. *The New Food Lover's Companion: Comprehensive Definitions of Nearly 6, 000 Food, Drink, and Culinary Terms.* Barron's Cooking Guide. Hauppauge, NY: Barron's Educational Series. ISBN 0764112589.

Kester D.F., T.M. Gradziel and C. Grasselly, 1991, Almond (*Prunus*), In: Genetic Resources of Temperate Fruits And Nut Crops. J.N. Moore and J.R. Ballington Eds. Int. Soc. Hort. Sci. pp. 701–758.Google Scholar.

Kislev, M.E., D. Nadel and I. Carmi, 1992 Epipalacolilhic (19, 000 BP) cereal and fruit diet at Ohalo II, Sea of. Galilee, Israel. Rev. Palaeobot. Palynology 73: 161–166.Google Scholar.

P.D. Drogoudi, G. Pantelidis, L. Bacchetta, D. DeGiorgio, H. Duval, I. Metzidakis and D. Spera. 2012. Protein and mineral nutrient contents in kernels from 72 sweet almond cultivars and accessions grown in France, Greece and Italy. *International Journal of Food Sciences and Nutrition.*

Puri, H. S. 2003. *Rasayana Ayurvedic Herbs for Longevity and Rejuvenation.* London: Taylor and Francis. ISBN 0203216563.

S. Prats-Moya, N. Grané-Teruel, V. Berenguer-Navarro and M.L. Martín-Carratalá. 1997. Inductively coupled plasma application for the classification of 19 almond cultivars using inorganic element composition. *Journal of Agricultural and Food Chemistry,* 45: 2093-2097.

S.K. Sathe, N.P. Seeram, H.H. Kshirsagar, D. Heber, K. Lapsley. 2008. Fatty acid composition of California grown almonds. *Journal of Food Science,* 73(9): C607-C614.

Spiller, G. A., D. A. J. Jenkins, O. Bosello, J. E. Gates, L. N. Cragen, and B. Bruce. 1998. Nuts and plasma lipids: An almond-based diet lowers LDL-C while preserving HDL-C. *Journal of the American College of Nutrition* 17(3): 285-290. Retrieved August 26, 2007.

T.M. Amrein, H. Lukac, L. Andres, R. Perren, F. Escher and R. Amadò. 2005. Acrylamide in roasted almonds and hazelnuts. *Journal of Agricultural and Food Chemistry*, 53: 7819-7825.

W.M. Cort, T.S. Vicente, E.H. Waysek and B.D. Williams. 2012. Vitamin E content of feedstuffs determined by high-performance liquid chromatographic. fluorescence. *Journal of Agricultural and Food Chemistry*, 31 (1983), pp. 1330-1333.

Ward, A. 1911. *The Grocer's Encyclopedia.* New York: The James Kempster printing company.

Zohary, D., and M. Hopf. 2000. *Domestication of Plants in the Old World: The Origin and Spread of Cultivated Plants in West Asia, Europe, and the Nile Valley.* Oxford: Oxford University Press. ISBN 0198503571.

Chapter 2

Apple

Introduction

Post-harvest handling encapsulates the many management decisions and processes that are involved in harvesting, handling, storage, packaging and transport of apple fruit necessary to provide the consumer with an acceptable product. Most of the characteristics that influence acceptance of fruit in the market place are present at the time of harvest, and include size and shape, colors and freedom from blemishes. Internal characteristics at harvest include the presence of or the potential to develop acceptable varietals flavour and texture. Some important changes that contribute to fruit acceptability occur after harvest, *e.g.* conversion of starch to sugars. However, post-harvest handling is largely concerned with maintaining, rather than improving quality. Consequently, the management tools used by apple growers, storage operators, packers and shippers are focused on the following

1. Reducing metabolic rates of processes that result in undesirable changes in colors, composition, texture, flavour and nutritional status.
2. Reducing water loss that can result in loss of marketable weight, shriveling, softening and loss of crispness.
3. Minimizing bruising, friction damage and other mechanical injuries.
4. Preventing the development of physiological and pathological disorders.

These objectives are met by harvesting the fruit at the maturity stage that will meet the quality standards required by the market either at harvest time or other storage, by handling fruit carefully and rapidly to avoid mechanical injury and to minimize deterioration and by applying protective chemical preservatives (antioxidants, fungicides). Fruit should be cooled quickly to remove field heat, and proper refrigeration should be applied during storage. High relative humidity should be maintained and the storage atmosphere should be controlled. Protective containers and packaging should be used during transport and marketing, along with refrigeration. Good sanitation practices must also be maintained through out these processes, as well as during harvesting, as outbreaks of food-borne illness can have a devastating effect in the market.

Harvesting of any horticultural product involves the imposition of stress – that is, an interruption, restriction or acceleration of normal metabolic processes. However, many recommended post-harvest conditions also causes stress, making the term in a post-harvest context somewhat nebulous. During post-harvest storage and handling, stress is an external factor that will result in undesirable changes in quality if the fruit is exposed to it for a sufficient duration or at a sufficient intensity. However, while recommended storage conditions for apple represent stress to the fruit, to the post-harvest physiologist they represent the optimum conditions for maintenance of product quality.

Responses of fruit to post-harvest treatments are greatly affected by these factors. Cultivar, for instance, it can influence post-harvest management through effects on ripening rates and physiology. Early maturing apple cultivars usually produce much higher levels of ethylene than later maturing ones and have the shortest storage life. Cultivars with lower ethylene production rates generally have longer storage potential. Responses of fruit to storage environment are influenced by physiological features, such as diffusion characteristic of the skin. Golden Delicious, for example, tends to shrivel faster than other cultivars because of breaks in the cuticle, and the Marshall McIntosh strain is less tolerant to low oxygen in controlled atmosphere (CA) storage than other McIntosh strains because of higher of higher resistance of the skin to gas exchange. Orchard management practices and climatic conditions have an impact not only on fruit quality at harvest but also on tolerances of fruit to post-harvest storage conditions.

Origin and Distribution

Primary centre of origin of apple (*Malus domestica*) is Southern Western Asia, in the Caucasus region near Gilan in Turkestan and domesticated by Greeks and Romans and few centuries BC in Middle-East and South-Eastern Europe as a result of their travels and invasions. The movement of apple to Western Europe was helped by the Christian settlers. Romans are believed to have been responsible for its introduction in France and England. There are still natural forests of apple on the caucasus mountains of Southern Asia. The cultivated apple usually referred to a Malus domestica Borkh is considered to be the parent of most of our cultivated apples. By the early 20th century, the USA and Canada were the two largest apple producing nations. Later in the century, the USSR also become important. By the beginning of 21 century, China has become the largest apple producers, with a large production of crop being exported as concentrated Juice. In India of all commercial varieties of Apple, Ambri is considered indigenous to Kashmir, whereas, rootstock species of *Malus baccate* are found growing in semi wild conditions in Kashmir and Himachal Pradesh. Jahangir the emperor has written about apples in his book 'Turk-e-Jahangir' which gives proof of the apple cultivation during the sixteenth century. In Jammu and Kashmir, apples have been grown as early as 2000BC. M.

Ermens, formely Head Gardeners of Public Works in Paris, who came to Kashmir during 1865 made a thorough investigation of the soil, climate, rainfall and other weather conditions in Kashmir. Being a gardener, Ermens brought with him a number of fruit plants which he believed would thrive in Kashmir together with implements of starting in experimental agriculture farm. These fruit trees were planted in Chashma Shahi, near Srinagar, J&K, in 1875. The J&K state established a Department of Agriculture and Horticulture in 1907. In Himachal Pradesh the first apple orchard was established at Bandrole in Kullu by Captain A. A. Lee around 1860. Therefore, many orchards were established at Manali, Raison and Nagar by the Englishmen. The cultivars such as Topred, Vance Delicious, Hardimen and spur type namely Red Spur, Golden Spur, Red Chief, Oregon Spur, Starkrimson, Siliver Spur and many other types were introduced between 1950-1985 at the Regional Horticulture Research Station Mashobra, H.P. which are now fast replacing the Delicious varieties.

In Uttarakhand apple was introduced in Kumaan and Garhwal by Britishers in the latter half of 19th century. Apple cultivation is a new introduction to a few isolated hilly tracts of Bihar, Darjeeling. Sikkim, Arunachal Pradesh, Meghalaya, Manipur, Nagaland and Assam. The effort of a regular cultivation in these isolated hill areas were mainly initiated after independence.

Species, Commercial Varieties and Rootstocks

Species

A wide range of genetic diversity in Malus is known and the exact number of Malus species is not yet confirmed. Rehder's Manual of Cultivated Trees and Shrubs (1940) enlist 25 Malus species while various authors have listed 74 different species of Malus in Knight's Bibliography of Apple Breeding and Genetics. Ponomarenko in the then USSR probably has made a more thorough study of Malus taxonomy than anyone else in the world. In 1988, he prepared a list of 122 apple species, of the total 30 odd species of genus Malus; only *M. sikkimensis* and *M. baccata* occur wild in India and adjoining areas of the neighboring countries in the North. However, in addition to *M. baccata* var. Himalaica and *M. sikkimensis* are reported from the Himalayas.

Commercial Varieties

The Delicious and its strains have gained popularity all over the world. They are cultivated on a large scale in India. Indian consumers have shown high preference for Red Delicious cultivar because of its high desert quality and attractive appearance.

1. **Red Delicious**: Red Delicious is the most popular variety of India. It was introduced in Himachal Pradesh by Satyanand Stokes around 1918 and soon its cultivation spread to Kashmir and Uttarakhand. The trees are vigorous and form spur freely. Fruit large sized and oblong conical in shape. Ground colour greenish yellow with red streaks. Flesh is creamish, juicy, aromatic and sweet in taste. It ripens in the third week of August and can be stored for 3-4 months.

2. **Starking Delicious (Royal Delicious)**: Starking Delicious was introduced to India by S.N. Stokes and named it Royal Delicious because he felt that Indian consumers will accept Royal far more rapidly than Starking.

   The trees are vigorous, form spur freely. Fruit large sized and oblong conical in shape. Ground colour greenish yellow covered with dark red stripes all over the fruits. The fruits ripe in second week of August. Flesh creamish, juicy, aromatic, sweet in taste with excellent quality. The fruits cannot be stored for a long time.

3. **Richared**: It is also a Delicious strain. The trees are vigorous and form spur freely. Fruit conical, skin with red blush. Fruits are sweet and juicy, matures in third week of August.

Colour Strains

Many coloured strains found in USA have been introduced to India. Most of them have better colour and shape and are early ripening.

1. **Vance Delicious**: It is a bud mutant of Delicious. Trees are vigorous and spreading. Fruit medium to large sized, conical, striped red. Flesh greenish, firm, juicy, aromatic and sweet in taste. It matures in the second week of August. It is more productive than starking delicious.
2. **Top Red**: This variety is the bud sport of Shotwell Delicious. Trees are moderately vigorous, spreading, frequently form spurs, more productive than Starking Delicious. The fruits are long, large size with red streakes. Pulp is yellow, sweet and juicy. The fruits ripen 7-10 days before Starking Delicious. It develops good colour at high altitudes.
3. **Hardeman**: Trees are vigorous, spreading forms spurs frequently, more productive than starking Delicious. Fruits medium to large, conical, striped red. Flesh creamish yellow, firm, juicy, aromatic and sweet in taste.
4. **Red Chief**: This cultivar is bud sport of Delicious. Tree size small, compact and forms more number of spurs. Fruits are similar to Delicious with prominent red streaks. The maturity is earlier by 15-20 days than Starking Delicious. Fruits are attractive bright in colour, pulp yellow, soft, sweet and juicy with excellent flavour.
5. **Red Spur**: Trees have close internodal growth. Fruit is medium to large in size and conical in shape. Fruits dark red in colour. Flesh is creamish yellow, firm, juicy and sweet in taste.
6. **Oregon Spur**: It is a bud support of Delicious, trees are small is structure but with large number of spur bearing branches. Fruit may be uniformly coloured. Flesh creamish yellow, firm, juicy and sweet in taste. Fruit ripens a few days earlier than starking delicious.
7. **Starkrimson**: It is a spur type of Delicious, trees are dwarf; fruits are oblong, deep red in colour, hard in texture, sweet and juicy. Flesh creamish yellow, firm, juicy, aromatic and sweet in taste. It ripens two weeks before starking delicious.

Other Varieties

1. **Ambri:** This is the only indigenous variety grown in Kashmir, originated as chance seedling in Kashmir. The trees are vigorous. It is an attractive apple with good keeping quality. The fruit is medium to large in size and oblong in shape. It has red streaks over a greenish yellow background. The pulp is white, crisp and sweet. The fruits ripen in the last week of September. The fruit has long keeping quality.
2. **Baldwin**: Trees vigorous, large sized fruits, reddish in colour, round to conical in shape and sour in taste. The fruit matures by the beginning of August. It is a triploid variety.
3. **Golden Delicious**: Tree moderately vigorous. Fruit medium to large, oblong, skin golden yellow with russetted prominent small dots scattered all over. Flesh creamy white, fine crisp, sweet with a blend of acidity. It is used as a pollinizer variety. Fruit matures in the second week of September.
4. **Black Ban Davis**: Grown in Kulu valley. Popularly known as Khali Devi. Trees vigorous, fruits are medium to large in size striped dark red colour, fruit shape is round to conical. Flesh is creamish white, firm, juicy, aromatic, sweet and mildly sub acid.
5. **Granny Smith**: Trees are moderately vigorous, upright, spreading, producing spurs freely. Fruits round in shape having green colour, light yellow in ripening. Flesh very firm, crisp greenish yellow to whitish. It ripens during the second week of October.
6. **Jonathan**: The trees are moderately vigorous, producing spurs freely. The fruits are medium sized and round to round conical shape, ground colour is pale greenish white. The flesh is white, has slightly acidic blend. It also acts as a pollinizer for Delicious varieties.
7. **King of Pippins**: The tree is moderately vigorous and upright. The fruits are medium sized and oblong conical in shape, the flesh is yellowish white, firm, crisp and sub acidic in taste. It is an early ripening variety.
8. **Maharaji (White dotted red):** It is widely grown variety in Kashmir. It is known as Maharaji because earlier trees were grown in the gardens of Maharaja of Kashmir. The trees are vigorous, the fruits mature in October-November. Medium to large size, its fruits are roundish to flattened in shape. The skin is smooth with white dots. Its flesh is yellowish white, crisp, tender juicy and sub acidic.
9. **McIntosh**: The trees are vigorous, hardy and very productive. The fruits are medium to large in size and flat round to round in shape. Flesh is white, soft, fine textures, very juicy and sweet in taste.
10. **Mollie's Delicious**: Fruits resemble with those of Delicious. It is very large and irregular. About half of the skin is red, the flesh is fairly dry with medium flavor, and it is a precocious cultivar. It ripens early than Delicious.

11. **Red Gold**: Tree is fairly vigorous. It has proved to be a superior pollinizer over Golden Delicious, the fruits are medium in size, round conical in shape with dark dull colour, the flesh is white, juicy and sub acid in taste. It is harvested in the second week of August.
12. **Red June**: Trees are moderately vigorous. An early maturing variety, the fruits are small in size and round to flat in shape. The flesh is whitish in colour. It is a good polinizer for the Delicious varieties. Being one of the earliest apple to came into the market, it fetches a fairly good price.
13. **Tydeman's Early Worcester:** Trees are moderately vigorous, spreading, producing, spurs freely. Fruits medium in size and round to round conical in shape. An early maturing variety, its fruits ripen in the second week of July, this is a very good pollinizer for Delicious apple.
14. **Fuji:** Trees are medium to high in vigour, tend to develop good spur system and are precocious in bearing. Fruit is medium to large in size and round to conical in shape. Flesh is yellowish white, firm and sweet with good aroma. At ordinary room temperature, it has good shelf life.
15. **Gala**: Its trees are vigorous and upright similar to those of Golden Delicious. Fruits are small to medium sized and oval round in shape. Flesh is creamish yellow, fine, firm and crisp with aromatic flavor, it is precocious and regular bearer.
16. **Jonagold:** Its trees are vigorous, medium sized strong wide crotches like Golden Delicious. This is a tropical variety. The fruits are large in shape. The flesh is moderately firm, texture is crisp and slightly coarse, juicy and sub acidic. It ripens along with Golden Delicious.
17. **Empire**: Its trees are vigorous, upright, spreading, having strong crotches, easy to prune regular bearer, the fruits are medium sized round in shape with dark red colour. Flesh is creamish, crisp, juicy, aromatic and excellent in quality, it ripens after McIntosh.

Scab Resistant Cultivars

1. **Red Free**: Fruit is round to flat in shape, deep red strips, medium sized, sub acidic in taste. Fruit matures during the second week of July in mild hills and during first week of August in the high hill region.
2. **CO-OP12**: Fruits are conical in shape, deep red in colour, sub acidic in taste, matures in 84 days after full bloom.
3. Amb Red: This is a hybrid between Red Delicious and Ambri. Fruits are conical, bright and striped over yellow ground, glossy, flesh whitish, crisp firm, aromatic and juicy.
4. **Amb Rich**: It is a hybrid between Rich-a-Red and Ambri. Fruit is medium to large conical in shape. Fruit skin is deep red in colour, flesh light yellow to white, slightly acidic and juicy. Fruit matures in the last week of September.

Root Stock

Apple cultivars do not come true from their seeds since apple is a cross pollinated and highly heterozygous plant. Propagation by seed leads to immense variability; therefore, its trees are raised by vegetative methods. Budding or grafting the scion cultivars into suitable rootstocks is essential. Thus rootstock is a very vital component for apple cultivation. The rootstock affects the performance of scion cultivars grafted on them. The rootstocks exert influence on scionic compatibility, nutrient uptake, water balance, resistance to pests and diseases, and tolerance to soil and environmental stress, the yield and fruit quality of the scion cultivars is markedly affected by the rootstocks. The performance of apple trees also depends on the use of right rootstock. Therefore, specific characteristics of rootstocks must be known.

Two types of rootstocks i,e Seedling rootstocks and colonel rootstocks are employed in the propagation of apple. The seedling rootstocks are propagated by seeds and clonal rootstocks propagated by vegetative methods.

Seedling Rootstocks

Seedling rootstocks are most commonly and widely used in India. However, there is a great diversity in type of material used for raising rootstock seedlings. Seeds of crab apple or any commercial cultivars of apple are used for raising seedlings. Diploid cultivars like Golden Delicious, Granny Smith, Wealthy, Yellow Newton, McIntosh, Red Gold *etc.* should be preferred, whereas, a triploid cultivars like Rhode Island Greening, Gravenstein and Stayman winesap could also be used. In Kashmir seeds of wild indigenous crab apple is used as root stock. Seeds of this apple do not germinate until given seed treatment. Some preserving treatments like chemical and growth regulators are affective in stimulating seed germination and seedling growth. Apple seeds treated with 200-1000 ppm ethrel gave maximum percentage of germination obtained maximum germination when seeds were treated with 50 ppm GA followed by 25 ppm kinetin.

Clonal Rootstocks

Clonal rootstocks, propagated asexually are usually multiplied by stooling as it is the most successful method. The first successful effort to test and evaluate rootstocks was initiated at East Maling Research Station, Kent, England in 1912, which resulted in the development of Malling series of clonal rootstocks.

In India, clonal rootstock trails were initiated on M and MM series of apple. These rootstock trails generated very useful information on the performance of different clonal rootstocks under different agro climatic conditions. Many rootstocks have been recommended for commercial use, these are M2, M4, M7 and M9 in Jammu and Kashmir and M7, M9, M26, MM106, MM111 in Mimachal Predesh.

The following rootstocks are recommended for planting:

M2: This rootstock is vigorous, precocious and productive. It is used in most of soil types but is not suitable for heavy soils. It is resistant to collar rot.

M4: Semi vigorous rootstock, slightly smaller than M2. It is more vigorous than M9 and promotes early and heavy bearing. Trees on this rootstock requires support during the early years, it is early to propagate through mound layering.

M7: A semi dwarfing rootstock, it is adaptable to a wide range of soil types and climates. It can tolerate temporary drought as well as water lodging. It is susceptible to root rot

M9: A dwarfing rootstock mostly used throughout the world. It is suitable for high density planting. Its roots are shallow and brittle and the trees require support. Does not tolerate draught, it is resistant to coller rot.

M13: Vigorous rootstock, produces a full size tree similar to apple seedlings. It is tolerant to wet soil conditions. It roots easily and is useful in shallow soils.

M26: Dwarf rootstock, takes 3-5 years to come into bearing. It can be planted in all type of soils, is resistant to collar rot and has spreading habbit.

M27: It is most dwarfing rootstock tested at Dr. Y.S. Parmar University of Horticulture and Forestry, Solan, H.P.

MM106: It is a semi dwarf rootstock. Trees on this rootstock are very productive. Fruit bearing starts early. The trees on rootstock MM106 gives good crop constantly. It is resistant to wooly aphids. It is easily propagated through hard and soft wood cuttings.

MM104: This is one of the most winter hardy clonal rootstock. Roots of trees on this rootstock are well anchored. It is susceptible to crown rot.

MM111: It is a vigorous rootstock. It is adapted to a wide range of soils and climate conditions. It is highly tolerant to drought.

Chemical Composition

Carbohydrate

Fresh apples are considered a food of moderate energy valve, whereas processed apple products are either comparable to fresh apples in energy valve or higher because of concentration, dehydration, or the addition of sugars during processing. Chemical composition of apple is affected by many factors, including cultivar, growing region, climate, maturity, cultural practices, and processing. The approximate composition of apples and some apples products as affected by processing methods is summarized in Table 2.1. Carbohydrates are the principal food constituents in apple. With starch and sugars the available carbohydrate and pectin, cellulose, and hemi cellulose the unavailable fractions. Total carbohydrates in fresh apples account for about 15 per cent comprising 0.89-5.58 per cent each of fructose and glucose; and 0.88-5.62 per cent sucrose.

Table 2.1: Approximate Composition of Apples and Apple products (per cent) (Young and How 1986; Gebhardt *et al.*, 1982)

Product	*Water*	*Energy*	*Protein*	*Lipid*	*Carbo-hydrate*	*Fiber*	*Ash*
Fresh apple with skin	83.9	59	0.19	0.36	15.3	0.77	0.26
Apple without skin, cooked	85.5	53	0.26	0.36	13.6	0.54	0.28
Canned apple, sweetened	82.4	67	0.18	0.14	16.7	0.54	0.27
Dehydrated apples (low moisture)	3.0	346	1.32	0.58	93.5	4.09	1.57
Dehydrated apples	31.8	243	0.93	0.32	65.9	2.87	1.10
Dehydrated apples sulfured and cooked without added sugar	84.1	57	0.22	0.07	15.3	0.67	0.26
Frozen apples	86.9	48	0.28	0.32	12.3	0.54	0.24
Canned apple juice	87.9	47	0.06	0.11	11.7	0.21	0.22
Apple juice frozen concentrated, undiluted	57.0	166	0.51	0.37	41.0	-	1.12
Applesauce, unsweetened	88.4	43	0.17	0.05	11.3	0.53	0.15
Applesauce, sweetened	79.6	76	0.18	0.18	19.9	0.46	0.14

Table 2.2: Sugar and Pectin Content (per cent) of some Apple Cultivars

Cultivar	*Total Sugars*	*Reducing Sugars*	*Sucrose*	*Pectin*
Delicious	11.79	8.81	2.89	0.42
Golden Delicious	12.39	7.89	3.78	0.64
Jonathan	11.45	8.29	2.79	0.59
Jubilee	12.60	8.02	3.90	0.61
McIntosh	10.89	8.30	2.61	0.52
Newton	11.67	7.50	4.18	0.60
Spartan	11.32	9.00	1.95	0.54
Stay man	11.69	7.05	5.02	0.61
Wine sap	12.82	10.67	3.14	0.75
Northern Spy	12.05	9.15	2.69	0.63
Stirling	11.98	7.94	3.16	0.60
Rome Beauty	10.65	7.16	3.49	0.56
York Imperial	12.38	8.07	4.31	0.53
Lowry	13.23	8.64	3.59	0.32

Organic Acids

Organic acids are among the most important constituents in apples. The primary acid in the fruit is malic, although others such as citric, lactic, and oxalic are also present. Various organic acids found in apple peel and pulp, though in smaller fractions, are presented in Table 2.3.

Table 2.3: Organic Acids found In Apple Fruits

Whole Fruit or Juice	*Peel*	*Pulp*
Malic	Glyoxylic	Pyruvic
	Isocitric	
Quinic	malic	Malic
Glycolic	Citric	Citric
Succinic	Quinic	Quinic
Galacturonic	Glyceric	Citramalic
Citramalic	Oxoglutaric	Glyceric
Mucic	Pyruvic	-oxoglutaric

The acidity in the fruit is of interest because it affects eating and cooking quality. The total acidity in apple juice in different varities ranges from 0.22 to 0.78 per cent with an average of 0.42 per cent as malic acid. Similarly, the ph variation among cultivars ranged between 3.36 and 4.25.

Proteins

Fresh apples with skin contain about 0.19 per cent protein, and as such are regarded as a poor source of this important nutrient. The proportion of different amino acids (Table 2.4) shows that aspartic and glucamic acid are the predominant amino acids in apples, followed by lysine and leucine.

Table 2.4: Amino Acid and Mineral Content of Fresh Apple

Amino Acid	*Per cent*	*Amino Acid*	*Per cent*	*Mineral*	*(ppm)*	*Mineral*	*(ppm)*
Alanine	0.007	Lysine	0.012	Calcium	7.0	Chloride	4.2-6.2
Arginine	0.006	Methio-nine	0.002	Iron	1.8	Chromium	0.03
Aspartic Acid	0.034	Pheny-lalanine	0.005	Mag-nesium	50.0	Cobalt	`0.10
Cystine	0.003	Proline	0.002	Phos-phorus	70.0	Copper	0.45
Glutamic cid	0.020	Serine	0.006	Potassium	1150.0	Iodine	0.02
Glycine	0.008	Threonine	0.007	Zinc	0.4	Moly-bdenum	0.30
Histidine	0.003	Trypto-pham	0.002	Copper	0.4	Selenium	0.9-1.6
Isoleucine	0.008	Tyrosine	0.004	Manga-nese	0.4	Sodium	8.9-9.2
Leucine	0.012	Valine	0.009				

Minerals

Fresh apples contain 0.26 per cent ash contents, while in dehydrated apples it is 4-5 times higher, owing mainly to the effect of concentration. Some variations in ash content among apples have been reported from different geographic regions. This variability is assumed to be due to the availability of different minerals in the soils of different regions. Potassium constitutes the main portion of the total mineral contents of apples, and it accounts for than 40 per cent of the total ash. Phosphorus and calcium are the next most prevalent minerals in the apples fruit (Table 2.4).

Vitamin

The vitamin content of fresh apples are presented in Table 2.5. The average ascorbic acid content is about 5 mg/100g of apple. In comparison with the recommended daily intake of vitamins, the proportion of all other vitamins except vitamin C in apple was found to be insignificant.

Table 2.5: Vitamin Contents of the Fresh Apple per 100 g of Tissue

Vitamins	*Concentration*
Ascorbic acid (mg)	5.7
Thiamine (mg)	0.017
Riboflavin (mg)	0.014
Niacin (mg)	0.077
Panthothenic acid (mg)	0.061
Vitamin B 6 (mg)	0.048
Folacin (mcg)	2.8
Vitamin A (retinol equelent)	5.3

Phenolic Compounds

Apple contains several classes of Phenolic compounds, including hydroxycinnamic derivatives, flavonols, anthocyanins, dihydrochalcones, monomeric flavan-3-ols, and tannins. There is wide variability in the total Phenolic contents of fruits. Total Phenolic content of ripe apple fruit and reported Phenolic content is ranging from 0.15 to 2.5 per cent. The variation in the values reported in the literature is due mainly to the method used for estimation of tannins, variety, stage of maturity, and environmental factor. The major phenolics identified in apple fruits are quinic acid, epicatechin, quercetin-3-0-β-D-galactopyranoside, phloretin-2-0-glucoside-2-0-glucoside, and cyaniding-3-galactopyranoside. The phenolics are located mainly in the vacuoles. The epidermal and sub epidermal layers have higher contents of phenolics than the internal tissues. The concentrations of Phenolic compounds are very high in young fruits and then rapidly decrease during fruit development. These Phenolic compounds are involved in enzymatic browning of apple products.

Harvesting

Fruit Maturity and Ripening

The apple is classified as a climacteric fruit. The term respiratory climacteric describes the rise in respiration rate that accompanies the maturation and ripening phase in apple, and this rise is associated with increase in internal concentrations of carbon dioxide and ethylene, respiration and autocatalytic ethylene production. In a climacteric fruit such as the apple, ethylene advances the timing of the climacteric, autocatalytic production continues after removal of ethylene and, in contrast to a non climacteric fruit, the management of the respiratory rise in independent of the concentration of applied ethylene. Thus timing of the climacteric and ripening of apple fruit is advanced by exposure to ethylene.

Ripening of apple fruit involves many physiological and biochemical changes. From an applied perspective, the most important of these are softening, the change of background or ground colour from green to yellow, loss of acidity, conversion of starch to sugars, formation of cultivar waxes and synthesis of aromatic compounds. The metabolism involved in these changes has been described bymany researchers. In the natural world, these changes result in a desirable product for seed dispersal by the activity of animals and/or breakdown and decay of the fruit. Many of these changes are at least partly desirable for human consumption, and the objective of apple industries is to harvest fruit at the appropriate maturity and apply post-harvest technologies to control the rates of these changes in order to provide the consumer with an acceptable product.

Harvest Timing and Fruit Quality

The harvest date within the maturation and ripening period has a profound effect on the storage quality of fruit. As quality factors such as flavour and aroma of the fruit increase, the storage potential of the fruit decrease and therefore harvest decisions are a compromise between the quality and storability of fruit. The increases in fruit quality are typically associated with the climacteric. The length of storage of apples can usually be increased by harvesting fruit before they are fully mature, but quality characteristics such as colour and varietal flavour develop less in these fruit, and they can be more susceptible to physiological disorders, such as bitter pit and superficial scald. Fruit harvested over mature tend to be softer and more easily damaged and may have water core and be more susceptible to diseases and physiological disorders, such as senescent breakdown.

Superimposed on the relationship between quality and storability during ripening is that between the harvest period and acceptable storage length. The harvest period over which quality of the fruit will be acceptable is relatively wide for fruit stored for short periods (1-2 months), but decrease as storage length increase. In mature fruit stored for only a month or two, acceptable flavour can develop in early harvested fruit, while, in those fruit harvested later, flavour is good and fruit are marketable if texture is maintained. In contrast the harvest window for a fruit that is marketed after long term CA storage for 9-12 months is much narrower, being limited by failure to develop flavour, loss of texture and development of storage disorders. The actual harvest periods involved may be affected by cultivar. For example, the optimum harvest period for an early season apple, such as 'McIntosh', is much shorter than that for a late harvested apple, such as 'Rome Beauty'. Therefore, the x-axis might be 1-2 weeks for the former cultivar and 3-4 weeks for the latter one. Regional differences can also be important. For example, the 'Cox's orange pippin' apple is a short term storage apple when grown in New Zealand, but a long term storage apple in England.

The importance of determining the optimum harvest date for storability will be related to how fruits are utilized and therefore the expectations of the consumer for the product. These expectations vary greatly, depending on the market. Continental Europe, for example, prefers apples at a more advanced stage of ripeness than does the UK.Greater Flavour, associated with tree ripened fruit, is likely to be a premium factor in 'pick-your-own' or gate sales during autumn. In contrast, for long term stored fruit, texture is more likely to be critical acceptance factor, although acceptable flavour must be present.The requirement for a maturity programme generally increases as an industry becomes larger and more complex, especially if it is reliant on long term storage technology.

Ethylene production of fruit can be manipulated before harvest, either by use of ethephon application to advance ripening, or by use of aminoethoxyvinylglycine (AVG), available commercially as ReTain, to reduce ethylene production and delay ripening. Harvest dates must be adjusted accordingly.

Harvest Indices

Many methods have been proposed as indices of apple maturity (Table 2.6). Several of these are indicators of quality, rather than maturity *per se*. Moreover, harvest decisions have to be based not only on physio- logical maturity, but also on market requirements, which include factors such as blush, background and usually the absence of water core. Therefore, the term harvest index is a more accurate terminology for the factors used in making harvest decisions. The most important harvest indices are as follows.

a) Ethylene Production or Internal Ethylene Concentration (IEC)

Ethylene, which is an important hormone that affects fruit ripening, can be measured relatively easily by gas chromatography. Since a rise in ethylene production is associated with initiation of ripening, it has been suggested that ethylene production or IEC should be a major determinant of harvest decisions. However the importance of ethylene in making harvest decisions is not straight forward. Relationships between ethylene productions and optimum harvest dates can be poor. Also, the timing, or presence, of increased ethylene production is a function of cultivar and, within a cultivar, is greatly affected by factors such as growing region, orchard within a region, cultivar strain, growing season conditions and nutrition. Ethylene production may not be relevant for determining the harvest of some cultivars such as golden delicious because it does not increase during the harvest period. Ethylene production may be a better indicator of when to complete the harvest especially in cultivars such as McIntosh, where autocatalytic ethylene production precedes preharvest drop.

Table 2.6: Example of Methods Evaluated as Indices for Determining Harvest Dates of Apple Fruit

Harvest Index	*Reference*
Ethylene production or internal ethylene Concentration	Dilley,1980; Lau,1985; Blanpied,1986; Chu, 1988 Knee *et al.*, 1989
Respiration	Wilkinson and Sharples,1967: Blanpied,1969
Starch Pattern index	Reid *et al.*, 1982; Knee *et al.*, 1989; Blanpied and Silsby 1992
Starch Content	Knee *et al.*, 1989
Firmness	Lau,1985; Knee *et al.*, 1989
Soluble solids concentrations	Lau,1985
Firmness and soluble solids concentrations (sliding scale)	Blanpied,1974
Firmness/Soluble solids concentration X starch index	Streif,1983; Delong *et al.*, 1999
Background colour	Lau,1985; Plotto *et al.*, 1995
Calendar date	Blanpied,1964; Fidler,1973
Days from full bloom	Haller, 1942; Truter and Hurndall,1988
Heat unit accumulation	Eggert,1960
Days from full bloom and temperature records	Luton and Hamer,1983; Blanpied and Silsby, 1992; Beaudry *et al.*, 1993
Water core incidence	Lau,1985; Blanpied and Silsby,1992
T stage	Stoll, 1968; Faragher *et al.*, 1984
Flesh Colour	Lau,1985
Seed Colour	Lau,1985
Free sugars and phosphorylated intermediates of carbohydrates metabolism	Knee *et al.*, 1989
Enzymes	Knee *et al.*, 1989
Peel pigments	Knee *et al.*, 1989
NADPH fluorescence	Cavalieri *et al.*, 1988
Fruit Size	Lau1985
Loss of bitter flavour	Blanpied and Silsby, 1992
Separation force	Smock, 1948; Lau, 1985
Titratable acidity	Lau, 1985
Visible/near-infrared spectroscopy	Peirs *et al.*, 2000

Starch Test

The hydrolysis of starch to sugars as fruit ripen can be estimated by staining starch with iodine solution. The resulting patterns, which reflect the extent of starch hydrolysis, are rated numerically, using starch charts. Many starch charts are available, either specific to cultivar, or generic. The starch test has become popular because of its ease of use. Optimum starch indices are available for many cultivars and when the change of indices is linear, the test can be used to predict optimum harvest dates. Indices developed for a cultivar in one region cannot necessarily be applied to other regions. Also, caution is required in interpreting starch patterns, as complications can occur under certain growing conditions. For example. Starch indices may appear more advanced than other harvest indices because initial starch accumulation in the fruit was not complete. Therefore, absence of 100 per cent starch staining may suggest incorrectly that starch hydrolysis has already occurred.

Flesh Firmness

Apples soften after reaching full size and as they mature and ripen. Flesh firmness has been used as a maturity index, but it is affected by many preharvest factors, including season, orchard location, nutrition and exposure to sunlight, which are independent of fruit maturity. However, as an indicator of internal quality, it can provide information that is important to fruit performance in storage. It can directly affect consumer satisfaction, and firmness is being used as a quality criterion by wholesalers, especially in Europe, rather than as a maturity index.

Soluble Solids Concentration

The soluble solids concentrations of apples generally increases as fruit mature and ripen, principally by the conversion of starch to sugars. It is also a quality index, rather than a maturity index, being affected by many preharvest factors, and levels do not necessarily reflect fruit maturity. As with firmness, soluble solids concentrations are increasingly being used as a quality criterion by wholesalers. Acceptability of apples may be affected by an interaction of firmness and soluble solids concentrations, *e.g.* a softer fruit with high soluble solids concentrations may be acceptable, whereas the same fruit with low soluble solids is not. Sliding scales incorporating both factors have been developed.

Titratable acidity

Malic acid is the predominant acid contributing to the Titratable acidity (Ta) of apple fruit. TA decreases during maturation and ripening, but optimum values vary by cultivar and season

Background or Ground Colour

As fruit mature and ripen, the background colour changes from green to yellow, reflecting the loss of chlorophyll. Ground colour has been used as a maturity indicator, but early harvest of fruit for long term storage may pre-empt the colour changes. Preharvest factors, especially those that affect nitrogen content, can markedly influence chlorophyll concentrations, independent of maturity changes. Also, in apples it is sometimes difficult to separate the initial decline of chlorophyll on a surface area resulting from fruit expansion from the subsequent loss resulting from chlorophyll breakdown. Nevertheless, assessment of background colour is used commonly for bicoloured cultivars, such as 'Gala', 'Brae burn' and 'Fuji'. Commercially, colour is usually assessed with colour charts, but it is commonly measured experimentally using chromameters.

Calendar Date, Days from Full Bloom, and Temperature Records

The usefulness of full bloom dates and days after full bloom with and without temperature records varies greatly by cultivar and growing region. Calendar dates alone have limited value in regions where temperature variations result in wide differences in bloom dates, but, in more consistent growing regions, days from full bloom can be the most reliable harvest index for several cultivars. The temperatures during the first 4-6 weeks after full bloom have been useful in predicting a harvest window for McIntosh.

Of the many other indices that have been tested, their usefulness as harvest indices has been variable, depending on the cultivar and growing regions in which they have been utilized. Seed colour change, for example, is specific to 'McIntosh', while flesh colour changes are used to determine harvest dates for 'Tydeman's Red and Spartan.

T-Test

Another criterion which has been related to the time of maturation is the date when the fruits reach the T-Stage. The T-Stage is the point at which the angle formed by the fruit receptacle and the pedicel reaches 90° and its timing is obtained by interpolation. The angle between pedicel and receptacle is greater than 90° early in the growing season and less than 90° late in the season. The number of days between the T-Stage and the date of optimal harvesting time of apple showed a remarkable consistency for a given location. It was determined that the number of days from the T-Stage to harvest for apple varied from 73 for cv. Discovery to 130 for cvs. Gloster and Idared in the Wedenswil region of Switzerland. It was reported that the time between this date and optimum harvest was similar from year-to-year within rootstock. However, others did not find this index to be reliable.

Harvesting Methods

The ultimate use of the apple fruits decides the methods to be used in harvesting. The Most commonly used harvest method for fresh-market and processing apples is by hand. However mechanical harvesting of fruit, particularly of relatively low-quality fruit for immediate processing, has now begun to make inroads.

Hand Harvest

The correct way to pick fruits is by lifting up with a slight twisting motion rather pulling down straight away from the spur. Pulling down results in many fruits being removed without stems, and it is more difficult to pick the fruit. Proper picking generally require careful handling at every step to prevent bruising.

Proper pickling and handling will prevent stem pulls, skin punctures, and bruising. Soft fruits require more careful handling than firm ones to avoid bruising, but firm fruit tend to get skin punctures more readily than soft types.

Mechanical Harvest

There are two major mechanisms for mechanical harvesting, *i.e.*, shake and catch, and shake and orchard floor sweep. These methods are not recommended commercially, as considerably physical damage occurs to fruits. However, for immediate consumption and processing, these quick methods may be applied.

The harvesting mechanism uses direct contact to push the fruit from the canopy. In this method, about 96 per cent of the apples are harvested with low bruise percentage and minimum mechanical damage. Robotic harvesting of apples has also been tried in France. The robot consists of a telescopic arm, a line –scan camera, and a microcomputer. A Pneumatic picker, located at the end of the hydrostatic manipulator arm, detaches the fruit and transfers it to a bin.

Grading, Packing and Transport

Grading

Apples are graded for size and quality under groups A,B and C, depending on their color, regular shape, freedom from injuries, blemishes, disease spots, *etc.* A and B grades are sent to market; C- grade fruits are not generally marketed through the fresh fruit trade. Grades A and B are further sub divided by size, the grades depending on the equatorial diameter of the fruit. Fruits meant for export are further designated as extra fancy, fancy class I, and fancy class II. Mechanical graders are also used to provide uniform standards of size grades. However, mechanical grading needs to be supplemented with visual screening for color, diseases, and uniform shape.

Packaging

The primary role of packing lies in the safe delivery of the produce from the production center to the ultimate consumer in prime form and fresh condition. Packaging is one of the major factors which influence the quality of fruit when it reaches the consumer.

Wooden Boxes

Like other temperate fruits, apples are generally packed in wooden boxes for marketing. The box is line inside with old newspaper sheets, keeping the margins for overhanging the flaps. The fruits are initially padded with wood wool pine needles at the bottom and later between intervening layers. Paper wrapped fruits in each layers; the top layer is covered with paper by bringing together the overhanging flaps. The top is then nailed on. The box is further reinforced externally by clamping with a tight 14-16 gauge steel wire for distant markets. Apple packing in wooden boxes results in high bruising losses and shows maximum loss in weight, which may be due to water absorption by the timber from the fruit and the subsequent loss of moisture to the atmosphere over and above the inherent drying of the timber during transportation.

Trays

Apples are also packed in paper pulp trays, with fruits in the trays often individually wrapped or unwrapped. These trays are then placed in boxes/cartons.

Corrugated Fiber Board Cartons

Corrugated fiber board (CFB) cartons have already replaced wooden boxes for packaging of most fruits and vegetables in the horticulturally advanced countries. They were introduced into the apple trade in India in the 1980.CFB cartons are capable of withstanding various transportation hazards both on mule back and by trucks. They cause minimum bruise damage (3.2-3.4 per cent) in comparison to wooden boxes (25.2-29.7 per cent), besides reducing loss in fruit weight. Further, CFB cartons are attractive, light in weight, and offer better printability, which helps in efficient marketing. They also reduce pressure on our already denuded forests. Plastic crates have also recently been recommended and introduced in the apple trade in Himachal Pradesh and J&K, India, to a limited extent, for field boxes for collection of fruit from orchards, stacking in cold storage, carriage to nearby markets, and to supply fruit to processing plants.

Transportation

Full telescopic corrugated containers are used as shipping containers for apples. Considerable reduction in bruising damage in tray-packed apples during transportation as compared to traditional packs has been reported. About 20-35 per cent of fruit is bruised in conventional wooden boxes during transportation. Safe transportation of apple packed wooden boxes along with CFB cartons during transportation has been reported. Adoption of modified atmosphere packaging (with retailer) for marketing and distribution of apples provides benefits in flexibility, stock control, quality maintenance, and reduced wastage.

Post-harvest Treatments

Superficial Scald Inhibitors and Fungicides

Superficial scald is a physiological disorder of apples that develops during cold storage. It was the major cause of fruit loss during storage until the commercial development of diphenylamine (DPA) as a scald inhibitor. Susceptibility to the disorder can be affected by cultivar and can decline with later harvest date. Several major cultivars, such as Delicious and Granny Smith are highly susceptible during normal commercial periods. DPA is typically applied as a drench, often at the storage receiving point while bins of fruit are still on trucks, or by bin dipping. The concentrations necessary for scald control vary by cultivar, and label rates on proprietary DPA products from commercial suppliers should be followed. Cultivars also vary in sensitivity to DPA induced skin injury. Lower DPA application rates may be used if scald risk is lower, for example because of climate. DPA use can be avoided where cultivars, such as Empire and Gala have a low scald risk.

The use of DPA is widespread, although it is not registered for use in some countries. Adequate control of scald may be obtained by rapid cooling and rapid CA storage, especially low oxygen concentrations, but efficacy can vary according to cultivar and growing region. Alternative means of controlling scald development, for example, stress treatments have also been investigated because of concern that DPA may become unavailable for this purpose due to consumer concerns about food safety relating to the use of such chemicals.

Calcium

Treatments of fruit with calcium can reduce development of bitter pit, senescent breakdown and rots and may slow softening. Depending on growing region and cultivar, storage operators will therefore apply calcium, usually as food grade flake calcium chloride or proprietary formulations. Cultivars vary in susceptibility to calcium induced skin burn, and product label application rates should be followed to minimize risk.

Fruit should be treated with calcium immediately after harvest, and it is applied either alone or with DPA (and sometimes with a fungicide). Application is usually by drenching or dipping. Calcium treatment is only commercially effective where disorders incidence is low to moderate, but effectiveness may be improved by CA storage.

Waxes

Apples are commonly waxed with shellac or carnauba shellac based coatings during the packing line operation, although some markets prefer line unwaxed fruit. Also, because shellac and carnauba shellac are associated with non food uses, such as floor and car waxes, alternative coatings are being tested.

The main advantage of waxing is to improve fruit appearance. Wax coatings can extend shelf life of fruit, by reducing water loss, respiration rates and ripening. The detergent wash before wax application can also remove dust, dirt and contamination arising from washing in dirty water, which can occur in non-chlorinated hydro handlers.

Methylcyclopropene (1-MCP)

Controlling ethylene production and action is a primary goal in the post-harvest management of apples. However, a new compound, 1-MCP, is now available for many apple industries, under the commercial name Smart Fresh, registration for 1-MCP was obtained in the USA and other countries in 2002 and European registration is pending. 1-MCP is structurally related to ethylene, has a no –toxic mode of action and is applied at very low dose levels, with low measurable residues in food commodities. The US Environmental Protection Agency has classified 1-MCP as a plant growth regulator.

1-MCP is thought to blind to the ethylene receptor, competing with ethylene for the available binding sites and thereby preventing ethylene action. Whereas ethylene diffuses rapidly from the binding site after ethylene treatment, this compound remains bound for long periods and formation of an active complex is prevented. Inhibition of ethylene action may eventually be overcome by the production of new receptors.

Softening, yellowing, respiration, loss of TA and sometimes a reduction in soluble solid concentration, as well as the development of several physiological disorders, are delayed or inhibited by 1-MCP application. Responses of fruit to 1-MCP may be affected by cultivar and fruit maturity. Volatiles production by apples is also inhibited by 1-MCP.These results are constant with the view that volatile production is regulated by ethylene, and consumer studies on the acceptability of 1-MCP- treated fruit will be required to ensure is not unacceptably compromised.

Storage

Depending on cultivar and growing region, some fruit are stored and marketed for up to a year, when the following year's crop becomes available. Therefore, unless fruit are sold for immediate consumption, the major objective of post-harvest management is to decrease the rates of metabolism and thereby the rates of deterioration and quality loss. Commonly, respiratory activity of harvested apple is used as an indicator of metabolic rates. The principal mechanisms available to slow down metabolism are temperature and control of storage atmospheres. Control of relative humidity is also an important component. The interaction of the three factors must be considered in developing a storage regime for any cultivar.

Temperature

Generally, low respiration rates and longer storage periods of apples are directly related to lowered storage temperatures. The lowest temperatures for storage must be above freezing and those at which chilling injury will develop. Maximizing quality maintenance of fruit requires attention to temperature not only immediately after harvest and during storage, but also during packing, transport and retail display. This combination of events is sometimes described as the 'cold chain', highlighting the importance of maintaining the links from harvest to consumer.

Effects of Temperature on Metabolism

Respiration

The respiration rate of fruit is directly affected by temperature and the respiratory climacteric is suppressed by storage temperatures below 10°C. However, respiration of chilling-injury-susceptible fruit may be stimulated by colder storage temperatures, *e.g.* 'Idared' fruit stored at 0°C initially had lower respiration rates than those stored at 2°C and 4°C but the respiration rate of these fruit increased as low-temperature breakdown developed. This effect was exacerbated by lowered oxygen concentrations in the atmosphere. Increased risk of chilling injuries and of the development of alcoholic off-flavours forms the basis of recommendations for raising storage temperatures when oxygen concentrations are lowered.

Ethylene

Low storage temperatures can slow down the onset of ethylene production in apples, but chilling temperatures can also enhance ethylene production in some cultivars. Accumulation of 1-aminocyclopropane carboxylic acid (ACC), a key intermediate in the ethylene biosynthetic pathway, and ethylene production of 'Granny Smith' apples at 20°C are markedly increased by exposure to 0°C for at least 8 days but this dose not occur bin 'Royal Gala' or 'Delicious'. Cold treatment appears to reduce resistance to ripening in 'Granny Smith', and the ripening physiology of this cultivar may be more analogous to that of winter pears.

Chilling Injuries

Some fruit types, typically tropical or subtropical in origin, develop chilling injury when exposed to storage temperatures below 10-15°C. Temperate fruit, though resistant to chilling injury are not immune to it. Most apple cultivars are not sensitive to chilling temperatures and maximum storage life is obtained by storing them as close to 0°C as possible. However, some cultivars are susceptible to the development of chilling-related disorders below 2-4°C. These disorders specifically include low-temperature breakdown and soft scald, as well as disorders that are aggravated by such temperatures.

Cooling after Harvest/Precooloing

Because the respiration rate of apple fruit is affected by temperature, the rate of fruit cooling after harvest can markedly influence quality maintenance. Fruit should be removed from exposure to radiant energy in the orchard to refrigeration, or at least shade, as quickly as possible after harvest. However, the importance of rapid cooling varies depending on cultivar, harvest maturity, nutritional status of the fruit and potential storage performance. Early-season cultivars tend to soften more rapidly than those that mature in the later part of the harvest season. Within

Table 2.7: Ranges of Atmospheric and Temperature Conditions for Selected Apple Cultivars (Watkins *et al.*, 2003). References should be consulted for specific atmosphere and temperature combination for each growing region.

Cultivar	*Oxygen (Per cent)*	*Carbon Dioxide (Per cent)*	*Temperature (°C)*
Braeburn	1-3	<0.5-1.2	0-1.5
Bramley's Seedling	1	5	4
Cortland	2-3	2-3	0
	2-3	2-3 for 1 month, then 5	2
Cox's Orange Pippin	1.2-2	0.7-4	3-4
Delicious	0.7-2.5	0-4.5	0-1.1
Elstar	1.5-2.5	1-4.5	0-1.5
Empire	1.5-3	0.5-3a	1-3
Fuji	0.7-2.5	1-2a	0-1
Gala	1-3	2-3	0-3
Golden Delicious	1-3	2-3	-0.5-2
Granny Smith	0.8-2.5	0.5	-0.5-2
Idared	1.3-3	0.5-4	0-4
Jonagold	1-2.5	2-3	0-3
McIntosh	3	2-3 for 1 month, then 5	2
	2	2-3 for 1 month, then 5	3
Marshall McIntosh	4-4.5	2-3 for 1 month, then 5	2
Must	1.5-2.5	1-3	0-1
Pink Lady®	1.5-2	1	0

a cultivar, apples tend to soften more rapidly at later stages of maturity than at earliest stages and in fruit with high nitrogen and low calcium contents. Rapid cooling appears to be more critical for fruit that are more likely to ripen quickly. Also, the negative effects of slow cooling on firmness and colour are magnified as storage length increases. Therefore, inadequate investment of resources at harvest to ensure rapid fruit cooling may not be apparent until late in the storage period, when fruit may not meet minimum firmness standards for marketing.

The predominant method of cooling apple fruit involves exposing fruit to normal air flow in a refrigerated room. However, this method is slow and inefficient because air flows around, rather than through, bins or cartons of fruit. Slow cooling can result in loss of quality when rooms are filled rapidly and refrigeration capacity cannot cope with large fruit loads. The conversion of many industries to rapid CA storage has also resulted in increased pressure to improve fruit cooling rates.

Faster room cooling of fruit can be achieved by separating and loading bins or cartons into separate rooms for pre-cooling before they are moved into the long-term storage room (modified room cooling) or by loading only the quantity that can be handled by the existing refrigeration system.

Modified room cooling involves increased movement of fruit and logistical planning, but can be extremely useful early in the harvest period when harvest of the faster-respiring early- season apples coincides with the availability of still-empty storage rooms that will be used later. In such situations, where fruit volumes exceed the refrigeration capacity of the cold storage, improved fruit cooling can be achieved by placing no more than two stacks of bins across the width of each CA storage room each day. In cases where the room is not down to 0°C by the next morning, the number of stacks should be reduced even further to one-deep. Faster cooling will be obtained if bins are placed in the downstream discharge of the evaporator with pallet runners oriented in the same direction as the air flow. Additional bins of fruit should be stacked, no more than two-high, in unfilled refrigeration rooms to cool overnight before loading into the CA room the next morning. These stacks should be placed randomly throughout the unfilled room to maximize air exchange with the fruit. The capacity to cool fruit is dependent upon the refrigeration capacity and room design.

Forced-air cooling and hydrocooling are less commonly used than room cooling for apples, although forced-air cooling is being used increasingly for cooling of fruit in cartons. In forced-air cooling, bins or cartons are stacked in patterns so that cooling air is forced through, rather than around, the individual container. A slight pressure gradient, usually developed with a fan, forces cold air through the containers, which contain ventholes placed in the direction that the air will move. Also, minimal packing material that will interfere with free movement of air through the container is used. Direct contact of cold air with the product can cool the fruit 0.1-0.25 times faster than room cooling. The cooling rate is controlled by the volume of air passing over the product, and the economics of cooling are affected by the fan speed and the number of containers being cooled. Water loss from the product is less than that found for room cooling.

In hydrocooling, apple temperatures are reduced by cold water flowing over the fruit surface, by either flooding, spraying or immersion of the fruit in chilled water. The rate of internal cooling is related to the size and shape and the thermal conductivity of the fruit being cooled. The method is simple, economical and effective and avoids water loss, although it is obviously limited to bins rather than cartons of apples.

Storage Temperatures

The recommended conditions for commercial storage of apples are -0.5°C to 4°C, the desired temperature being a function of sensitivity to low-temperature-associated injuries. However, these disorders typically develop over long-term storage and sometimes temperatures closer to 0°C are used for short-term(1-2 months) storage of chilling sensitive cultivars to maximize firmness retention. Additional factors in selecting storage temperatures include the interaction between temperature and low oxygen and/or high carbon dioxide concentration. Also, temperature affects humidity requirements. It is easier to maintain relatively humidity above 90 per cent at 1°C than at 0°C.

Temperature should be monitored throughout the storage period by measurements at different points throughout the storage room with thermocouples. It is dangerous to rely on a single thermometer at the door, as temperatures within stacks and throughout the room may be lower or higher than indicated by such readings.

Post-storage Temperature Control

While fruit are kept at optimum temperatures during storage, the cold chain is most frequently broken after storage during packing, transport and retail display. When storage rooms are opened for marketing, fruit can be exposed for extended periods to temperatures above optimal due to operations required for movement of fruit for packaging. Fruit temperatures can increase during this time, as well as during packing. After fruit have been packed into plastic bags or trays and placed into cartons, it is important to remove this heat. Passive cooling may take several days because cardboard cartons are good insulators and offer poor air circulation. Moreover, fruit respiration contributes to temperature increases within cartons. Failure to remove heat accumulated during packing processes may be responsible for firmness losses during subsequent storage and transport and have a negative impact on fruit quality at the consumer level. Forced-air cooling of packed fruit, with modification of carton design to improve air flow over fruit resulting in faster cooling, is becoming more common. A downside of using vented cartons, however, is that fruit can also warm up faster if good refrigeration is not maintained.

Transport requirements for apples can range from simple, involving short distances to a nearby local market, to sophisticated, involving long distances internationally by land and sea. Most transit vehicles have sufficient refrigeration capacity to maintain, rather than cool, fruit. Thus apples should be cooled properly before loading into the container or other transit vehicle. To minimize warming of the crop and therefore avoid losing the benefits of cooling, it is necessary to ensure fast and efficient loading, under ideal conditions the loading dock and load assembly areas are refrigerated. Pallets contains cartons should be loaded into transit vehicles in such a way as to allow an air flow that is adequate for temperature control.

Relative Humidity

Relative humidity is a major factor in preventing moisture loss of apples and therefore maintaining their quality. Relative humidity of 90-95 per cent is recommended for apples to prevent shrivel. The major causes of dehydration of fruit in cool stores are the small surface areas of refrigeration coils and/or frequent defrosting. Therefore, CA rooms should have the largest coil size feasible for the room. The number of defrost cycles can also be reduced to optimize relative humidity in the room, and some operators reduce the oxygen concentration to the minimum safe level and then raise the temperature to 1-2°C to minimize the need to defrost. Storage rooms can be cut fitted with high-pressure water-vapour systems that can add moisture to the room and which are suited for operation at temperatures around 0°C. the air-distribution system should be designed to prevent condensation of water droplets on fruit in order to prevent decay. The use of plastic rather than wooden bins or of bins-liners inside wooden bins can also help minimize shrivel—for example, with 'Golden Delicious'.

Controlled-Atmosphere Storage

The apple is the predominant horticultural commodity stored under CA conditions. The objective of CA storage is to lower oxygen and increase carbon dioxide concentrations to levels that will maintain fruit quality by decreasing respiratory metabolism and reducing ethylene production and action, but not to levels that induce fermentation or other damaging events.

Effects of CA on Metabolism

Respiration

The metabolic rates of apple fruit are reduced by decreasing oxygen and/or increasing carbon dioxide concentrations in the storage atmosphere. Low oxygen concentrations reduce oxygen uptake and carbon dioxide production proportionally, and therefore the respiratory quotient remains constant. High carbon dioxide concentrations reduce carbon dioxide production more than oxygen uptake, and the respiratory quotient therefore declines under these conditions. Decreasing oxygen concentrations delays the onset of ethylene production in a ventilated system in which low ethylene concentration in the atmosphere are maintained, but not in a product-generated atmosphere where ethylene can accumulate in the room. Decreasing oxygen concentration also decrease the rates of chlorophyll loss and softening and the associated release of soluble cell- wall component, such as uronides.

CA may affect respiratory metabolism via effects on glycolysis, fermentative metabolism, the tricarboxylic acid (TCA) cycle or the electron-transport chain, but the extent to which respiration is affected directly or whether other cellular processed whose energy demand affects respiratory rates are slowed is uncertain. Severe stresses resulting from injurious levels of either oxygen or carbon dioxide induce accumulation of acetaldehyde, ethanol and ethyl acetate. In addition, succinate accumulation is a feature of fruit treated with high carbon dioxide concentrations, probably because of inhibition of succinate dehydrogenase under these conditions.

Ethylene

CA storage cab also affect ethylene biosynthesis and action, and therefore the effects of CA are not solely due to effects of these atmospheres on respiration. Ethylene production is reduced by as much as 50 per cent in a 3 per cent oxygen atmosphere. It has been reported that low oxygen concentrations impeded ethylene binding, but the primary effects of low oxygen are likely to be directly on ethylene biosynthesis, since ACC oxidase, a key enzyme in ethylene biosynthesis, has an absoluter oxygen requirement for activity.

The action of carbon dioxide concentrations on ethylene biosynthesis is less clear because:

1. It is difficult to separate the direct effects of elevated carbon dioxide from those of low oxygen concentrations on ethylene biosynthesis.
2. Solubilization of carbon dioxide produces H^+ and HCO_3, which may affect the pH of the cytoplasm and consequently enzyme activities.
3. Maximal activity of ACC oxidase in vitro requires carbon dioxide and it is uncertain to what extent enzyme activity is regulated in vitro by the gas. It has been proposed that carbon dioxide displaced ethylene from ethylene receptor sites, but carbon dioxide effects are separable from those associated with ethylene perception.

In apples, low oxygen concentrations inhibit only the ACC-to-ethylene step, while high carbon dioxide also inhibits formation of ACC. The fruit have lower internal ethylene concentrations and ACC accumulation in a 2.5 per cent oxygen atmosphere than in air. Patterns of delayed and decreased ethylene production of preclimacteric apple fruit in 2 per cent oxygen, 2 per cent oxygen plus 5 per cent carbon dioxide of air plus 5 per cent carbon dioxide, compared with air alone, have been associated with reduced expression of genes encoding ACC synthase and ACC oxidase and of ACC-synthase activity ACC-oxidase activity in air plus 5 per cent carbon dioxide, however, was not different from that in air alone. Primary effects of carbon dioxide appeared to be on ACC-synthase transcription and on accumulation of active ACC-oxidase protein. Fungistatic levels of 20 per cent carbon dioxide or 0.25 per cent oxygen on preclimacteric and climacteric apples have also indicated that ACC- synthase transcript abundance and enzyme activity may be a key step in the inhibition of ethylene production.

Factors that Affect Tolerances of Fruit to CA

1. Two factors affect the tolerance of apple fruit to low oxygen and elevated carbon dioxide concentrations:
2. The resistance of the skin to gas exchange between the atmospheres outside and inside the fruit.
3. Metabolic sensitivity to the gases.

The effect of these factors is illustrated by comparing the effects of oxygen on ethanol concentration (used as an indicator of fermentation) in the fruit of 'Delicious' and two 'McIntosh' strains. The 'Delicious' apple is very tolerant to low oxygen, whereas the 'Marshall McIntosh' accumulates ethanol at higher external oxygen concentrations. 'Red Max' is somewhat intermediate between the others. The skin resistance of the 'Marshall' strain is much greater than that of the 'Red Max' strain, and thus the internal oxygen concentrations in the two strains will be similar at external oxygen concentrations of 2 per cent and 0.8 per cent, respectively. However, differences in tissue sensitivity to low oxygen concentrations between strains also exist, as, even at the same internal oxygen concentrations, ethanol accumulations very.

The effects of oxygen and carbon dioxide can interact, with fruit sensitivity to low oxygen increasing as carbon dioxide concentrations increase. Sensitivity to both gases is increased by lower storage temperatures and is dependent on cultivar and, within cultivars, on factors such as harvest date. Injuries associated with oxygen and elevated carbon dioxide.

Recommendations for CA Storage

The recommended gas composition and temperature conditions for CA storage are specific to cultivar, growing region and sophistication of the equipment available for monitoring and controlling the atmospheres. The wide range of recommended atmospheres, even for the same cultivar, reflects these factors, as well as different storage strategies employed in different growing regions. Thus, a CA recommendation for a cultivar may not apply everywhere. Regulations on CA storage, covering both the safe operation and the use of the legal definition of 'controlled atmosphere' for stored apples may exist and differ in different states or countries. These regulations include the rate of establishment of CA conditions, the maximum concentration of oxygen permitted and the minimum length of time the fruit can be in CA.

CA technology has undergoes major changes in the last 30 years. Old protocols, where CA rooms were loaded over 8-10 days and where fruit respiration then lowered oxygen to 2.5-3 per cent over an additional 15-20 days, are obsolete. 'Rapid CA' in which the oxygen concentration in CA rooms is reduced to less than 2.5 per cent by flushing the atmosphere with nitrogen within 4-7 days from the time of harvest to when CA conditions are applied, is now standard practice for many industries. Some cultivars must be cooled before rapid CA to avoid injury.

The use of rapid CA is critical for long-term (> 6 months) storage. Longer periods between harvest and CA storage can still result in a superior product to that obtained in air storage, but the benefits of CA storage for fruit quality may be lost over tine. Even when rooms are filled over extended periods, oxygen concentrations in the CA rooms are usually lowered by nitrogen flushing. It is becoming more common to open rooms briefly to remove some of the fruit for marketing, and nitrogen flushing is used to re-establish atmospheres in resealed rooms.

CA storage regimes fall into one of three categories, depending on the sophistication of equipment and technology involved.

Standard CA

Standard CA involves application of conservative atmosphere conditions with a minimum risk of gas-related injuries. Carbon dioxide concentrations are typically in the 2-3 per cent range and provide additional quality benefits to those obtained by using 2-3 per cent oxygen alone. Control of these atmospheres may be manual by daily reading and adjustment or via computer-controlled equipment. The margin of safety is large enough when oxygen concentrations are around 2 per cent for fluctuations in gas concentrations found in manually adjusted storage rooms not to cause fruit injury.

Low-Oxygen CA Storage

In low-oxygen CA storage, fruit are kept at oxygen concentrations below 2 per cent but above the concentrations at which fermentation will occur. In Europe, the term ultra-low oxygen (ULO) is used to describe these atmospheres. The safe oxygen concentration varies by cultivar and growing region. 'Delicious' apples from British Columbia, Canada, for example, can be stored safely at 0.7 per cent oxygen concentrations, allowing control of superficial scald without the use of DPA. Fruit of the same cultivar from other growing regions may show injury when stored at these low oxygen levels. Strains within a cultivar can also vary in sensitivity. In general, it is necessary to increase storage temperatures when low-oxygen CA storage is used. The carbon dioxide concentrations under low-oxygen storage are usually much lower than under standard CA, because additional benefits of carbon dioxide for firmness are not observed.

Low-Ethylene CA Storage

Low-ethylene ($<1\mu$ 11^{-1}) CA storage has been evaluated as a method for reducing superficial scald, as a method for reducing superficial scald, as a safe substitute for low-oxygen CA storage and for retarding flesh softening and other forms of senescence. Ethylene is removed by absorption on to potassium permanganate, by oxidation or by catalytic burners.

Results of this technology for fruit quality have been variable, as low- ethylene CA storage does not maintain quality if concentrations of ethylene gas within the fruit cannot be controlled, it is possible to have high concentrations of ethylene within the fruit, with consequent effects on ripening, even when concentrations in the storage atmosphere are low. Therefore, the technology is limited to naturally low-ethylene-producing cultivars such as 'Empire', and/or to early-harvested fruit to ensure that only preclimacteric fruit are stored. Loss of firmness, but not colour, may be delayed by low-ethylene CA storage. The return on investment in this technology for maintaining fruit quality has not been adequate. However, reduced superficial-scald incidence by lowering ethylene, even when concentrations are above 1μ 11$^{-1,}$ in CA storage could be useful if DPA is no longer available to the apple industry.

CA technologies, as well as associated marketing strategies, will continue to develop. For example, decreasing periods of acceptability for air-stored fruit, especially of softer cultivars, such as 'Cortland' and 'McIntosh', are resulting in increasing use of short-term CA. this type of storage is occurring even though storage time of fruit under CA does not meet regulatory requirements for labeling the fruit as being 'CA-stored'. CA can also be used during shipping, either using modules that maintain atmospheres in ships' holds or by use of CA-fitted containers. Apples are a relatively low-value commodity, however, and the utilization of new and existing technologies is influenced greatly by cost-effectiveness. In some regions, sophisticated technologies are not required to meet market demands, whereas, in others, investment in these technologies is the cost of staying in the business.

Modified-atmosphere (MA) storage, in which atmospheres around the fruit are modified passively by polymeric-film bags, has been tested for both transport and wholesale purposes, but have been used on only a limited scale commercially. The major impediments to their use include additional cost of the films, as well as those associated with the logistical requirements of training the packers, sealing of bags and preventing damage to bags.

Stress Treatments

There has been increasing interest in short-term stress treatments, such as low oxygen and high carbon dioxide concentrations that are outside the normal range for storage, and heat, for maintenance of quality and disease control and to meet quarantine requirements. This research has been prompted, for example, by the search for non-chemical substitutes for chemicals, such as DPA and fungicides. These treatments, which may be applied immediately after harvest or intermittently during storage, are suitable only if the tissue recovers without damage after removal from storage.

Low-Oxygen Stress

initial low-stress treatments of 0-0.5 per cent oxygen for 5-10 days control the incidence of superficial scald and the losses in firmness and colour of 'Granny Smith'. Inhibition of superficial scald development by initial oxygen stress has also been shown for other cultivars, and it may be possible to devise treatment strategies appropriate to the cultivar and length of storage period. The limitation to commercial adoption of initial low-oxygen stress treatments may be seasonal variations in cultivar response, and the associated risk of fermentation.

Heat Treatments

Heat treatments, applied either as hot air (*e.g.* 40-50°C for up to an hour), inhibit softening of apples and delay development of superficial scald during storage. However, heat treatments can accelerate the rate of degreening. While hot water treatments may be suitable for quarantine treatments, commercial implementation for apple fruit may be hindered by the relatively small margin of safe temperatures that do not result in damage, expressed as skin browning and internal breakdown.

Physiological and Pathological Disorders

Physiological Disorders

Given the extended storage periods and the many cultivars that exist, it is not surprising that a wide variety of physiological disorders have been identified in apple fruit. Susceptibility to disorders varies by cultivar, preharvest factors and post-harvest conditions.

Disorder can be considered in three categories:

1. Disorders that develop only on the tree. The most important of these is water core, in which intercellular air spaces in the core and cortical tissues become filled with liquid, predominantly sorbitol. Usually the occurrence of water core is associated with advancing fruit maturity and low night temperatures prior to harvest, but a variant of the disorders can occur as a result of heat stress. In Fuji, its occurrence and type of development can be affected by growing region. Presence of water core in fruit at harvest creates problems in certain cultivars, such as Delicious, because fruit with moderate or severe water core can later develop breakdown during storage. Therefore, occurrence of water core in delicious is an indicator of the end of the harvest period. In contrast, water core is desirable in the Fuji apple because of the sweetness it imparts to the fruit, and mild or moderate water core does not appear to result in the development of breakdown. Grade standards for Fuji have recently been modified so that water core is not a grade defect for the cultivar in the USA or Canada
2. Disorders that develop on the tree and during storage. Bitter pit is a disorder characterized by development of discrete pitting of the cortical flesh, the pits being brown and becoming desiccated with time. The pits may occur predominantly near the surface or deep in the cortical tissue. An associated disorder, known as lenticel's blotch, is also observed in some cultivars, such as Braeburn and Delicious. Sun-scald another disorders, occurs in fruit on the tree, but browning blackening of the skin develops in storage.
3. Disorders that develop during storage. These can be divided into senescent breakdown disorders, chilling disorders and disorders associated with inappropriate atmospheres during storage. The most common disorders associated with temperature and atmospheres are superficial scald, soft scald, low-temperature breakdown, brown core, internal browning, low oxygen injury and high carbon dioxide. Several other disorders have been described. Because of their commercial importance, factors associated with the development of common physiological disorders are briefly summarized.

Bitter Pit

The incidence and severity of bitter pit are affected by cultivar, but within a cultivar bitter pit is related to harvest date and climate. In susceptible cultivars, harvest of less mature fruit can result in higher bitter pit incidence, as can excessive pruning or high temperatures and/or water stress conditions during the growing season. Influences of climatic conditions are at least partly related to their effects on calcium concentrations in the fruit.

Development of bitter pit during storage results in financial loss, and a number of strategies have been employed to prevent its occurrence. Preharvest calcium sprays are commonly applied to reduce bitter pit development. Methods to predict bitter pit susceptibility risk based on mineral concentrations at harvest or infusion of magnesium have been developed. Rapid cooling and CA storage may also reduce its development during storage, but most focus has been on post-harvest application of calcium to fruit. Calcium drenches are commonly applied to bitter pit susceptible cultivars and in the USA are routinely applied with DPA.Vacuum infiltration of fruit with calcium was used commercially in New Zealand for several years but has now been phased out. Neither drenches nor vacuum infiltration will totally control bitter pit in highly susceptible apples, and good preharvest management practices that reduce risk of subsequent disorder development should be stressed Recommended rates for pre and post-harvest application of calcium vary by cultivar and region.

Senescent Breakdown

Senescent breakdown incidence is related to the harvesting of over mature fruit and/or fruit with low calcium concentration. It can be exacerbated by storing fruit at higher than optimal temperatures. Fruit that are susceptible to breakdown because of low calcium are commonly drenched with calcium salts before storage, but the incidence of senescent breakdown can also be reduced by harvesting fruit at a less mature stage, rapid cooling and reducing the duration of storage. Increased senescent-breakdown incidence was shown to be related to the climacteric in one study, but not in others. However, because the occurrence of breakdown is highly dependent on calcium concentrations in the fruit, the ability to defect relationships between the disorder and the climacteric may be confounded. A fruit with high calcium concentrations may never develop senescent breakdown, while one with low calcium concentrations may show increased disorder incidence with later harvest dates.

Superficial Scald (Syn. Storage Scald)

Superficial scald is a physiological disorder associated with long-term storage. It was the major cause of apple-fruit loss until the advent of Post-harvest treatment with the antioxidant DPA. Cultivar, climate and harvest date affect susceptibility of fruit to the disorder.

DPA is usually applied with a fungicide to reduce decay incidence, and calcium salts may also be included at the same time to reduce bitter pit or senescent breakdown. Both DPA use and DPA residues on imported fruit are prohibited in some countries. Another antioxidant, ethoxyquin, is no longer permitted for use on apples.

Alternative ways of controlling superficial scald are being investigated, and storage operators are reducing the use of DPA where possible. Low concentrations of oxygen in CA storage reduce the risk of scald developing and may also permit the use of lower DPA concentrations. Low-oxygen and low-ethylene CA storage also reduce scald incidence. In British Columbia, Canada, 0.7 per cent oxygen storage is used as a substitute for DPA treatment. This technique cannot be used universally because fruit grown in other regions may be susceptible to low-oxygen injury.

Low-Temperature Disorders: Soft Scald, Low-Temperature Breakdown, Brown Core and Internal Browning

Soft scald is characterized by irregular but sharply defined areas of soft, light brown tissue, which may extend into the cortex. Susceptibility of fruit to soft scald is cultivar and climate-related, and disorder incidence may be aggravated by harvesting over mature fruit and delays between harvest and cooling. Storing fruit at 3°C rather than at lower temperatures can sometimes control the disorder, and DPA, used for the control of superficial scald, may also reduce the incidence of soft scald. Storage at a lower temperature following prompt cooling can reduce the incidence of soft scald on 'Golden Delicious' fruit.

Low-temperature breakdown, brown core and internal browning are affected by cultivar sensitivity to low temperatures and generally these disorders increase in incidence and severity as the length of storage is increased. Climate affects the sensitivity of fruit to the disorders, with more problems occurring after colder, wetter growing seasons. Low-temperature breakdown is characterized by markedly brown vascular bundles, browning of flesh and a clear halo of unaffected tissue underneath the skin. In contrast to senescent breakdown, the affected tissues are more likely to be firmer, more moist and darker in colour. Brown core (syn. Core flush) involves browning of the flesh, initially in the core area and later in the cortex, where it becomes difficult to distinguish from low-temperature breakdown. Internal browning does not involve the breakdown of the flesh but rather a graying of the flesh apparent when the apple is cut. Internal browning and core flush are often associated with higher concentrations of carbon dioxide, since both can occur in CA when the carbon dioxide concentration is higher than of oxygen.

Low-oxygen injury affects fruit in a number of ways. The first indication of injury is loss of flavour, followed by fermentation odours. These odours may disappear if storage problems are identified soon enough and severe injury has not occurred. Injury symptoms range from purpling or browning of the skin in red-coloured cultivars, development of brown soft patches resembling soft scald and abnormal softening and splitting of fruit. As discussed earlier, cultivars vary greatly in response to low oxygen, and susceptibility to injury is influenced by number of pre and Post-harvest factors.

Carbon dioxide injury may be external or internal. The external form consists of wrinkled, depressed colourless or coloured areas restricted to the skin surface and usually on the greener side of the fruit. Internal forms are expressed as brown heart and/or cavities in the flesh. Occurring of the disorders is cultivar-specific, and generally external carbon dioxide injury is associated with early harvest, while that of internal injury is associated with later harvest. The growing region also affects susceptibility of fruit to carbon dioxide injury.

Susceptibility to carbon dioxide injury has long been a problem for certain apple cultivars. It is the basis, for example, for the commercial CA recommendations for 'McIntosh', in which carbon dioxide levels in the storage atmosphere are kept to 2-3 per cent for the first 4-6 weeks and then allowed to increase to 5 per cent. Several newer apple cultivars, including 'Empire', 'Fuji' and 'Braeburn', appear to be susceptible to carbon dioxide, resulting in commercial losses. Carbon dioxide injuries have occurred in air-stored fruit under conditions where carbon dioxide can accumulate (*e.g.* warm fruit in sealed cartons).

At least in the case of 'Empire', losses due to carbon dioxide injury have occurred when DPA usage has been stopped because the cultivar is not susceptible to superficial scald. It was not previously recognized that DPA reduced the incidence of both external and internal carbon dioxide injuries. These observations highlight the point that chemicals that are technically applied for one reason may also have other effects that may not be appreciated until the chemical is withdrawn from use. Non-chemical strategies to reduce carbon dioxide injury include acclimatization of fruit in air before CA storage, if this can be done without loss of quality, and initially maintaining low carbon dioxide concentrations in the storage environment.

Pathological Disorders

The main Post-harvest diseases of apples that develop in storage are blue mould, caused by Penicillium species, and grey mould, caused by Botrytis cinerea. Mucor pyriformis Fischer can cause severe losses in pears but is less common in apples. Botrytis and Penicillium species enter fruit primarily through cuts, stem punctures and bruises, but Penicillium expansum Link can invade some apple cultivars via the stem during long-term CA storage. Other pathogens of stored apples include the Colletotrichum species that cause bitter rot, the Botryosphaeria species that cause black rot and white rot and Pezicula malicorticis (Jackson) Nannf., the cause of bull's-eye rot. Decays caused by Collectotrichum, Botryosphaerias and Pezicula are initiated in the field and must be controlled by using fungicides or other disease-management strategies during the growing season.

Blue mould and grey mould are usually controlled during long-term storage by Post-harvest application of the benzimidazole fungicide, thiabendazole (2-(4-thiazolyl) benzi-midazole (TBZ), applied in combination with DPA. TBZ may be applied a second time as a line spray or in wax as apples are packed. Benzimidazole-resistant strains of *P. expansum* and *B. cinerea* were discovered in apple storage during the mid-to late 1970s, but TBZ plus DPA continued to control blue and grey moulds, because most benzimidazole-resistant strains of the pathogen were highly sensitive tom DPA. During the mid-1990s, the incidence of blue mould bergan to increase in some apple packing-houses where the predominant strains of *P. expansum* had developed resistance to the benzimidazole-DPA combination. Grey mould is still controlled by the benzimidazole-DPA combination, presumably because this pathogen does not recycle on field bins as readily as does *P. expansum*, and it has therefore been subjected to less selection pressure for fungicide resistance. Captan (N-trichloromethylthio-4-cyclohexene-1,2-dicarbondioxide; Captan 50w, Captan 80w, Captan 4L) has a post-harvest registration but has proved to be only moderately effective for controlling *P. expansum* and *B. cinerea*. Captan residues are not acceptable in some markets.

Much effort has been devoted to the development of biocontrols for Post-harvest diseases of apples, but, while many of the biocontrol agents selected and developed to date have proved to be very effective in controlled tests, commercialization of biocontrols has been slow. Reasons may include limited markets compared with field crops, liabilities associated with the value of the stored crop, a high public profile for apples in debates relating to food-safety tissues and difficulties in devising shelf-stable formulations of biocontrol agents. Moreover, biocontrols generally cannot currently provide eradicant activity against established infections. Using combinations of biocontrols and reduced rates of TBZ may be more effective than using either product alone. When such combinations are used, the chemical fungicide may provide eradicant and short-term protestant activity necessary to prevent decay until the biocontrol agents become established in wounds or other infection sites.

Regardless of the Post-harvest fungicide or biocontrol, good sanitation will remain essential for reducing inoculum on contaminated field bins and in packing-houses and storage rooms. Badly contaminated bins should be cleaned and disinfested (steam-cleaned) before they are reused for a new crop. Plastic bins may carry less inoculum than wooden bins and also have the advantage of reducing bruising and abrasion where apples contact the sides of bins. Careful fruit handling, rapid cooling after harvest and storage at recommended temperatures also help to limit the development of Post-harvest decay.

Processing

Apples are processed into various products such as juice, concentrate, vinegar, sauce, butter, preserve, candy, jam, jellies, and canned products. Apples are also dried as rings, chops, or cubes. They are also used for making fermented beverage such as cider and wine. The waste from the apple processing industry, such as peel, core, or pomace, can be utilized for production of pectin and various edible products.

Juice

Apple juice is a popular drink. In earlier times, apple juice in the form of cider was a seasonal treat, but now it ranks a distant second to orange juice in fruit juice consumption. Apple juice contains a considerable proportion of the soluble components of the original apples, such as sugars, acids, and various other carbohydrates. Malic acid is the predominant acid in apple juice. Several distinct forms of apple juice available in the market include clarified apple, natural apple juice, pulpy apple juice, and apple juice blends with other juices/extracts.

Clarified Apple Juice

Preparation of clarified apple juice involves grating and pressing the apples, clarification with pectinol enzymes, filtration, and packaging. Traditional packaging involves pasteurization at 80-88°C, then filling and hermetically sealing the juice in glass containers or metal cans. This aseptic process, with the product packed in laminated flexible containers, has been successfully introduced in many countries of the world.

Natural Apple Juice

The characteristics of a natural apple juice are considered to be very close to the juice which comes directly from the press. Commercially, this is accomplished through the addition of ascorbic acid or through heating the pressed juice to flocculate unstable compounds. Ascorbic acid helps in preserving the very light color of the juice by reserving the oxidation of juice constituents. The juice is then immediately pasteurized to inactivate the oxidizing enzymes occurring naturally in apple juice.

Another process for making natural apple juice utilizes heat to flocculent the unstable compounds in the juice. In this process, the juice from the press is heated to 95-97°C to include flocculation, followed by cooling to 18-20°C until bottling. Plate or tubular heat exchangers are used. The juice is centrifuged to remove the flocculent and nonsoluble solids and heated to 88°C, a lower temperature than the initial heating, and bottled.

This juice from both the processes is higher in viscosity than clarified apple juice. The product from the ascorbic acid process is bright cream to bright yellow in color, with a stable suspension of small solid particles, and may have light sediment. The heat-treated products is light in color and may have a slight haze. For best utilization of apple hybrids, *viz.*, Ambrich, Ambred, and Red Delicious X Ambri-51.

Pulpy Apple Juice

Pulpy (crushed) apple juice has a light color and a high pulp content of fine cells. In its production, washed apples are coarsely ground and passed through a pulper with a fine screen. The pulped juice is then deaerated by passing it through a vacuum chamber, which helps in minimizing oxidation, and then homogenized, pasteurized at 88°C, and filled into containers. It is a continuous process with very little time elapsing between the grinding of the apples and final sealing of the containers.

Apple Juice Blends

Apple juice and apple juice concentrates are used as the base4 for blended fruit juices and fruit juice drinks. Apple-cranberry and apple-pear are favorite blends. Combinations of apple and tropical fruits are available, as are blends with citrus juices. Several of these blends are sold as frozen concentrate as well as in single-strength forms. Apple juice blends with citrus juice and ginger extract has been developed as an apple appetizer. Efforts to improve the nutritional qualities of apple juice by blending with either egg yolk or soya bean proteins have also been successfully made.

Concentrate

Apple juice and other fluid foods are concentrated in order to reduce their volume and weight, which results in lower costs of packaging, storage, and transportation. The principal methods applicable to apple juice concentration include evaporation, reserve osmosis, and freezing. In preparation of apple juice concentrate, the clarified juice is concentrated to sixfold and the concentrate is cut back to fourfold (42°Brix) with fresh juice. However, concentrate prepared by stripping of the juice of voltaic flavor constituents prior to concentration and then adding back the voltaic flavor constituents to the concentrated juice was reported to be superior in flavor and aroma. Prepared concentrates are frozen and stored at -18°C. further, the apple juice concentrate prepared with adding back of voltaic flavor constituents was found to be stable without any loss in sensory qualities for approximately 2 years, 1 year, and 2-4 months at -18°C, -12°C and -6.6°C, respectively. The clear juice is passed through a filter press using diatomaceous earth as a filter aid to ensure complete removal of small particles. The filtered juice is pasteurized at 80-87°C for 30 s in flash pasteurizers. The hot juice is filled into sterile bottles and sealed. The juice is aloes canned is lacquered enamel cans, and sometimes it is also fortified with vitamin C.

Canned Apples

Canning of apple rings is not practiced commercially due to some inherited problems such as the presence of high volume of gases (29., 5 per cent) in the fruit tissues, difficulty of their removal during exhausting, less drained weight, mashy texture, *etc.* there are a few reports pertaining to canning of apple slices in which firming agents such as calcium chloride for the improvement of texture have been tried on a laboratory scale.

Canned apples, which are usually available in larger-size cans, are generally used in pies. The varieties commonly employed for canning are Yellow Newton, Pippin, Spitzenberg, Winesap, Baldwin, Russet, Jonathan, Delicious, and Rome Beauty. The fruits are first washed in warm dilute hydrochloric acid to remove any lead or arsenic spray residue and then rinsed in cold water. They are then peeled by hand or by machine and cut into slices, 0.31-0.63 cm thick. The slices are placed in 2-3 per cent common salt solution to prevent their darkening due to enzyme action. They are then blanched at 71-80°C for 3-4 min in plain boiling water or in 3 per cent boiling brine. Blanching is essential to remove oxygen from the tissues and thus prevent pinholing in the cans during storage. The blanched slices are filled into cans, covered with either hot water or dilute sugar syrup, exhausted and processed. Pinholing of cans during prolonged storage, especially in warmer climates, is a serious problem in the case of canned apples. In an effort to develop a relatively new technique of osmocanning of apple rings with the application of osmosis, Pretreatment of apple rings in 70 per cent sugar solution at 50°C for half an hour prior to canning improved the physicochemical and sensory characteristics of the canned product. The application of osmotic technique resulted in products of desired drained weight, color, appearance, texture, and sugar/acid blend in comparison to those canned by using conventional canning technology.

Frozen Products

For freezing, apple slices, after treating with 3 per cent brine solution, are subjected to vacuum to remove air, which is responsible for enzymatic browning. They are reimmersed in salt solution, washed, and filled with sugar in a proportion of 4:1. Alternatively, apple slices are frozen by subjecting them to a high vacuum, treating with salt solution, blanching the brined slices in free-flowing steam, cooling in water, and packing in slipover cans. Slices can also be prepared for freezing by immersing them in 0.2 per cent SO_2 solution or in bisulfite solution containing citric acid for 1 min. the slices are kept under refrigeration for several hours to allow proper penetration of SO_2 into the slices, which are then filled into slipover cans with sugar (5:1) and frozen at 6.0°C or below.

Dried Products

Apples can be preserved by drying. The peeled and cored apples are prepared as rings, segments, chops, or cubes and treated with a weak solution or citric acid and a bisulfate dip. The latter provides SO_2, which inhibits enzymatic browning. The sulfured slices are dried at 60-70°C for 6-8 h. Different apple varieties can be dried as rings and found that Golden Delicious to be best with respect to yield, appreance, and taste. Among different treatments, a 2500 ppm SO_2, 1-h dip of apple rings resulted in best dried product on sun drying and dehydration as well as after 180 days storage. The dried products are packed in moisture proof containers. A freeze-dried product based on apple and milk can be prepared by using 50 per cent apple, 3.5 per cent each of sugar and lemon juice, and milk in various proportions (0-43 per cent).

Cider

Alcoholic beverage from apples is generally called cider or wine, depending on the alcohol content in the final product. There are two types of apple ciders, dry and sweet ciders. The fruits are crushed or grated and juice is obtained by hydraulic press, then sugar is added to the juice to raise the brix to 22°. In addition, 100 ppm of SO_2 and a pure culture of wine yeast, Saccharomyces Cerevisae strain ellipsoideus, are added. After fermentation, at 20-25°C, the cider is racked and filtered. Before bottling, the cider is made sparkling clear. During the aging process, most of the suspended material settles down, leaving a major portion of the liquid clear. The fermented liquid is further clarified by using bentonite, casein, gelatin, or filtering through pulp filters. After aging and clarification, the cider is pasteurized to prevent spoilage. A process for making cider from apple juice concentrate (72° Brix) has also been standardized. Must prepared by direct dilution of concentrate to 20o Brix gave cider of better chemical and sensory quality. The addition of pectinolytic enzyme to the must improved the fermentability and made available the minerals and increased the color appeal of the product.

Vinegar

Vinegar made from fermented apple juice by acetic acid fermentation is called apple cider vinegar or cider vinegar. Apples are grated and pressed to get juice. Even after this, the pomace contains a large percentage of juice, which is rather difficult to extract. To extract this residual juice, the pomace is ground finely, and actively fermenting cider is added in order to promote yeast fermentation. The pomace is allowed to ferment for 2-3 days and then pressed. By this method, a larger yield of juice is obtained than by simple grinding and pressing. The juice extracted by this method is of inferior quality and is used for production of cider vinegar. Apple juice is fermented with wine yeast. When fermentation is complete., the yeast and fruit pulp settle to form a compact mass at the bottom of the tank, from which fermented liquid is separated. The clear liquid is stored in vessels. The acetic acid fermentation is brought about by acetic acid bacteria (*Acetobacter* sp.) for acetic acid fermentation, the fermented liquid is adjusted to 7-8 per cent alcohol content. Mother vinegar containing acetic acid bacteria is then added to hasten the process and inhibit the growth of undersirable microorganisms. The vinegar is prepared by the Orins slow process or the German quick process. Once the process is complete, the fermented liquid is allowed to age to improve the flavor. Acetic acid may also react with alcohol.

Other Products

Apple Butter

Apple butter is similar to apple jam except that it is made from finely sieved apple pulp to which small quantities of spices consisting of nutmeg, cinnamon, clove. *etc.*, are added. The pulp: sugar ratio is generally 1:3/4. On account of its mild, spicy taste and flavor, apple butter is popular among a large number of consumers. Butter can be prpared with apple of excellent organoleptic qualities with 30.2 per cent moisture, 13.6 per cent sugar by mixing unsalted cream butter with apple puree (25 per cent), skim milk (10 per cent), and granulated sugar (8 per cent)

Chutney

To make chutney, apple slices are cooked along with other ingredients (sugar, salt, and spices) until they become thick. The product is bottled hot.

Apple Sauce

Apple sauce is made from peeled, cored, and sliced apple which are cooked in stream and passed through a pulper. The pulp is mixed with sugar, spices, and heated under stream at about 85°C. Acetic acid is added to adjust the acidity in the product. the hot mixture is filled into glass bottles and then heat processed and cooled prior to storage.

Pickles

Pickles are made by adding apple slices to a boiling mixture of vinegar, sugar, and spices and continuing boiling for 5 min. The mass is then simmered until the pieces become soft. The product is then packed into jars. The vinegar and sugar mixture is reboiled to a syrupy consistency and poured on the slices and filled in the jar. If desired, spices are also added to the jars.

Jam and Jelly

In the production of apple jam, good-quality fruits are selected and washed in cold water. The fruits are peeled and the skin and seeds are removed. The peeled fruits are cut into small pieces. The fruit pieces are cooled and crushed with a paddle and made into a fine pulp by sieving. to 1 kg of pulp, an equal quality of sugar and 2.5 g of citric acid are added and the mixture is mixed thoroughly. The mixture is cooked slowly with occasional stirring until it passed a sheeting or drop test. The final weight of jam is in the range of 1.5 times the sugar added. The hot jam is filled into clean glass jars. Similarly, apple jelly can be prepared from apple using apple juice or apple pectin extract obtained by boiling unpeeled apple pieces in water for 25-30 min and filtering through muslin cloth.

Preserves

Apples for preserves are peeled without removing the stem, pricked with a fork, and kept in 2-3 per cent sodium chloride solution to prevent browning. This is transferred to 2 per cent lime water and kept there for some time. Alum solution and a pinch of sodium bisulfite are added to bleach the colour. Fruits are blanched for 2-3 min. An equal quality of sugar is required for good- quality apple preserves. Apples are placed in layers of sugar (half quality only) in a vessel and left undisturbed for 24h. During this period, sugar absorbs the water and syrup may be formed. The mass is heated to boiling for a minute and sugar is added to raise the total soluble solids to 59-60°Brix. A small quantity of citric acid is added and the mass is boiled for about 5-8 min and then kept undisturbed for another 24h. On the third day, the strength of the sugar syrup is raised to 70°Brix and the product is allowed to stand for a week.

Baked Apple Product

Symmetrical- shaped firm apples of Rhode Island, Northwest, Gravenstein and stayman, winesap varieties are used for making baked apple products. Apples are washed and cored. Cans are filled with two or three apples, and then spiced and acidified hot syrup (40-50° Brix) is poured into the can at about 71° C. baking of apples occurs in the cans during processing,which takes about 30 min in boiling water.

Waste Utilization

Flavor Compounds

Peels and cores from apple canneries and apple dries can be utilized to produce vinegar and for jelly juice stock.Apple pomace obtained after extraction of juice can be used to produce natural flavoring compounds. These compounds can be obtained by extracting with liquid CO_2, which is fractionated at two different temperatures to obtain a flavorless fraction. This procedure gave a broader flavor spectrum than did those prepared by distillation. Apple processing waste can be used as fuel source or animal feed.

Pectin

Pectin can be obtained from apple processing waste. To obtain pectin, the dried pomace is boiled in water for half an hour. Protopectin is hydrolyzed to pectin by heating and acid hydrolysis and extracted by the alcohol extraction method. The extract obtained is filtered, bottled, may be pasteurized as such or may be spray dried to 5 per cent moisture and used after dispersion with water as an additive. Low-methoxyl pectin is produced by treating a solution of pectin with pectin methyl esterase, which removes a methyl group from the ester unit of galacturonic acid. Low – methoxyl pectin forms gels in the presence of a comparatively low concentration of soluble solids and high pH (6.5) if a calcium source is present. Extraction and evaluation of different apple cultivars for pectin has been reported. It was found that, of four different apple varieties tried, Golden Delicious pomace was promising with respect to pectin yield, jelly grade, and other qualities. The harvesting period of cultivars did have noticeable influence on the jelly grade.

Animal Feed

Apple pomace can be used as animal feed by feeding either as fresh or as dried pomace. However, pregnant cows fed with apple pomace supplemented with nonprotein nitrogen have been found to give birth to dead or weak calves.

Citric Acid

Citric acid can be produced from apple pomace by growing *Aspergillus niger* under controlled conditions. More than 250g of citric per kilogram of pomace solids can be produced.

Charcoal

Charcoal briquents can be prepared from pomace by heating the dried apple pomace at 160-200°C followed by grinding the pyrolyzate to pass a 40-mesh sieve and molding the particles. Apple pomace charcoal can also be used for water purification in place of commercial charcoal.

Microbial Biomass Production

Protein-enriched product can be obtained by using batch and fed-batch processes utilizing *Saccharomycopsis lipolytica* and *Trichoderma reesei*. The product can be used for cattle feeding. Dried apple pomace can also be used for the preparation of some bakery products.

REFERENCES

Adams, D. O. and Yang, S. F. 1979. Ethylene biosynthesis: identification of 1-aminocyclopropane-1-carboxylic acid as an intermediate in the conversion of methionine. *Proceedings of the National Academy of Sciences USA* 76, 170-174.

Alleyne, V. and Hagenmaier, R.D. 2000. Candelilla-shellac: an alternative formulation for coating apples. *HortScience* 35, 691-693.

Amiot, M., S. Aubert, J. Nicholas, P. Goupy, and P. Aparicio. 1992. Phenolic composition and susceptibility of various apple and pear cultivars at maturity, *Bull, Liaison Groupe Polyphenols 16:* 48.

Autio, W.R. and Bramlage, W.J. 1982. Effects of AVG on maturation, ripening, and storage of apples. *Journal of the American Society for Horticultural Science* 107, 1074-1077.

Bartsch, J. A. and Blanpied, G. D. 1990. *Refrigeration and Controlled Atmosphere Storage for Horticultural Crops.* Bulletin 22, Northeast Regional Agricultural Engineering Service (NRAES), 45 pp.

Beaudry, R., Schwallier, P. and Lennington, M. 1993. Apple maturity prediction: an extension tool to aid fruit storage decisions. *HortTechnology* 3, 233-239.

Bhardwaj, J. C., and B. B. Lal. 1990. A study on drying behaviour of rings from different apple cvs. Of Himachal Pradesh, *J. Food Sci. Technol. 27*(3): 144.

Bishop, D. 1996. *Controlled Atmosphere Storage. A Practical Guide.* David Bishop Design Consultants, Heathfield, UK, 58 pp.

Blankenship, S. M. and Unrath, C.R. 1987. Use of ethylene production for harvest-date prediction of apples for immediate fresh market. *HortScience* 22, 1298-1300.

Blanpied, G. D. 1974. A study of indices for earliest acceptable harvest of 'Delicious' apples. *Journal of the American Society for Horticultural Science* 99, 537-539.

Blanpied, G. D. 1986. A study of the relationship between fruit internal ethylene concentration at harvest and post-harvest quality of cv. Empire apples. *Journal of Hortcultural Science* 62, 465-470.

Blanpied, G. D. 1990. Controlled atmosphere storage of apples and pears. In: Calderon, M. and Barkai-Golan, R. (eds) *Food Preservation by Modified Atmospheres.* CRC Press, Boca Raton, Florida, pp. 265-299.

Blanpied, G. D. and Jozwiak, Z. 1993. A study of some orchard and storage factors that influence the oxygen threshold for ethanol accumulation in stored apples. **In**: *Proceedings of the Sixth International Controlled Atmosphere Research Conference, Vol.1. cornell University, Ithaca, New York*, pp. 78-86.

Blanpied, G. D. and Silsby, K. J. 1992. *Predicting Harvest Date Windows for Apple.* Information Bulletin 221, Cornell Cooperative Extension Publication, Ithaca, New York, 12 pp.

Blanpied, G. D., Johnson, D, S., Lau, O. L., Lidster, E. C. and Porritt, S. W. 1999. Apples. **In**: Lidster, P. D., Blanpied, G. D. and Prange, R. K. (eds) *Controlled atmosphere Disorders of Commercial Fruit and Vegetables.* Publication 1847E, Agriculture *Canada, Ottawa*, pp. 7-22.

Bown, A. W. 1985. CO_2 and intracellular pH. *Plant Cell and Environment* 8, 459-465.

Bramlage, W.J.and Meir, S. 1990. Overview of chilling injury of horticultural crops. **In**: Wang, C. Y. (ed.) *Chilling Injury of Horticultural Crops.* CRC Press, Boca Raton, Florida, pp. 37-49.

Brookfield, P., Murphy, P., Harker, R. and MacRae, E. 1997. Starch degradation and starch pattern indices: interpretation and relationship to maturity. *Post-harvest Biology and Technology* 11, 23-30.

Burmeister, D. M. and Dilley, D. R. 1995. A scald-like controlled atmosphere storage disorder of Empire apples-a chilling injury induced by CO_2. *Post-harvest Biology and technology* 6, 1-7.

Burda, S., W. Oleszek, and C. Y. Lee. 1980. Phenolic compounds and their changes in apples during maturation and cold storage, *J. Agr. Food Chem. 38*: 945.

Bump. V. L., Apple pressing and juice extraction, *Processed Apple Products* (D. L. Downing, ed.), AVI. Varr Nostrand Reinhold, New York, 1981, p. 53.

Cavalieri, R. P., Stecker, T. D. and Fellman, J. K. 1988. *Reduced Pyridine Nucleotides as a Measure of Fruit Maturity*. Paper No. 88-6565, Amrican Society of Engineers, St Louis, Michigan.

Chang, Y. L., and L. R. Mattick, 1989. Composition and Nutritive value of apple products, *Processed Apple Products* (D. L. Downing, ed.). AVI, Van Nostrand Reinhold, New York, p. 303.

Chu, C. L. 1988. Internal ethylene concentration of 'Mclntosh', 'Northern Spy', 'Empire', 'Mutsu', and 'IdaRed' apples during the harvest season. *Journal of the American Society for Horticultural Science* 113, 226-229.

Chauhan, S. K., B. B. Lal, and V.K. Joshi 1993. Development of protein rich apple beverage, *Res. and Ind.* 38: 227.

Couey, H. M. and Olsen, K. L. .1975. Storage response of 'Golden Delicious' apples after high-carbon dioxide treatment. *Journal of the American Society for Horticultural Science* 100, 148-150.

DeLong, J. M., Prange, R. K., Harrison, P. A., Schofied, R. A. and DeEII, J. R. 1999. Using the Streif Index as a final harvest window for controlled atmosphere storage of apples. *HortScience* 34, 1251-1255.

Dilley, D. R. 1980. Assessing fruit maturity and ripening techniques to delay ripening in storage. In: *110th Annual Report.* Michigan State Horticultural Society, Morrice, pp. 132-146.

Downing, D. L. 1989. Apple cider, *Processed apple Products.* AVI, New York, p. 169.

Elgar, H.J., Watkins, C.B. and Lallu, N. 1999. Harvest date and crop load effects on a carbon dioxide related storage injury of 'Braeburn' apple. *HortScience* 34, 305-309.

Elgar, H. J., Burmeister, D. M. And Watkins, C. B. 1998. Storage and handling effects on a carbon dioxide related internal browning disorder of 'Braeburn' apples. *Hort-Science* 33, 719-722.

Emonger, V.E., Murr, D.P. and Lougheed, E. C. (1994) Preharvest factors that predispose apples to superficial scald. *Post-harvest Biology and Technology* 4, 289-300.

Fan, X., Matthesis, J.P and Blankenship, S. M. 1999b. Development of apple superficial scald, soft scald, core flush, and greasiness is reduced by MCP. *Journal of Agricultural and Food Chemistry* 47, 3063-3068.

Fan. X., and Mattheis, J. P. 1999. Impact of 1-methylcyclopropene and methyl jasmonate on apple volatile production. *Journal of Agricultural and food Chemistry 47*, 2847-2853.

Faragher, J.D., Brohier, R.L, little, C.R. and Peggie, I.D. 1984. Measurement and prediction of harvest maturity of jonathan apples for storage. *Australian Journal of Experimental Agriculture and Animal Husbandry* 24, 290-296.

Ferguson, I.B and Watkins, C.B. 1989. bitter pit in apple fruit. *Horticultural Reviews* 11, 289-355.

Fernandez-Trujillo, J. P., Nock, J.F. and Watkins, C.B. 2001. Superficial scald, carbondioxide injury, and changes of fermentation products and organic acids in 'Cortland' and 'Law Rome' apple fruit after high carbon dioxide stress treatment. *Journal of the American Society for Horticultural Science* 126, 235-241.

Fidler, J.C. .1973. Conditions of storage. **In**: Fiddler, J.C. Wilkinson, B.G., Edney, K.L.and Sharples, R.O. (eds) *The biology of Apple and Pear Storage*. Common wealth Bureau of Horticultural and Plantation Crops, East Malling, UK, pp.1-61.

Gebhardt, S. E., R. Cutrufelli, and R. H. Matthews. 1982. Composition of foods. Fruits and fruit juices, *U. S. Dept. Agr. Bull. 8*

Gorny, J.R. and Kader, A.A. 1996a. Regulation of ethylene biosythesis in climacteric apple fruit by elevated CO_2 and reduced O_2 atmosphere. *Post-harvest Biology and Technology* 9, 311-323.

Gorny J.R. and Kader, A.A 1996b. Controlled-atmosphere suppression of ACC synthase and ACC oxidase in Golden delicious' apples during long-term cold storage.*Journal of the American Society for Horticultural Science* 121, 751-755.

Gorny, J.R. and Kader, A.A. 1997. Low oxygen and elevated carbon dioxide atmospheres inhibit ethylene biosynthesis in preclimacteric and climacteric apple fruit. *Journal of the American Society for Horticultural; Science* 122, 542-546.

Gussaman, C.D., Goffredz, J.C. and Gianfagana, T.J. 1993. Ethylene production and fruit-softening rates in several apples fruit ripening variants. *HortScience* 28, 135-137

Harker, F.R., Redgwell, R.J., Hallett, I.C., Murray, S.H. and Carter, G. 1997. Texture of fresh fruit.*Horticultural Reviews* 20, 121-224.

Hardenburg, R. E. and Spalding, D. H. 1972. Post-harvest benomyl and thiabendazole treatments, alone and with scald inhibitors, to control blue and gray mould in wounded apples. *Journal of the American Society for Horticultural Science 97,* 154-158.

Hewett, E.W. and Thompson, C.J. 1989. Modified atmospheres during storage and transport for bitter pit reduction in 'Cox's Orange pippin' apple. *New Zealand Journal of Crop and Horticultural Science* 17, 275-282.

Hewwett, E.W. and Watkins, C.B. 1991. Bitter pit control by sprays and vacuum infiltration of calcium in 'Cox's Orange pippin' apples. *Hortscience* 26, 284-286.

Hours, R. A., A. E. Massucco, and R. G. Ertola. 1985. Microbial biomass product from apple pomace in batch and fed batch cultivars, *Appl. Microbial. Biotechnol. 23*(1): 33.

Hulme, A.C. 1956. Carbon dioxide and the presence of succinic acid in apples. *Nature* 178, 218-219.

Ingle, M. and D'Souza, M. C. 1989. Physiology and control of superficial scald: a review. *HortScience* 24, 28-31.

Janisiewicz, W.J. 1998. Biocontrol of Post-harvest diseases of temperature fruits challenges and opportunities. In: boland, J. and Kuykendall, L. D. (eds) *Plant-*

Microbe Interactions and Bbiological Control. Marcel Dekker, New York, pp.171-198.

Joshi, V. K., B. B. Lal, and R. Sharma. 1988. A study of preparation of apple cubes, IFCON, Central Food Technological Research Institute, Mysore, India, abstr. FRD 32.

Joshi, V. K., B. B. Lal, and K. L. Kakkar. 1990. Updating the technique of apple chops making and its utilization, *Beverage and Food World 16*: 21).

Joshi, V. K., D. K. Sandhu, B. L. Attri, and R. K. Walia. 1991. Cider preparation from apple juice concentrate and its consumer acceptability, *Indian J. Hort.48*(4): 321.

Jobling, J., Mc Glasson W.B. and Dilley, D.R. 1991. Induction of ethylene synthesizing competency in Granny Smith apples by to low temperature in air. *Post-harvest Biology and Technology* 1, 111-118.

John, P. 1997 Ethylene biosynthesis: the role of 1- aminocyclopropane-1-carboxylate (ACC) oxidase, and its possible evolutionary origin. *Physiologia Plantarum* 100, 585-592.

Johnson, D.S. and Ertan, U. 1983. Interaction of temperature and oxygen level on the respiration rate and storage quality of Idared apples. *Journal of Horticultural Science* 58, 527-533.

Kaushal, B. B., and J. C. Anand 1986. Recent trials on grading and packaging of apples, *Indian Food Packer 40*(2): 29.

Kays, S.J. 1997. *Post-harvest Physiology of Perishable Plant Products*. Exon Press, Athens, Georgia, 532pp.

Klein, J.D. 1987. Relationship of harvest date, storage conditions, and characteristics to bruise susceptibility of apple. *Journal of the American Society for Horticultural Science* 112, 113-118.

Knee, M. 1993. Pome fruits. In: Seymour, C.B., J.E. and Tucker, G.A. (eds) *Biochemistry of Fruit Ripening*. Chapman and Hall, London, pp.325-346.

Knee, M and Hatfield, S.G.S. 1981. Benefits of ethylene removal during apple storage. *Annals of Applied Biology* 98, 157-165.

Knee. M., Looney, N.E., Hatfield, S.G.S. and Smith, S.M 1983. Initiation of rapid ethylene synthesis by apple and pear fruits in relation to storage temperature. *Journal of Experimental Botany* 34, 1207-1212.

Knee, M., Hatfield, S.G.S. and Smith, S.M 1989. Evaluation of various indicators of maturity for harvest of apple fruit intended for long- term storage.*Journal of Horticultural Science* 64, 403-411.

Knee, M. 1980. Physiological responses of apple fruits to oxygen concentrations. *Annals of Applied Biology* 96, 243-253.

Kupferman, E. 1994. Report to the industry on fruit quality and packing practices for Washington grown apples. *Trees Fruit Post-harvest Journal* 5 (2), 3-26.

Larrigaudiere, C., Pinto, E. and Vendrell, M. 1996. Differential effects of ethephon and seniphos on color development of 'Strarking Delicious' apple. *Journal of the American Society for Horticultural Science* 121, 746-750

Larrigaudiere, C., Graell, J. Salar, and Vendrell, M. 1997. Cultivar difference in the influences of a short period of cold storage on ethylene biosynthesis in apples. *Post-harvest biology and Technology* 10, 21-27.

Lau, O.L 1985. Harvest indices for BC apples. *British Columbia Orchardist* 7(7), 1A-20A.

Lau, O. L. 1997. The effectiveness of 0.7 per cent O_2 to attenuate scald symptoms in 'Delicious' apples is influenced by harvest maturity and cultivar strain. *Journal of the American Society for Horticultural Science* 122, 691-697.

Lal, B. B., R. S. Rana, H. L. Kochhar, T. R. Chandha, and S. B. Maini., 1988. Packaging and transportation of apple-A study on commercial aspects, *Production and Conservation of Forestry* (P. K. Khosla, D. K. Khurana, and Atal, eds.), Indian Society of Tree Scientists, Solan, India, p. 226.

Li, Z., Liu, Dong, J., Xu, R. and Zhu, M. 1983. Effect of low oxygen and high carbon dioxide on the levels of ethylene and 1-aminocyclopropane-1-carboxylc acid in ripening apple fruits. *Journal of Plant Growth Regulation* 2, 81-87.

Lal, B. B., V. K. Joshi, P. C. Sharma, and R. Sharma. 1992. Development of apples based appetizers. Golden Jubilee national Semina on Emerging Trends in Temperate Fruit Production in India, held at Y. S. Parmar, UHF, Nauni, Solan, abstr. 127.

Little, C.R. and Holmes, R.J. 2000. *Storage Technology for Apples and Pears*. Department of Natural Resources and Environment, Knoxfied, Victoria, Australia, 528 pp.

Little, C.R., Faragher, J.D. and Taylor, H.J. 1982. Effects of initial oxygen stress treatments in low oxygen modified atmosphere storage of 'Granny Smith' apples. *Journal of the American Society for Horticultural Science* 107, 320-323.

Lal, G., G. S. Siddappa, and G. L. Tondon. 1986. *Preservation of Fruits and Vegetables*, Indian Council for Agricultural Research, New Delhi.,

Liu, F.W. and King, M.M. 1978. Consumer evaluations of 'Mclntosh' apple firmness. *HortScience* 13, 162-163.

Lurie, S. 1998. Post-harvest heat treatments. *Post-harvest Biology and Technology* 14, 257-269.

Luton, M.T. and Hamoer, P.J.C. 1983. Predicting the optimum harvest dates for apples using temperature and full-bloom records. *Journal of Horticultural Science* 58, 37-44.

Maini, S.B., B. Diwan, B.B. Lal, and J.C. Anand. 1982. Packaging transport and storage of apples in wooden containers, *Indian Food Packer* 36(3): 34).

Marlow, G.C. and Loescher, W.H. (1984) Watercore. *Horticultural Reviews* 6, 189-251.

Marmo, C.A., Bramlage, W.J. and Weis, S.A. 1985. Effects of fruit maturity, size and mineral concentrations on predicting the storage life of 'Mclntosh' apples. *Journal of the American Society for Horticultural Science* 110, 499-502.

Massey, L.M. 1989. Harvesting, storing and handling apples. *Processed Apple Products* (D. L. Downing, ed.), AVI, Van Nostrand Reinhold, New York,

Mattick, K. R., and J. C. Moyer. 1983. Composition of apple juice, *J. Assoc. Off. Anal. Chem. 66: 1251*

Nicholas, J. J., F. C. Richard-Forget, P. M. 1994. Goupy. M. J. Amiot, and S. Y. Aubert, Enzymatic browning reactions in apple and apple products, *Crit. Rev. Food Sci. Nutr. 34*: 109.

Park, Y.M., Blanpied, G.D., JOzwiak, Z. and Liu, F.W. 1993. Post-harvest studies of resistance to gas diffusionin Mclntosh apples. *Postharvst Biology and Technology* 2, 329-39.

Peirs, A., Lammeetyn, J., Ooms, K. and Nicolai, B.M. 2000. Prediction of the optimal picking date of different apple cultivars by means of VIS/NIR-spectroscopy. *Post-harvest Biology and Technology* 21, 189-199.

Plotto, A., Azarenko, A.N., Mattheis, J.P. and McDaniel, M.R. 1995. 'Gala', 'Braeburn', and 'Fuji' apples: maturity indices and quality after storage. *Fruit Varieties Journal* 49, 133-142.

Poovaiah, B.W., Glenn, G.M. and Reddy, A.S.N. 1988. Calcium and fruit softening: physiology and biochemistry. *Horticultural Reviews* 10, 107-152.

Peterson, D. L., and T.S. Kornecki. 1987. Mechanical apple harvester for T-trellis canopies, *Am. Soc. Agr. Eng*. 30(3): 597.

Reid, M.S., Padfield, C.A.S., Watkins, C.B. and Harman, J.E. 1982. Starch iodine pattern as a maturity index for Granny Smith apples. 1. Comparison with flesh firmness and soluble solids content. *New Zealand Journal of Agricultural Research* 25, 239-243.

Rosenberger, D. A. 1990. Post-harvest diseses. In: Jones, A. L. and Aldwinckle, H. S. (eds) *Compendium of Apple and Pear Disease.* APS Press, St Paul, Minnesota, pp. 53-54.

Rosenberger, D. A. and Meyer, F. W. 1985. Negatively correlated cross-resistance to diphenylamine in benomyl-resistant *Penicillium expansum. Phytopathology* 75, 74-79.

Rupasinghe, H. P. V., Murr, D. P., Paliyath, G. and Skog, L. 2000 Inhibitory effect of 1-MCP on ripening and superficial scald development in 'Mclntosh' and 'Delicious' apples. *Journal of Horticultural Science and Biotechnology* 75, 271-276.

Saftner, R.A., Conway, W.S. and Sams, C.E. 1998. Effects of post-harvest calcium and fruit coating treatments on post-harvest life, quality maintenance, and fruit-surface injury in 'Golden Delicious' apples. *Journal of the American Society for Horticultural Science* 123, 294-298.

Saltveit, M.E. and Morris, L.L. 1990. Chilling injury of crops of temperate origin. In: Wang, C.Y. (ed.) *Chilling injury of Horticultural Crops.* CRC Press, Boca Raton, Florida, pp.3-15.

Sharma, T. R., B. B. Lal, S. Kumar, and A. K. Goswami. 1985. Pectin from different varieties of Himachal Pradesh apples, *Indian Food Packer 39*(4): 53.

Sharma, R. C., V. K. Joshi, S. K. Chauhan, S. K. Chopra, and B. B. Lal 1991. Application of osmosis, osmocanning of apple rings, *J. Food Sci. Technol. 28*(2): 86.

Sisler, E.C., Serek, M. and Dupille, E. 1996. Comparison of cyclopropene, 1-methylcyclopropene and 3, 3- dimethylcyclopropene as ethylene antagonists in plants. *Plant Growth Regulation* 18, 169-174.

Snowdon, A.L. 1990. *A Color Atlas of Post-harvest Diseases and Disorders of Fruits and Vegetables.* Vol. 1, *General Introduction and Fruits.* CRC Press, Boca Raton, Florida, 302 pp.

Stover, E., Fargione, M. J., Watkins, C.B. and lungerman, K.I. 2003. Interactions of ReTain^{TM} (AVG) with ethephon and summer pruning on preharvest drop and fruit quality of Marshall 'Mclntosh' apples. *HortScience* 38 (in press).

Stow, J.R., Dover, C.J. and Genge, P.M. 2000. Control of ethylene biosynthesis and softening in 'Cox's Orange Pippin' apples during low-ethylene, low oxygen storage. *Post-harvest Biology and Technology* 18, 215-225.

Streif, J. 1983. Der optimale erntetermin beim apfel. I. Qualitatsentwicklung und reife. *Gartenbauwissenschaft* 48, 154-159.

Sharma, P. C. 1982. *Studies on packaging Systems and their effects on shelf-life and quality of apples,* M.Sc. thesis, Himachal Pradesh Krishi Vishvavidalaya, Palampur, India.,

Sunako, T., Sakuraba, W., Senda, M., Akada, S., Ishikawa, R., Niizeki, M. and Harada, T. 1999. An allele of the ripening-specific 1-aminocyclopropane-1-carboxylic acid synthase gene (ACSI) in apple fruit with a long storage life. *Plant Physiology* 119, 1297-1304.

Smith, K.C. and Lay-Yee, M 2000. Response of Royal Gala apples to hot water treatment for insect control. *Post-harvest Biology and Technology* 19, 111-122.

Thompson. J.P. 1985. Harvesting systems, *Post-harvest* Technology of Horticultural *Crops* (A.A.Kader, N. E. Sommr, J. F. Thompson, F.G. Mitchell, and M.S. Reid, eds.), Division of Agriculture and Natural Resources, University of California, Berkeley.,

Truter, A.B. and Hurndall, R, F. 1988. New findings on determining maturuity of 'Starking' 'Topred' and 'Starkrimson' apples. *Deciduous Fruit Grower* 38, 26-29.

Volz, R.K., Biasi, W.V. and Mitcham, E.J. 1998. Fermantation volatile production in relation to carbon dioxide-induced flesh browning in 'Fuji' apple. *HortScience* 33, 1231-1234.

Vyas, K. K., and V. K. Joshi 1982. Applegtone-A new fortified beverage from apple, *Indian FoodPacker 36*(3): 66.

Wang, Z. and Dilley, D.R. 2000. Initial low oxygen stress controls superficial scald of apples. *Post-harvest Biology and Technology* 18, 201-213.

Watkins, C.B. 1999. Maintaining firmness of apples: effects of packing, cooling, and transport. In: CA *Storage: Meeting the Market Requirements* Bulletin 136, Northeast Regional Agricultural Engineering Service (NRAES), Ithaca, New York, pp.65-71.

Watkins, C.B. 2002. Ethylene biosynthesis mode of action, consequences and control. In: Knee, M. (ed.) *Fruit Quality and its Biological Basis*. Sheffield Academic Press, Sheffield, pp. 180-224.

Watkins, C. B. and miller, W.B. 2003. A summary of physiological processes or disorders in fruits, vegetables and ornamental products that are delayed or decreased, increased, or unaffected by applications of 1-melhylcyclopropene (1-MCP).

http: / /www.hort.cornell.edu / department / faculty / watkins / ethylene / index.htm

Watkins, C.B., Hewett, E.W., Bateup, C., Gunson, A. and Triggs, C.M. 1989a. Relationships between maturity and storage disorders in 'Cox's Orange Pippin' apples as influenced by preharvest calcium or ethephon sprays. *New Zealand Journal of Crop and Horticultural Science* 17, 283-292.

Watkins, C.B., Bowen., J.H. and Walker, V.J. 1989b. Assessment of ethylene production by apple cultivars in relation to commercial harvest dates. *New Zealand Journal of Crop and Horticultural Science* 17, 327-333.

Watkins, C.B., Silsby, K.J. and Goffinet, M.C. 1997b. Controlled atmosphere and antioxidant effects on external CO_2 injury of 'Empire' apples. *HortScience* 32, 1242-1246.

Watkins, C.B., Nock, J.F. and Whitaker, B.D. 2000. Responses of early, mid and late season apple cultivars to post-harvest application of 1-methylcyclopropene(1-MCP) under air and controlled atmosphere storage conditions. *Post-harvest Biology and Technology* 19, 17-32.

Watkins, C.B., Kupferman, E. and Rosenberger, D.A. 2003. Apple: post-harvest quality maintenance guidelines. In: Gross, K., Wang, C.Y. and Saltveit, M.E. (eds.)*The Commercial Storage of Fruits, Vegetables, and Florist and Nursery Stocks*. Agriculture Handbook No. 66(revised), United States Department of Agriculture, Washington, DC (in press).

Wilkinson, B.G. and Fidler, J.C. 1973. Injuries caused by incorrect concentrations of carbon dioxide and or oxygen. In: Fidler, J.C., Wilkinson, B.G., Edney, K.L. and Sharples, R.O. (eds) *The Biology of Apple and Pear Storage.* Commonwealth Bureau of Horticultural and Plantation Crops, East Malling, UK, pp.81-87.

Watkins, C.B., Brookfield, P.L., Elgar, H.J. and Mcleod, S.P. 1997a. Development of a modified atmosphere package for export of apple fruit. In: *Proceedings of the International Congress on the uses of plastics in Agriculture*. Laser Pages Publishing, Jerusalem, Israel, pp.586-592.

Wills, R.B.H., Hopkirk, G. and Scott, K.J. 1981. Reduction of soft scald in apples with antioxidants *Journal of the American Society for Horticultural Science 106, 569-571.*

Young, C. T., and J. S. L. How. 1986. Composition and nutritive value of raw and processed fruits, *Commercial Fruit Processing. AVI,* New York.,

Yamaki, S. 1984. Isolation of vacuoles from immature apple fruit flesh and compartmentation of sugars, organic acids, phenolic compounds and amino-acids, *Plant Cell Physiol. 25*: 151.

Chapter 3

Apricot

Introduction

Apricot (*Prunus armeniaca* L.) is the species of *Prunus* classified with prunoidae sub family Rosaseae, family of Rosales order. The apricot is an attractive, delicious and highly nutritious fruit with pleasing flavor. Apricot fruit has a high content of beta-carotene which is converted to vitamin A in the body. Other vitamins and nutrients present in this fruit are vitamin C, iron and fiber contents. The apricot probably originated in central or western China, rather than in Armenia as its name suggests. The apricot requires cool weather to break dormancy and dry sunny weather or spring and a warm summer for fruit maturity. Hot weather with temperatures above 38°C, however, injures the fruit. It is cultivated in temperate climate of all the continents of the world. But Asia and Europe are the largest producers and the major apricot producing nations are China, USSR, Turkey, Italy, Spain, Greece, France, and Morocco. It is consumed fresh or in processed form- canned, dried or frozen. In India apricot is grown in J&K, Himachal Pradesh and Uttrakhand and to a limited extent in North Eastern Hills. In India it ranks second to plum among the stone fruits in area, production and popularity. It is a hardy, drought resistant rather less susceptible to pests and diseases. Apricots are the first deciduous fruit trees to produce flowers in the spring after almond and blooms approximately for a period of 1-2 weeks, depending upon the weather conditions and produce more flowers than are needed to ensure the production of adequate crop and often requires removal of small fruits (thinning) to ensure optimum fruit size. The market demand for high qualitative standards and having new attractive appearance and good flavour cultivars are needed to satisfy consumers demand for table use and as value added products. Apricot is a climacteric fruit with high respiratory and metabolic rate and highest ethylene emission. Rapid post-harvest softening is one of the major problems of apricot which limits its availability and marketing. Owing to these features, shelf life of apricot is extremely short

(1-2 weeks) at 0° C temperature and 90 per cent relative humidity. In most of the countries in addition to consumption of fresh apricots, there is huge demand of its dried products. Availability of fresh fruit in the market is from May end to August while dried fruits may be available round the year. Apricot fruit is used in many ways. Tree ripe fruit is excellent dessert fruit. Because of its perishable nature, the fruit is canned, candied, frozen and dried. The fruit is also processed in to number of products such as jams, nectar and fruit bars *etc.* which are appealing, nutritive and has excellent flavour and aroma. Apricots are also consumed as dry fruit. Drying is one of the oldest food processing methods known. Dried apricots are produced from plump and fully ripe fresh fruits. The essential feature of the drying is to reduce moisture content of the fruit below the level at which enzymatic and microbial damage at minimum. The special flavour and texture are winning over and new generation of consumers who uses dried apricot as snacks because of the flavour and nutritional value of the fruit.

Botany

The apricot belongs to the family Rosaeceae and most cultivated apricots belongs to the species *Prunus armeniaca* L. Closely related species include *P. mume* Sidb and Zucc., the Japanese apricot, *P. dasycarpa* Ehrh, the black apricot *P. brigantiaca* Vill, the *Briancon apricot* from the French Alps, *P. ansu* Komar, *P. sibirica* L. and *P. mandshurica* (maxim). The apricot is diploid ($2n = 16, x=8$). Flowers are borne singly or doubly at a node on very short stems (peduncles). Most commercial cultivars are self-fertile, but Perfection (Gold Beck) and Riland are examples of self-incompatible cultivars. Trees generally produce vigorous upright growth, but not as upright as plum. Floral initiation occurs in summer (late April-May) and most of the flowers which set into fruit are produced on spurs. Spurs are productive for 305 years. The highest-quality fruit is borne on younger spurs. Apricots are the first deciduous fruit tree to produce flowers in the spring after almond and because of that are subject to frost damage. Trees bloom over a period of approximately 1-2 weeks, depending on weather conditions. Flowering is followed by the appearance of leaves, which are simple, alternate and serrated, round-ovate to avota and sharp-pointed. Apricot cultivars available require approximately 300-1200 h of chilling temperatures (below 7.2°C). Vegetative buds require less chilling than reproductive (flower) buds. In these years with insufficient winter (December and January) chilling, bud drop can occur. Apricots produce more flowers than are needed to ensure the production of an adequate crop and often require the removal of small fruit (thinning) to assure adequate fruit size. The apricot is a stone fruit (drupe). Botanically, a drupe fruit is a fleshy, one-seeded fruit that does not split open by itself, with the seed enclosed in a stony endocarp, called a pit. The apricot fruit consist of a stony endocarp, a fleshy mesocarp and an outer exocarp (skin).

Cultivars Suitable for Table and Processing

Apricot cultivars and selections with promise for commercial production are described briefly, listing general characteristics each cultivar.

Exotic Varieties

Goldrich

The fruits are large, bright, attractive orange, hang well at maturity and do not split after heavy rains. Flesh quality is fair.

Harcot

An early apricot with an attractive red blush. It has good fruit quality markets and direct fruit sales, but unsuitable for shipping or processing. Harcot has a sweet kernel.

Harglow

This cultivar blooms slightly later than most commercial apricots. The fruits are medium in size and bright solid orange in colour. The fruit is firm and flavourful. This cultivar is best suited for the fresh market.

Hargrand

The fruits are dull orange in colour and the size is excellent when the crop has been thinned. The flesh is juicy and tasty. Hargrand is suitable for the fresh market.

Harlayne

Ripens late in July and has proven to be very hardy at Harrow. The fruit have a red blush and are bright and attractive in the basket. This cultivar requires careful thinning to attain size. It is best suited to the fresh market but is also suitable for home processing.

Harval

A late-maturing cultivar. It has large, attractive fruit that are bright orange with a red blush.

Veecot

Mid-season variety bearing highly attractive, smooth finish and a deep, dark orange ground colour. It hangs well at maturity, but because of its intense colour, must be picked carefully for optimum maturity. Bacterial spot can be a problem for some early years, but tree health is excellent. Recommended for the fresh market and home canning.

Blenheim

This is an old variety, dating back more than 160 years in England. It comprises nearly 30 per cent of the apricots grown in California and is used for canning and drying. Blenheim bloom is relatively late and heavy. The flesh was light orange, firm, juicy, and aromatic with a nice balance of sugar and acid. Flavor was very good to excellent. Pre-harvest drop may be severe some years.

Chinese

A variety from the South Haven Experiment Station in Michigan. Flower buds are reported above average in cold hardiness. Trees bloom in mid-season and produce only light to moderate amounts of flowers. Harvest season is early. Ten to 50 per cent of the fruit surface was covered with a red-orange blush with red spots over an orange background. The suture was fairly pronounced, the suture ratio was 1.07, and the length: diameter ratio was 1.04. Attractiveness was rated fair. The flesh was light orange, soft, and fairly dry. Flavor was similar to a plum and was rated good.

Perfection

This early-blooming variety blooms heavily. Harvest season is early. The flesh is soft and juicy and the mild flavor rated fair.

Wilson Delicious

Trees have moderately heavy bloom during the mid-season. The orange flesh is soft, dry, aromatic, and sweet with some acid. Flavor rated good.

Tilton

The second leading variety grown for canning, drying, and fresh market. Trees bloomed heavy and late. Fruits were moderately attractive. The flesh is fairly dry, mealy, and aromatic, and the flavor was sweet and rated good.

Goldcot

Trees bloom early and very heavy and fruit ripens in mid-season. The flesh is moderately soft and dry, with a mild, almost bland flavor that rated fair.

New Castle

The fruit of the apricot tree resembles a small, yellow peach and is used for drying, desserts, preserves and canning. They can grow up to 25 feet in height and require full sun and medium water with good drainage. Brown rot, fruit bark beetle, peach borer, plum curculio and San Jose scale can all be a problem. This cultivar is best suited for southern California.

Nugget Apricot

Large, flavorful yellow freestone. Attractive orange skin blushed with red. Vigorous, productive tree. Originated in Ontario, CA. Introduced in 1956. Chilling requirement 500 hours or less.

Royal Rosa Apricot

Extremely vigorous - more disease tolerant than other apricots. Bears young and heavy. Especially nice fruit: sweet, low acid, fine flavor. Very early harvest (late May in Central CA). Excellent backyard apricot.

In India, apricot is grown on mid hills to high hills having variable climatic condition. Varieties which are suitable for mid hills are not suitable for high hills or dry temperature region. About 100 varieties of cultivated apricot are available in India. Most of them are of exotic origin. The promising varieties recommended for different regions are given in Table 3.1.

Figure 3.1a: Apricot Varieties being Grown at CITH Srinagar, J&K.

Figure 3.1b: Apricot Varieties being Grown at CITH Srinagar, J&K.

Apricot Varieties Suitable for Drying

Variety CITH-AP-2, CITH-AP-3, Afghani, Turkey and Erani were found good for dehydration purpose and stored maximum with good quality, colour, texture and flavour.

Table 3.1: Apricot Varieties Recommended for Cultivation in different States of India

Himachal Pradesh	
Mid hills	New Castle, Early Shipley and Shakarpara
High hills	Kaisha, Nugget, Royal, Suffaida, Charmagaz and Nari
Dry temperate	Charmagaz, Suffaida, Shakarpara and Kaisha
Uttar Pradesh	Charmgaz, Kaisha, Moorpark, Turkey, Ambroise, Early Shipley, Chaubattia Alankar, Chaubattia Madhu, ChaubattiaKesri and Bebeco
Jammu and Kashmir	
Ladakh	Halman, Rakchakarpa, Tokpopa, Margulam, Narmu and Khante
Kashmir	Turkey, Australian, Charmagaz, Rogan and Shakarapara
New promising varieties for mid hills	Early maturing - Baiti,Beladi
	Late-maturing - Farmingdale, Alfred

Table 3.2: Apricot Varieties Released from ICAR-CITH, Srinagar, J&K

Variety	*Description*	*Photos*
CITH Apricot-1	Fruits are bigger in size (50-60g), round in shape, orange in color with reddish coloration on one side (25-30 per cent), high yielder (30-35 t/ha), low acidity, high TSS (14°Brix), suitable for table use and also for processing.	
CITH Apricot-2	Fruits are yellowish orange in colour, medium in size (40-50g), round in shape, low acidity, high TSS (14°Brix) and high yielding (25-30 t/ha). Suitable for table use and also for processing.	
CITH Apricot-3	Fruits are very attractive with bright colour (30-40 per cent area of fruit with orange back ground), medium in size (30-40g), low acidity, high TSS (16°Brix) and good yielder (20-25 t/ha), suitable for desert use.	

Afghani

Early sweet and juicy variety flowers during second to 4th week of March and ripens in first week of June. Fruit size is medium and colour of fruits is yellowish green, kernel taste sweet. Suitable for table purpose and drying.

Turkey

Fruit weighs about 24 round and creamy in colour, soft and free pit variety. Tastes sweet and suitable for drying. TSS 21 Brix.

Errani

Fruit weighs about 41 g, roundish, frim textured, TSS 16.0 B, good for during, having good TSS/acidity blend, rich in carotene.

Ladakhi Apricot and their Value Addition

Apricot (locally known as "Chulli") is one of the most nutritive delicious and commercially important fruit crops of Ladakh. It has a wide range of distribution in different parts of Ladakh with particularly abundance in Sharm areas (lower ladakh) including Dha-hanu, Garkhon, Skurbuchan, Domkhar, Wania, Khaltse and Timosgang. Apricot in Ladakh is believed to have been introduced a century back either from china or Central Asia. Since then, apricot has become one of the most preferred and commercially cultivated fruit crops of Ladakh and has become an integral part of the people there.

Apricot being a unique tolerant and highly stable plant can grow exuberantly in wide range of jagged sandy sill having low nutrient and moisture content in the cold desert of Ladakh. Luxuriously adapted in the extreme environment here in Ladakh, the apricot tree can attain a height of about 4-7m bearing heart shaped leaves and produces flowers in spring and fruits in summer. With the onset of breezy spring these trees overcome the long terrible winter dormancy and start producing young healthy leaf buds and by the month of April-May they produce beautiful white or pinkish flowers that not only ensures the continuity of their population but also give a unique look to the sandy desert of trans-Himalaya. By the month of august-September, they start producing yellow-orange, rounded r oval shaped fruits that are juicy, sweet taste with peculiar floavour.

Figure 3.2: Apricot Oil from Chuli Apricots in Leh.

There are many varieties of apricot grown in Ladakh, which differ from one another in taste (sweet, bitter, sour0, Size, shape and physical appearance. Some of these varieties include Halman, laktse-karpo, Safaida, Khanteh *etc.* Halman and Laktse-karpo are the most preferred one for commercial purpose. Both the fruit and kernel of apricot is believed to be highly nutritive and consumed as either fresh or dried. They are known to possess a good amount of vitamin-A, vitamin-C, potassium, calcium, iron, carbohydrate, amino acids and sugars. Apricot has been consumed by the people of Ladakh for decades. It has become an integral part of the traditional culture of people here. Local people serve dried or fresh apricot as an excellent dessert, particularly on traditional festival occasions. During the chilly winters, when people prefer to remain indoor dried apricot fruits make an excellent eatable that compensates the long cold entire, especially for children who use to fill their pocket with dried fruits and enjoy themselves.

Figure 3.3: Apricot Cultivation in Leh.

In entire ladakh, a farmer practices one of the best and oldest methods of fruit preservation and storage by open sun drying. The local people, particularly women and children, collect the fully ripe apricot in a large traditional basket 9locally known as Tsepo) and wash them under running water to remove the dusts, and then spread on the roof lop from drying under open sun light. The fruit are dried either as whole fruit (locally known as fating) or seed are separated before drying, and the dried fruit without seed are called Chuli skampo. During the sun drying process, the fruit losses its natural colour and turns dark brown. This is the major drawback of traditional method of sun drying in Ladakh. However at present the Ladakhi farmers have adapted several improved methods of drying including treatment with sulphurdoxide and use of polyhouse apricot drier. These methods are believed to reduce or prevent the browning of fruits.

From the commercial point of view, apricot has been the major source of income for many Ladakhis who are engaged in cultivation and marketing of this fruit. Halman and Laktse-karpo are the two prime varieties that have a good demand in the market and are profitably sold @ Rs 200-250 per kilogram. Besides the kernel is consumed as dry fruit and make a good market by locals. The seed with sweet kernel is consumed as dry fruit and make a good market price of Rs 100-150/kg while the seed with bitter kernel are used for oil extraction. The apricot oil 9locally called tseghumar) is multipurpose oil with a pecular apricot flavor and is sold at a remarkable price. Traditionally, the oil is extracted from the semi-roasted kernels by crushing them in a large wooden mortar (locally termed as Thom), followed by heating and compressing them with few drops of water on a flat stone (called as Tsigg). Besides, several other products such as apricot jam, squash, jelly and cake are being produced for commercial purposes.

This is worth mentioning here that Ladakh is one of the major producer of apricot in India, but at present almost 90 per cent of the fresh apricot being produced in this cold and arid region go waste and its market value stands abysmal. The prime reason for such debacle is the lack of proper network for processing and supplying apricot products in Ladakh and elsewhere in India. The local apricot growers though have the knowledge of cultivation and drying but they are devoid of any modern technical skills for proper preservation, storage, transportation and marketing of apricot. This results in a huge loss not only to the poor farmers but also to the economy of Ladakh general.

Composition and Uses

Apricot fruit has a high content of beta-carotene; beta-carotene which is found in apricots is converted to vitamin A in the body. Other Nutritional contained in fruits are vitamin C iron and fiber.

Table 3.3: Nutritional Composition of Apricot

Nutrient	*Unit*	*Value per 100g*
Proximates		
Water	g	86.35
Energy	kcal	48
Protein	g	1.48
Total lipid (fat)	g	0.39
Carbohydrate by difference	g	11.12
Fiber, total dietary	g	2.0
Sugars, total	g	9.24
Minerals		
Calcium, Ca	mg	13
Iron, Fa	mg	0.39
Magnesium, Mg	mg	10
Phosphorus, P	mg	23
Potassium, K	mg	259
Sodium, Na	mg	1
Zinc, Zn	mg	0.20
Vitamins		
Vitamin C, total ascorbic acid	mg	10.0
Thiamin	mg	0.030
Riboflavin	mg	0.040
Niacin	mg	0.600
Vitamin B-6	mg	0.054
Folate, DFE	µg	9
Vitamin B-12	µg	0.00
Vitamin A, RAE	µg	96
Vitamin A, IU	IU	1926

Health Benefits

Anemia

Anemia is the condition where the blood has lower concentration of red blood cells (RBCs) or RBCs having below average amount of hemoglobin. This condition is most commonly caused by iron deficiency. The presence of iron in apricot makes it an excellent food for anemia sufferers. The small but essential amount of copper in the fruit makes the iron available to the body. Liberal consumption of apricot can increase the production of hemoglobin in the body. This is ideal for women after their menstrual cycle, especially those with heavy flow.

Constipation

Apricots are rich in fiber and hence are good for smooth bowel movements. It is often recommended to patients who are regularly suffering from constipation due to its laxative natures. The cellulose and pectin content in apricot is a gentle laxative and are effective in the treatment of constipation. Cell membrane helps the intestines to work properly, and the latter helps in absorbing water in intestine. The pectin absorbs and retains water, thereby increasing bulk to stools, aiding in smooth bowel movement. Hence, consumption of 6 to 8 apricots per day should help to control constipation. Fiber also binds to cancer-causing chemicals in the colon, preventing them from damaging colon.This may be one reason why diets high in fiber-rich foods, such as apricots, are associated with a reduced risk of colon cancer.

Digestion

Apricot helps to clear entire acidity and digestion problem. Consumption of an apricot before meal to aid digestion, as it contains alkaline reaction in the digestive system.

Eyes/Vision

The large amount of vitamin A is essential to maintain or improve eye sight. Insufficiency of this vitamin can cause night blindness and impair sight. Apricot is highly recommended for eye because it contains Vitamin A which helps in improving vision for the person.

Fever

Apricot juice is given to the patients suffering from fever as it provides necessary vitamins, minerals, calories and water to the body. Some people also use steamed apricot during fevers. Blend some honey and apricots with some mineral water and drink to cool down fever. It quenches the thirst and eliminates the waste products effectively from body. Tone up eyes, stomach, liver, heart and nerves by supplying vitamins and minerals.

Skin Disease

Apricot oil is best for skin care. It quickly absorbs the skin and does not keep the skin oily when applied. Apricot not only useful for maintaining the skin smooth and shiny, also aids in treating number of skin diseases. It also binds the bile salts in the colon and efficiently flushes them out of your system.

Heart Disease

Fiber lowers high cholesterol levels, people at risk for atherosclerosis or diabetic heart disease. Since bile salts are made from cholesterol, the body must break down more cholesterol to make more bile, a substance that is also necessary for digestion. The end result is a lowering of cholesterol levels.eat the apricot whole for its precious fiber that is highly beneficial for your colon health.

Kidney Stones

Person with containing calcium oxalate as kidney stones should not consume too much of apricot fruit, because it contains small amount of oxalate and hence, it will cause severe damages to the kidney.

Asthma Problems

Sulfur compounds such as sulfur dioxide are rich in dried apricots. Patients who suffer from asthma strictly must not eat this fruit as the chemical compounds present in this may causes severe reactions in your body.

Respiratory Disease

Due to Excessive intake of amygdalin present inside the apricot fruit can cause cramps, vomiting, respiratory distress and in severe cases, nervous system depression and fatal respiratory failure. So persons who suffer from respiratory and nervous disease should avoid this fruit.

Diarrhea

Diarrhea can be caused due to infections, antibiotics, or teething effects in children. Loose motion will occur frequently. In this case, stool becomes watery. If one loose too much water, might suffer from dehydration. Diarrhea kills more children than malaria, AIDS and TB. In fact, diarrhea is the second leading cause of death in children under 5 years of age.

Other Uses for Apricots

Seed kernels can be pressed for apricot seed oil after grinding. The oil can be used in high quality soap (Amandine soap) and toiletries manufacture or as an ingredient in human and feed animal foods (as 'press-cake'). The aromatic essences can be extracted and used in foods and toiletries. Seeds or kernels of the apricot grown in central Asia and around the Mediterranean are so sweet that they may be substituted for almonds. The Italian liqueur Amaretto and amaretti biscotti are flavoured with extract of apricot kernels rather than almonds. Oil pressed from these cultivars has been used as cooking oil.

Seed hulls can be ground and used in:

- ✰ Toiletries (exfoliants)
- ✰ Abrasives in commercial applications: filler for polymers, in polishing-creams, abrasive for cleaning delicate wood and metal surfaces, dental abrasive
- ✰ Can be burned at high heat and used as fuel or activated charcoal.

Fruit Development

A double-sigmoid growth curve is characteristic of stone fruit including apricots. Phase 1 of the growth curve results from enlargement of all parts of the ovary with exception of the endosperm and embryo. Lignifications of the endocarp take place during phase II and growth is confined mainly to the endosperm and embryo. It is during phase III that expansion of the monocarp (edible portion) leading to the mature fruit is resumed. The induction of parthenocarpy by gibberellic acid (GA) has been reported in apricot.

Quality Characteristics and Criteria

Fruit size, shape, freedom from defects (including gel breakdown and pit burn), and freedom from decay are all important quality criteria. High consumer acceptance is attained for fruit with high (>10 per cent) SSC and moderate TA (0.7 to 1.0 per cent). Apricots with 2 to 3 lb-force (8.9 to 13.3 N) flesh firmness are ready to eat. Most cultivars soften very fast, making them susceptible to bruising and subsequent decay.

Maturity and Harvesting

In apricots, harvest date is determined by changes in skin ground color from green to yellow. The exact yellowish-green color depends on cultivar and shipping distance. Apricots should be picked when still firm because of their high bruising susceptibility when fully ripe and soft.

Apricot fruit attains a characteristic flavor if it is allowed to ripen on the tree. As the fruit is left on the tree longer, the soluble solids content increases, acids decrease, firmness decreases, and consumer acceptance increases. Gradual loss in soluble and insoluble proteins during ripening also takes place. Apricots used for drying are harvested fully ripened. Those used for canning should be less mature and firm enough to be pitted and processed. Fruit to be frozen is harvested riper than for canning, *i.e.*, before it is firm enough to undergo heat processing.

Since apricots do not mature uniformly, selective harvesting involving two or more pickings is desirable. Selected trees with the highest proportion of mature fruit may be harvested completely. Apricots are always harvested by hand, usually into picking bags or plastic totes.

Figure 3.4: Maturity and Ripeness Stages of Apricot.

Grades, Sizes, and Packaging

Apricots are generally handled in half bins or totes and hand-packed. In some cases, apricots are dry-dumped onto a padded packing line belt. Apricots are tray-packed in single and double layers or volume-filled (about 10 kg net). Apricots should be uniform in size, and not more than 5 per cent in each container may vary more than 6 mm when measured at the widest part of the cross section.

Optimum Storage Conditions

Apricots are seldom stored in large quantities, though they keep for 1 to 2 weeks (or even 3 to 4 weeks, depending on the cultivar) at -0.5 to 0°C with RH of 90 to 95 per cent. Susceptibility to freezing injury depends on SSC, which varies from 10 to 14 per cent. The highest freezing point is -1.0°C.

Controlled Atmosphere (CA) Considerations

The major benefits of CA during storage and shipment are retaining fruit firmness and ground color. CA conditions of 2 to 3 per cent O_2 + 2 to 3 per cent CO_2 are suggested for moderate commercial benefits; the extent of benefits depends on cultivar. Exposure to less than 1 per cent O_2 may result in development of off flavors, and higher than 5 per cent CO_2 for more than 2 weeks can cause flesh browning and loss of flavor. The addition of 5 to 10 per cent CO_2 as a fumigant at during transport (less than 2 weeks) may improve the potential for benefit from CA. Pre storage treatment with 20 per cent CO_2 for 2 days may reduce incidence of decay during subsequent transport and storage in CA or air.

Retail Outlet Display Considerations

Cold-table display is recommended because of apricots' fast ripening. Ripening before consumption should ideally be done at temperatures of 18 to 24°C.

Chilling Sensitivity

Chilling-sensitive cultivars develop and express chilling injury symptoms (gel breakdown, flesh browning, and loss of flavor) more rapidly at 5°C than at 0°C. Storage at 0°C is necessary to minimize incidence and severity of chilling injury on susceptible cultivars.

Rates of Ethylene Production and Sensitivity

Ethylene production rate increases with ripening and storage temperature from under 0.1 $\mu L\ kg^{-1}\ h^{-1}$ at 0°C (32°F) to 4 to 6 $\mu L\ kg^{-1}\ h^{-1}$ at 20°C (68°F) for firm-ripe apricots and higher for soft-ripe apricots. Exposure to ethylene hastens ripening (as indicated by softening and color changes from green to yellow). Also, ethylene may encourage growth of decay-causing fungi.

Respiration Rate

Temperature mg $CO_2\ kg^{-1}\ h^{-1}$

0°C 4 to 8

10°C 12 to 20

20°C 30 to 50

To get mL $CO_2\ kg^{-1}\ h^{-1}$, divide the mg $kg^{-1}\ h^{-1}$ rate by 2.0 at 0°C (32°F), 1.9 at 10°C (50°F), and 1.8 at 20°C (68°F). To calculate heat production, multiply mg $kg^{-1}\ h^{-1}$ by 220 to get BTU per ton per day or by 61 to get kcal per tonne per day.

Physiological Disorders

Gel Breakdown or Chilling Injury

Develops in cold storage, particularly at 2 to 7°C (35 to 45°F) when apricots are stored for a long time period. This physiological problem is characterized in early stages by the formation of water-soaked areas that subsequently turn brown. Breakdown of tissue is sometimes accompanied by sponginess and gel formation. Fruit stored at these temperatures have short market-life and lose flavor.

Pit Burn

Flesh tissue around the stone softens and turns brown when the apricots are exposed to temperatures above 38°C (100°F) before harvest. This heat injury increases with higher temperatures and longer durations of exposure.

Post-harvest Pathology

Brown rot is caused by *Monilia fructicola* and is the most important post-harvest disease of apricot. Infection begins during flowering. Fruit rots may occur before harvest, but often occur post-harvest. Orchard sanitation to minimize infection sources, preharvest fungicide application, and prompt cooling after harvest are control strategies.

Rhizopus rot, caused by *Rhizopus stolonifer,* occurs frequently in ripe or near-ripe fruit held at 20 to 25°C (68 to 77°F). Cooling fruit and holding below 5°C (41°F) are very effective for controlling this fungus.

Suitability as Fresh-Cut Product

Fresh-cut apricot wedges should be kept at 0°C (32°F) and 90 to 95 per cent RH to maintain quality for 2 to 5 days, depending on cultivar and ripeness.

Special Considerations

The greatest hazard in handling or shipping apricots is decay, mainly brown rot and rhizopus rot. Quick cooling of apricots to temperatures of 4°C (39°F) or lower and holding them as near to 0°C (32°F) as possible will retard ripening, softening, and decay

Processing

Most of the apricots produced are generally canned, dried or frozen. Fruits of suitable varieties for processing are picked by hand or to a lesser extent mechanically, are delivered to the canning plant and washed in a water bath containing chlorinated water. These are graded for size, and the small fruits which are unsuitable for canning are cooled and directed for eventual use in concentrate, fruit bars, dehydrated products and other nectar-type products.

Fruits and vegetables can lose their typical fresh appearance and characteristic in only a short time after harvesting. However, if they are quickly cooled, this rate of deterioration can be retarded. This rate of change is increased if the plant cells have been bruised or if they have been ruptured during harvesting. This physical rupturing of the cells causes an increase in respiration rate, which can lead to physiological and physical decay through oxidative and other adverse cut surface must be treated to minimize degradative chemical reaction. Also, respiration and chemical reaction rates can be lowering the product temperature.

The product discoloration that occurs in stored fresh fruits and vegetables is due to enzymatic browning. The main enzymes associated with fruit and vegetable browning is polyphenol oxidase. The reaction sequence is either hydroxylation (at the one hydroxyl group) or oxidation (at the two methyl groups). For browning to occur, enzymes must be able to react with subtracts such as chlorogenic acid, coffeic acid and catechol. An oxygen supply and a catalyst such as copper also are required. It is necessary to control the reactivity of the enzyme to control enzymatic browning. There are different chemical and physical ways to inactive enzymes. Ascorbic acid will be act as a reducing agent, antioxidants and as a metal-sequestering material, when used as a browning retardant. It also has nutritive value. Fresh apricots that had been halved and dipped in ascorbic acid solution followed by some experimental dips, drained, sealed in pouches and held in cold storage for 10 weeks remained fairly light during the storage time.

Physical methods such as heating can be used to inactive enzymes, but these results in changes in textural properties. A product temperature has to reach about 60°C before most enzymes start to loss their activity. Enzyme denaturation is also dependent on acidity. Heat does not seem to be a viable method for use in minimally processed product because of its effect on texture.

A. Canned Apricots

The canning process is typical of that used with other stone fruits. The washing process has several advantages, such as product cooling, removal of dust and other contaminants and reduction of microbial load. In addition, the quality and appearance of the product are improved. Washing is done in equipment designed to effectively direct high-pressure sprays so as to accomplish the desired effects without injuring the product. At this stage inferior and damaged fruits are removed. Size sorting is usually accomplished, by passing the product over screens containing different size holes or slits. The defective apricots are removed by passing the product over an inspection belt where trained personnel look for defects (immature, green and overripe) and undesired material (leaves, sticks and stones). The inspected and sized fruits are delivered to mechanical apricot cutters, which align the fruits so that they are cut along their sutures. The cut fruits are then opened and the loose pits drop through perforated stainless plates. Cut and pitted fruits are then delivered to the sizer and grader to be mechanically sorted, resulting in uniformly sized fruits. Consumers except a specific number of units in each can. Therefore, it is important to mix various sizes to result in a specified number of units per can. Once the can has been filled, a topping medium-for example, sugar, water, or fruit juice-is added and the cans are immediately seamed. Canned apricots are sealed in heavy syrup, light syrup, juice pack (apricot or pear), or water.

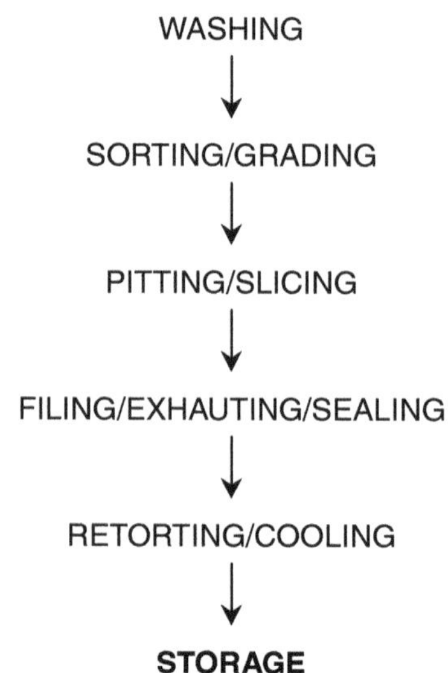

Figure 3.4: Flow Chart for Canning Process of Apricot Fruits.

The texture of canned apricot is influenced by many factors, including conditions of sterilization, maturity of fruits and storage. Softening of canned apricots is due to conversion of protopectins to water-soluble pectins. This occurs enzymatically during ripening. In case of canning, pectin materials gradually move the cell wall into the syrup, causing a gradual softening of texture. Higher acidity of the fresh fruit also results in complete breakdown of tissue. The higher concentration of hydrogen accelerates breakdown of the binding materials between cell walls during the heating process.

Dried Products

Dried apricots are produced from plump, fully ripe, fresh fruits. The fruits are picked and mostly sun dried. Common pre drying treatments include (a) selection and sorting of fresh fruits, (b) washing, (c) cutting into halves and removal of pits, (d) spreading of fruits on drying trays, (e) sulfuring with burning sulfur or gaseous SO_2 or potassium meta bisulfate (1.0 per cent) (f) placing of trays in poly tunnels for sun drying or cross flow cabinet dryers to reduce moisture content to 15-20 per cent and 9g grading. The dried fruits can be stored at 4°C and 75 per cent RH for at least 6-9 months. Drying is one of the oldest food processing methods known. The essential feature of drying is to reduce moisture content of the fruit below the level at which enzymatic damage occurs. Although one of the principal reasons for sun drying of fruit is preservation, other factors play an important part in the selection of this process. Drying reduces the bulk weight of the product, but most important is the development of the flavor and texture in the finished dried apricot.

The species flavor and texture are winning over a new generation of consumers who used dried apricot as snacks because of the fruits flavor and nutritional value.

Figure 3.5

Sulfur dioxide (SO_2) has been used for many years to preserve the color of dried fruits and is the only chemical added to dried apricots. SO_2 is generally recognized as safe for use and is approved for use by the Food and Drug Administration for food use, apricots prepared for drying as usually exposed to gaseous SO_2 before being put in the sun for drying. In addition to prevention of enzymatic browning. SO_2 treatment reduces degradation of carotene and ascorbic acid. The SO_2 treatment of apricots before drying retains their natural color if closely controlled so that enough SO_2 is present to maintain the physical and nutritional properties of the product throughout its life. Sulfured dried apricots contain SO_2 levels of 2500 to 3000 ppm. The amount of SO_2 must be controlled in dried product, since different countries permit different SO_2 levels in the fruit. After sulfuring, the trays are placed in poly tunnel dryers (72-80 hours) or cross flow cabinet dryers (12-14 hours) for dehydration. Drying is complete when the apricots have a moisture content of 15-20 per cent. Drying is dependent on condition of fruit, air moisture and consistency of sun exposure. After drying, the apricots are transferred to boxes to cure and to bring them to equilibrium moisture content. The dried apricots are then ready for grading and packaging. Dried apricots are sorted according to grades for sizes. The size grades do not refer to overall quality of product.

Relatively low moisture levels, high natural sugar levels and low pH of dried apricots product spoilage by microbial or enzymatic deterioration. Most commonly observed deterioration is darkening of the product due to gradual loss of SO2. This darkening is not spoilage in the true sense, but it does cause the product to become unappealing and undesirable. The loss of SO_2 cannot be eliminated completely but can be controlled. The storage temperature plays an important role in determining the storage life of dried apricots. Temperature is so critical that storage life is cut approximately in half for every 1°C increase. Dried apricots may be frozen with no adverse effects. However, dried frozen fruits should be allowed to defrost in a low-humidity area prior to use. The following conditions will help to obtain maximum storage life for dried fruits.

1. Store dried fruits at 4-5°C and 75 per cent RH for excellent keeping for at least 6-9 months (if washed and resulfured apricot)
2. Keep temperature and relative humidity constant.
3. Be sure that product is wrapped and not exposed to air.
4. Protect dried fruits from high-intensity direct light.

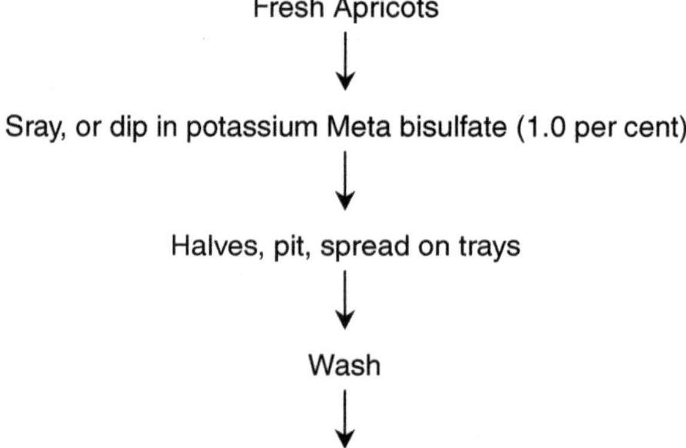

Drying in poly tunnel dryers in sun (72-80 hours) or cross flow cabinet dryer (12-14 hours)
(Final moisture 12-15 per cent) cross flow cabenet dryer (Dehydrate to final moisture)

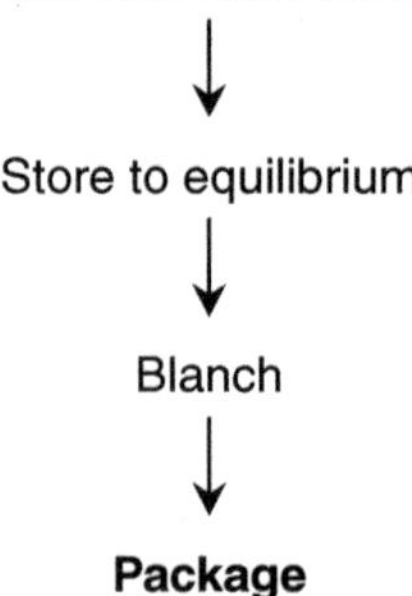

Figure 3.6: Flow Chart for Drying of Apricot.

Apricots are one of the first summer fruits. They are full of fiber, beta-carotene, and vitamins A and C. Drying, as a method of food preservation, has several benefits. Unlike frozen apricots, dried apricots can be stored at room temperature and do not rely on electricity to maintain their quality. Unlike canned apricots, dried apricots take up significantly less space and are processed more naturally, with less heat, resulting in a nutritionally superior, less altered apricot. Also, home drying your own apricots ensures they are sulfite free.

Osmo Air Dried Apricot Products

A technology was developed at ICAR-CITH Srinagar. J&K, for the preparation of quality osmo dehydrated product f apricot and cherry. Technology developed in the form of flow diagram is given in Figure 3.8.

Figure 3.7: Osmo Dehydrated Product of Apricot.

Nectar and Concentrate

Fruits that are unsuitable for canning are delivered to a thermal screw where they are heated to approximately 100°C and delivered to a pulping unit which

Apricot Nectar

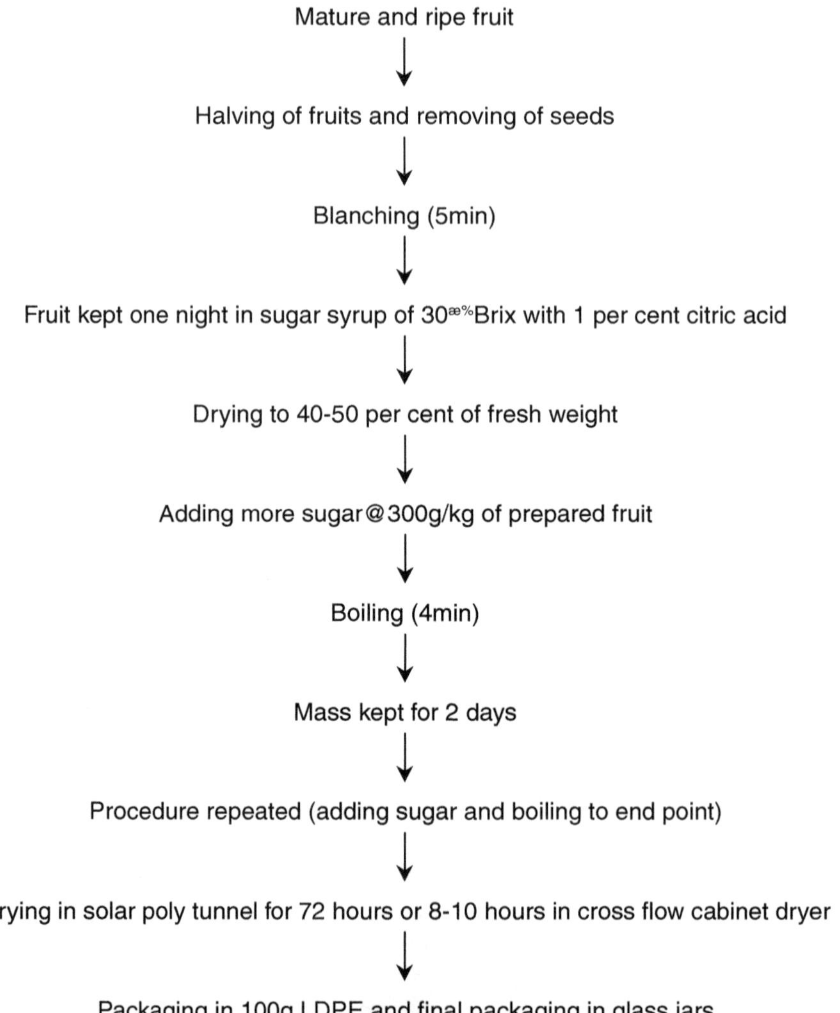

Figure 3.8: Flow Chart for Preparation of Osmo Air Dried Apricot.

removes the pits. The heated pulp is pumped through a series of finishers that remove some fibrous material such as the skins. The finished juice is then ready for the evaporation process or is used for preparation of nectar by mixing with sugar, water and citric acid. The nectar is canned and sealed similar to eh cut-fruit canning process.

The juice is concentrated to 32° Brix. The product is canned in 3.25-liter cans or aseptically filled in 250-liter drums. Also, juice is concentrated to dry powder for subsequent use in beverages, bakery products, jellies, puddings and deserts. In addition, apricot juice powder has the following advantages:

1. It saves weight and pace because more than 85 per cent of the water is removed; hence it is economical in transportation, storage and handling to distant markets.
2. Juice powder can be preserved for extended periods without refrigeration.
3. Ease in handling and convenience and rapid reconstitution make it suitable for domestic and military requirements, especially for overseas use.
4. It has concentrated nutritive values.
5. It is stable at elevated storage temperatures and for an extended period of time.
6. It has high quality and nutritive value at the time of consumption.

Concentrate can be processed further into different products.

Jam, Jellies and Preserves

Apricot jam is processed similar to other fruit jam except that 0.5 per cent more citric acid is added. Jellies consist of a ratio of 45 parts fruit juice to 55 parts sugar. Jellies and preserves contain whole chunks of fruit pieces and are permitted additional amounts of added pectic, citric, tartaric, malic or lactic acid, and a buffer salt. No added flavor or color is permitted with the exception of mint and cinnamon jellies. Fruit gets can be obtained with lower sugar levels by using low-methoxy pectin, which produces a gel in a product without requiring large amounts of sugar. The gelling action is initiated by addition of calcium salts, which cause pectin to crosslink, thus producing a structural matrix. The general procedure is to blend dried apricot with water to obtain a semisolid paste or to use a commercially prepared puree. Low-methoxyl pectin and low amounts of sugar are added and mixed well. Flavoring, orange peels and pineapple are added to give flavor and texture to the

Apricot Jelly

product. The product is poured into hot packed in containers and sealed. Upon cooling to room temperature, the product forms a flavored gel with good storability.

Apricot Fruit Bar

Processing Technology

For preparing best quality apricot fruit bar, fruits should have ideal colour, texture and flavour. Fruit should be ripe but not over ripe. Selected fruits are washed and treated with 100 ppm sodium hypochlorite for one minute. Stems, stones are removed. Bruised portion of fruits are discarded. Apricot halves are dipped in ascorbic acid (5 per cent Wv) and citric acid (5 per cent Wv) for 30 seconds. The stuff is then steam blanched for 5 minutes. Puree the fruit halves in a blender or processor until smooth slurry is produced. The puree/slurry is drained and passed through screen pulper. The final puree is concentrated with sugar up to 55±2 f Brix. The final concentrate is spread in food approved plastic or steel trays about 4 mm thickness in tunnel dryer (sun drying) for 22-24 sun shine hours (Temperature 48-50°C and relative humidity of 30 per cent) or cross flow cabinet dryer for 6-8 hours. The dried spread is uniformly cut into shape of small fruit bars in different layers and packed in 25 μ poly ethylene film or foods approved plastic cups and finally shrink wrapped.

Table 3.4: Composition of Apricot Fruit Bar

Moisture	12-14 per cent
TSS	55°Brix
Vitamin C (ascorbic acid)	6.8—7.00 mg/100 g
Carotenoids	400 μg/100 g

Organoleptic Evaluation

	0 days	*270 days*
Colour	8.5	8.0
Appearance	9.0	8.5
Taste	9.0	8.5
Aroma	8.5	7.5
Texture	9.0	8.5

Storage Studies

	90 days	*180 days*	*290 days*
TSS (°Brix)	52	50	48
Vitamin C (mg/100 g)	6.8	6.5	6.0
Carotenoids (μg/100 g)	380	360	330

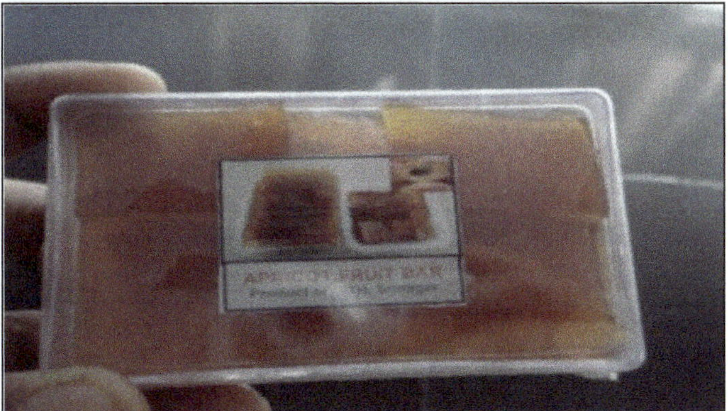

Figure 3.9: Quality Apricot Fruit Bar.

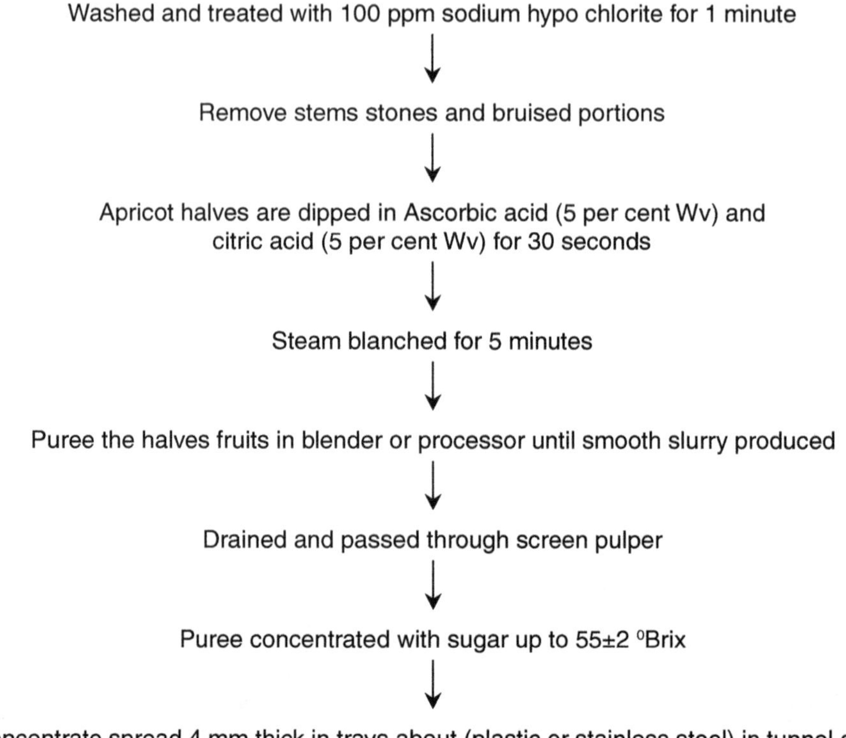

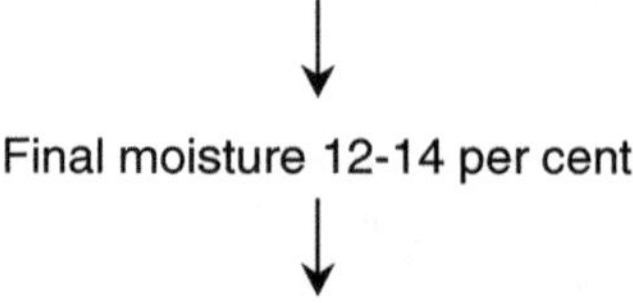

Figure 3.10: Flow Diagram for Preparation of Apricot Fruit Bar.

A process technology is developed at CITH, Srinagar for making ***Apricot Fruit Bar.*** *The final product i.e.* ***Apricot Fruit Bar*** *is having excellent texture, colour, aroma, taste, least browning, not sticking to gums and teeth, and can be stored up to nine months in cool and dry places without loss in quality, nutrition and appeal.*

REFERENCES

A.O.A.C 2000. Official methods of analysis, 16th edition, Association of Official Analytical Chemists, Washington, D.C.

Agar T and Polat A. 1995. Effect of different packaging material on the storage quality of some apricot varieties. *Acta Hort.* **384**: 625-632.

Antunes M D C and E M Sfakiotakis. 1997. The effect of controlled atmosphere and ultra low oxygen on storage ability and quality of 'Hayward' kiwifruit. *Acta Hort.* **444**: 613-618.

Arthey D and Philip R A. 2005. Fruit Processing Nutrition, Product and Quality Management, Ed 2nd Brijbasi Art Press Ltd. India, pp: 45.

Bhat MY, Padder B A, Wani I A, Banday F A, Ahsan Hafiza, Dar M A and Lone A A. 2013. Evaluation of apricot cultivars based on physic-chemical characteristics observed under temperate conditions. *International Journal of Agricultural Sciences* **3(5)**, pp. 535-537.

Carlos H and Adel K A. 1999. Apricot Post-harvest Quality Maintenance Guidelines, California Deptt. of Pom., pp: 1-5.

Chambroy Y, Souty M, Jacwuemin G, Gomez R M, Audergon J M. 1995. Research on the suitability of modified atmosphere packaging for shelf life and quality improvement of apricot fruit. *Acta Hort.* **384**: 633-638.

Chevallier A. 1996. The encyclopedia of medicinal plants. Dorling Kindersly, London.

Echeverna E and Valich J. 1989. Enzymes of sugar and acid metabolism in stored Valencia organs. *J. Am. Soc. Hort. Sci.* **114**: 445-449.

F.A.O. 2011. (http: //faostat3.fao.org).

Ghorpade V.M, Hanna M A and Kadam S S. 1995. Apricot In: Handbook of Fruit Science and Technology. Salunkhe D. K., S.S. Kadam (Eds).Marcel Dekker Inc., New York.

Harris S and Reid M S. 1981. Techniques for improving the storage life of kiwifruit (Actinidia chinensis). Publication G120, Auckland Indus. Develop. Division, DSIR, Auckland, New Zealand.

Haydar H, Ibrahim G, Mehmet O M and Bayram M. 2007. Post-harvest chemical and physical mechanical properties of some apricot varieties cultivated in Turkey. *J. Food Eng.* **79**: 364-373.

Hussain A, Yasmin A and Ali J. 2010. Comparative study of chemical composition of some dried apricot varieties grown in northern areas of Pakistan. *Pak. J. Bot.* **42**, 2497-2502.

Ishaq S, Rathore H A, Awan S S and Shah S Z A. 2009. The studies on the physic chemical and organoleptic characteristics of apricot (*Prunus armeniaca* L.) produced in Rawalakot, Jammu and Kashmir during storage. *Pak. J. Nutr.* **8(6)**: 856-860.

Joshi S.M, Adhikari K S, Seth J N and Divakar B L. (1990). Comparative study of biochemical changes in new apricot hybrid, ChaubattiaMadhu and in its parents during maturity and post-harvest storage. *Progressive Horticulture* **20(3-4)**: 246-252.

Kays S J. 1991. Post-harvest Physiology of Perishable Plant Products. Vas Nostrand Rein Hold Book, AVI Publishing Co., pp: 149-316.

Kazankaya A. 2002. Pomological traits of apricots (*Prunus armeniaca* L.) selected from Bitlis seedling population. *J. Am. Pomol. Soc.* **56(3):** 184-188.

Larmond E. 1977. Laboratory Methods for Sensory Evaluation of Foods Research Branch Canada. Deptt. *Agri. Publication*, pp: 1637.

Leccese Annamaria, Sylvie Bureau, Maryse Reich, Catherine Renard M G C, Jean-Marc Audergon, Carmelo Mennone, Susanna Bartolini and Raffaclla Viti. 2010. Pomological and Nutraceutical Properties in Apricot Fruit: Cultivation Systems and Cold Storage Fruit Management. *Plant Foods Hum Nutr.* **65**: 112-120.

Lee S K and Kader A A. 2000. Pre harvest and post-harvest factors influencing vitamin C content of horticultural crops. *Post-harvests Biol. Technol.* **20:** 207-220.

Lone Abid A. 2013. Evaluation of apricot cultivars based on physic chemical characteristics observed under temperate conditions. International Journal of Agricultural Sciences ISSN: 2167-0447 **3 (5)**, pp. 535-537.

Shahnawaz M, Khan T U and Tariq M. 2005. Performance evaluation of optimum doses of sulfuring and sulphiting in dried apricot to minimize the effect of stringency. *Sarhad J. Agric.* **21(2)**, 247-250.

Pramer C and Kaushal M K. 1982. Wild fruits of Sub Himayalyan Region. Kalyani Publishers, New Delhi, India.

Rai R D and Saxena S. 1988. Effect of storage temperature on vitamin C content of mushrooms (*Agaricus bisporus*). *Current Sci.* **57**: 434-435.

Rubio P and Infante. R. 2010. Preconditioning "Goldrich" and "Robada" apricot (*Prunus armeniaca* L.) harvested at two different maturity stages. *Acta Hort.* **862**: 605-626.

Salunkhe D K, Bolin H R and Reddy N R. 1991. Storage, Processing and Nutritional Quality of a fruits and Vegetables. **1**, Ed 2nd, CRC Press, Boca Raton Floride.

Sharma D P, Sharma N, Bawa R and Rajesh K. 2005. Potential of apricot growing in the arid-cold desert region of North-Western Himalayas. *Acta Hort.* **696:** 61-63.

Sharma K D, Kumar R and Kaushal B B. 2004. Mass transfer characteristics, yield and quality of five varieties of osmotically dehydrated apricot. *J. Food Sci. Technol.* **41:** 264-275.

Singh D B, Nazeer A, Kumar D and Lal S. 2014. Apricot fruit bar to enrich value added products. *Indian Horticulture*, **59**, No.4.

Souty M, Reich M, Breuils L, Chamboy Y, Jacquemin G and Audergon JM. 1995. Effect of post-harvest calcium treatments on shelf life and quality of apricot fruit. *Acta Hort*. **384**: 619-623.

Spayed S E, Proebsting E L and Hayrynen L D. 1986. Influence of crop load and maturity on quality and susceptibility to bruising of 'Bing' sweet cherries. *J. Am. Soc. Hortic. Sci.* **111(5)**: 678-682.

Vardzelashvili M G and Lebanidze VZ. 1974. Some biological and commercial properties of early sweet cherry cultivars. Sadovodstva, Vinogradarstva I Vinodeliya Gruzinskaya2: 5-17 (cf: Horticultural Abstracts **44 (9)**: 6361.

Wenkam N S. 1979. Nutritional Aspect of Tropical plant Foods. In: Inglett, G.E., Charalambous, G. (Eds.), Tropical Food: Chemistry and Nutrition, Academic Press, New York, USA, pp: 341-350.

Wills R H H.; F M. Scriven and H Green. 1983. Nutrient composition of stone fruit (*Prunus* spp.) cultivars: apricot, cherry, nectarine, peach and plum. *J. Sci. Food Agri.* **34:** 1383-1389.

Yildiz, F. 1994. New Technology in Apricot Processing. *J. Standard Apricot* Special Issue Ankara pp: 67-69.

Chapter 4

Berries

Introduction

Berries are considered soft fruit and include botanically different types of fruits such as blackberries, blueberries, strawberries, cranberries, gooseberries and currants and raspberries. These types are used as desserts as well as in processing. They are canned, frozen or made into jams, jellies or preserves. The juices are used in beverages and ice cream. Production figures for all berries are not available. However, they are produced mainly in the United States, European countries. North America, China and the, New Zealand is the largest producers of kiwi fruit, followed by Italy, Japan, France, the Unites States and Chile. Germany, Austria and Japan are the major importers, with high per-capita consumption. The rapid rise in kiwi production, including highly successful international trade and rapid spread of production makes this fruit the most successful new fruit crop of this century.

Botany

Blackberries are native to North America, they strong and erect stems, except for the dewberry (*Rubus procimbens*). Blackberries vary in color ranging from dark red to reddish black. Brison, Rosborough Womack Cheyenne and Hull Thornless are some of the cultivars of blackberries.

Of about 15 species of the blueberry, only three are of commercial importance. The high-bush blueberry (*Vaccinium corymbosum*) is the most important, followed by the low-bush (*V. angustifolium*) and rabbit-eye (*V. ashei*) species. While cultivation of the low-bush species is concentrated in the northern parts of North America, the rabbit-eye blueberry is grown widely in the southeastern United States.

Table 4.1: Commonly Grown Berries in the world

Common Name	*Botanical Name*
Blackberries and dewberries	*Rubus* sp.
Blue berries	
High-bush	*Vaccinium corymbosum*
Low-bush	*V. angustifolium*
Rabbit-eye	*V. ashei*
Cranberries	*V. macrocarpon*
Gooseberries and currants	*Ribes* sp.
Raspberries (red and black)	*Rubus* sp.

Heat-resistant hybrid species adaptable to southern regions of a warmer climate are being developed by crosses between high-bush and rabbit-eye species. June and Croafan are some of the commonly grown cultivars in the United States. The blue Chip cultivar of blueberry fruit have excellent color, firmness and a pleasant acid flavor. It is a mid-season cultivar with resistance to cane canker.

All commercial cultivars of cranberries are derived from *Vaccinium macrocarpanm* which is native to North America. The other commercially important cranberries are the small cranberry *V. oxycoccus* native to Europe and *V. idaea,* the low-bush cranberry or cowberry of eastern Canada.

Gooseberries are grown both in Europe and North America. The European species is classified as *Ribes grossularia. R. hirtellum* (smoothberry) and *R. cynosbati* (prickly berry) are the most important species of gooseberry native to North America. Currants are grown mainly in Europe and their cultivation is favored especially in Great Britain. The red currants are *R. rubrum;* the black currants are *R. nigrum,* the black currant is distinctly different in characters and originated from red and white currants. A large proportion of the gooseberry is used for processing. Colossal cultivar of the gooseberry has a large, round to oblong fruit of high quality.

Kiwi fruit (*Actindia chinesis*) was formerly called Chinese gooseberry. It is vine crop, having excellent keeping quality of the fruit. The fruit is also called *Ichang gooseberry,* monkey peach, kiwi and kiwi berry. The name commonly accepted in the North-American market is kiwi fruit. Internally, the fruit has a white central core running the length of the fruit and surrounded by a translucent inner pericarp. This region contains black seeds in locules radiating from the core. The inner pericarp is surrounded by an outer pericarp composed of thin walled parenchyma cells. Both the inner and outer pericarp contains chlorophyll, giving the fruits its unique internal green color. Early commercial cultivars included Abbott, Hayward, Bruno, Allison and Monty, but he superior keeping quality of Hayward resulted in the almost total exclusion of other varieties. Recently, a new selection, Dexter, with better local adaptation, has been introduced in Australia. Hayward is the most common commercially grown cultivar in the United States.

The commercial raspberry cultivars, ranging in color from red to black are selections from the wild species of Europe, namely, *Rubus idaeus* or *R. stratagems*, a red raspberry native of North America. All black raspberries (black caps) originated from *R. occidentails*, a species native to the southeastern United States. Yellow and purple (hybrid) raspberries of both *R. strigogus* and *R. occidentalis* are also available. The cultivars of raspberries include Black Knight, Black Treasure. Bountiful, Giant, Forever, Amber, Jet, Prestige, Sensation, Sparkling, Gem, Cumberland, Plum Farmer, Columbian, Royal Purple, Cuthbert, Herbert, Letham, Taylor and Indian Summer, Brooks and Olmo have described the genetics and characteristics of these cultivars.

Some of the popular strawberry cultivars grown in the United States are Cardinal, Sunrise, Delite, Atlas and Apollo. The Arking is a late-ripening, root-rot-resistant cultivar adapted to Arkansas conditions. Several new cultivars of strawberries like Allstar, Amazing, American Sweet Heart, Autumn Surprize. Candy Red, Deep Red, Honeoye, Prelude. Surprise, Temptation, Tribute, tyee, Universal Red, Aberdeen, Aroma, Blackmore, Catskill, Dorsett, Dresden, Fairmore, Fairfax, Gem, Green Moutain, Herbron, Mastodon, Maytime, Pathfinder, Redstar and Wayzata has also been reportd.

Production

Blackberries

Most cultivated blackberries have been developed from native species. The most suitable soil is a clay loam, well drained but receiving plenty of moisture. The upright blackberries may be propagated either by root cutting or by suckers. Trailing blackberries are usually propagated by up tip layering. Blackberries of an upright growing habit are usually planted in row 2 to 2.5 m apart, with the plant 1-2 m apart in the rows. Blackberries are pruned in such a way that they may be tied to wire trellises or stakes. Strongly growing varieties are usually pruned back to 1-1.2 m. trailing varieties are cut back, leaving 2.5-5 m according to the distance between the plants. Nitrogenous fertilizers are applied in the spring, with plenty of potash and phosphates later in the summer to help ripen the canes. A bright orange rust on the underside of the leaves and crown gell, a bacterial disease, cause damage to blackberries.

Blackberries are very perishable and thus have a short post-harvest life of 3-4 days. They should therefore have a medium of handling and after picking. Most blackberries used for processing are harvested mechanically. Under commercial conditions at harvest, mechanically harvested blackberries and both raw and processed quality comparable to hand-picked fruits regardless of berry temperature. The machine-harvested berries at higher temperature (30°C). However, deteriorated more rapidly during storage than the hand-picked berries at the same temperature. The anthocyanins, soluble solids and titratable acidity of blackberries are influenced by preharvest temperature. The anthocyanin and soluble solids content decrease when blackberries ripe at lower temperatures. Loss of acids during fruit ripening accelerates by increasing temperate.

Blueberries

Acidic soil with pH from 4.5 to 5.0 is most suitable for blue soil should be well drained, with plenty of moisture below the root level. The application of 300kg of superphosphate per acre is enough for blueberries; or one application of 180-h/acre of complete fertilizers (about 5:10:5) can be made on the spring when the buds are sprouting. Usually, very little pruning is necessary for the first 3 years after planting. General pruning may be done at any time from the fall of leaves in the autumn to the beginning of growth in the spring.

Blueberries picked early in the harvest season are commonly infected by *Alternaria tenuis*, whereas those harvested late are more commonly attacked by *Botrytis cinera* and *Glomerella cingulata*. Plump, firm and uniformly colored (light blue to blue black) blueberries that are free from injury and decay are harvested for the fresh market. While green or red color indicates unripeness, overripe berries are dull and usually soft textured.

Cranberries

The most suitable soil for cranberries is clay loam, well drained with plenty of moisture, the cranberry vines are reproduced by cuttings, and these are set out in rows 30-50 cm apart and 15-28 m apart in the row. Fertilizers are applied in the spring, after the last reflow has been removed. The application of 35 kg of sodium nitrate, 135 kg of rock phosphate, and 25 kg of sulfate of potash per acre is sufficient for cranberries. In cranberry growing, water is used extensively, but most for winter flooding. Enough water must be supplied to keep the vines in a healthy condition. The cranberry is attacked by several insects, such as leaf hoppers, fire worms, and blossom worms, while the important cranberry disease are blast, rot and false blossoms.

Cranberry fruit harvested for the fresh market should be plump, bright, firm, clean, uniform in color (light red to reddish black) and free from mechanical injury, sun scald, insect injury, or decay. The uniform maturity of cranberries permits growers to pick them in one lot. Berry size slight amount of green color are considered immature.

Gooseberries

Gooseberries are grown in any good, loamy sol, preferably in a dry, sunny spot, they often flourish on land which is wet during a large part of the year. Gooseberries are usually propagated from cutting or by layering. When planted in rows, they are placed 75-15 m apart in the rows, with the rows 1.5-2 m apart. Just after fruiting time, those stems which have sprouted from the base of the plant are pruned out. The important disease of gooseberries is white pine blister rust and cane blight. Most of the commercial production of gooseberry is used for processing. Both green and ripe fruits are harvested, depending on end use. Color and flesh firmness have been employed as indices of harvest maturity.

Kiwi Fruit

Deep, fertile, well-drained, sandy loam soils are recommended for cultivation of kiwi fruit. Waterlogged or hard soils are avoided. Kiwi fruits are best grown in a warm temperate climate, but they can tolerate a wide of conditions. Early winter frosts do not damage the fruit, but frost at any time during and early flowering may cause considerable injury. On the other hand, in climates that do not provide sufficient winter chilling, spring bud burst is erratic and cropping is poor. Both fruit and vines are easily damaged by winds; hence wind protection is usually required for kiwi crops. Soft or hardwood cuttings are ideal for economic and rapid multiplication. Shield budding and whip grafting are recommended for large scale propagation, for top-working on old economic vine, or for grafting male plants. The plants are transferred to permanent sites in the orchards when in a dormant condition and planted in rows at 7.5-9.0 m x 4.2-5.0 m spacing. Male and female flowers occur on different plants, it is essential to have an adequate number of male plants, which should be evenly spaced and surrounded by female plants to ensure the most efficient pollination. Hence, on average, the ratio of female to male plants is maintained at about 6:1. seed propagation is not recommended for kiwi fruit.

The fruit-bearing habit of kiwi fruit resembles that of the grapes and the plants are supported and trained either on single or multiple wire fences, or on pergolas. Fruit are generally borne the current-season's growth arising from the first three to six buds of the shoots of the previous year. Pruning is done twice a year, once in summer, when the growth of vines is very active to remove the undesirable vines and again in winter, when the vines are in a dormant condition, to obtain regular crops of good quality. Either one or both of the following systems of winter pruning are adopted: (a) the leaders, which bear the fruiting laterals, are removed and replaced by typing down well-grown leaders from the growth of current season and (b) the fruiting laterals may be shortened to two buds beyond where the fruit was borne during the previous season. The aim is to retain as many of the few leaders as possible and to remove those which have already fruited. This avoids overcrowding of leaders and laterals on the vines, which is essential for the production of high-quality fruits.

Kiwi vines have relatively few pests or diseases, but where crops are destined for export, quarantine requirements necessitate a high degree of control. Leaf roller insects and armored scale are the major pests, but occasionally mites can be important. In some locations, root-knot nematodes, particularly Meloidogyne spp. can reduce production. Armillaria root rot and Phytophthora crown rot are both capable of causing the death of vines. Leaf infection by bacteria, *Pseudomonas viridiflava*, or by a range of weakly pathogenic fungi, occurs mainly as a result of leaf damage caused by wind or frost. Such leaf infections generally do not affect vine vigor. Infection of unopened flowers by P. viridiflava, known as bacterial blossom blight, can be important in some orchards but generally occurs spasmodically.

Kiwi fruits are harvested while hard and unripe, but to optimize storage life and to ensure that they ripen to satisfactory eating quality, a minimum level of physiological development must be attained before harvest. In New Zealand the fruits are harvested when the soluble solids content is above 6.2 per cent. an indication of eating quality in ripe fruits can be obtained by measuring the soluble solids content. Fruits with a soluble solids content of less than 12.5 per cent are regarded as having an unacceptable eating quality (2.20). late-harvested fruits retain their flesh firmness during storage better then early-harvested fruits. Minimum soluble solid concentration cn also be adopted as a maturity standard. Kiwi fruit, if left on the vine, will ripen and abscise from the plant at slightly overripe stage. However, ripening is not synchronized, and individual fruits can ripen over several months. Once the orchard attains acceptable maturity, all the fruits can be harvested on one occasion. Fruits are picked by snapping them off at the calyx, so that the fruit stalk remains on the vine. The fruits are collected in small boxes and then transferred into 200-250 kg bins for transport to the packing shed. After quality and size grading, the fruits are stored, either packed in singly-layer trays ready for marketing or in larger bins. Packaging is done in single-layer trays with perforated polyethylene liners in wooden or corrugated cartons. The polyethylene maintains high relative humidity and allows a small percentage of CO_2 to build up, which aids in long-term storage.

Raspberries

Any type with food drainage and sufficient soil moisture content is suitable for raspberries. Upright raspberries may be propagated either by root cuttings or by suckers. Raspberries usually produce fruit on a cane of one season's growth. The canes necessary for the bearing of the crops are produced during one season. Flower or bear fruit the next and must then be removed. The chief diseases are anthrocnose, cane blight or wilt, root knot, orange rust and mosaic. American raspberry beetle, raspberry cane borer and sawfly are the important pests of raspberries.

Strawberries

Land for strawberries should be thoroughly cultivated and mannered. Any good garden soil is satisfactory. Strawberries are propagated by means of runner shoots which come out from the parent plant and root at alternate nodes. Thorough but shallow cultivation is required and removal of all runners until the end of the bear in of season is recommended. Heavy manuring is necessary in order to obtain a good crop and recommends 25 cartloads of well-rotted cattle manure per acre. The plants should be irrigated every third or fourth day. White grub, leaf roller, leaf beetle, weevil and cutworm are the important pests, while leaf spot is the important disease of strawberries.

Strawberries are generally harvested manually, owing to their tenderness and vulnerability to mechanical damage. Despite these factors, attempts have been made to develop prototype mechanical harvesters. Berries ready for harvest are fully red, or at least three-fourths of their surface is red or pink. A dull and shrunken appearance usually indicates over ripeness. The percentage of calyx-free fruit of the strawberry was generally higher early in the season and decreased rapidly as the season progressed. Euparen (dichlofluanid) spray combined with gibberellic acid (5 ppm) plus 0.5 per cent KH_2PO_4 applied just before harvest improved harvesting of calyx-free fruit. As the strawberry fruits matured on the plant, the weight, percentage of soluble solids, ascorbic acid and water-soluble pectin increases. The total solids, acidity, total phenols, cellulose, protopectin and activity of polyphenol oxidase decreased with increasing fruit maturity.

Cape Gooseberries

Ethnic names of cape gooseberry is Jam fruit, Peruvian cherry, Uchuva, whereas scientific name *Physalis peruviana* was derived from the Greek word physa meaning bladder, for the calyx covering the fruit and peruviana meaning of Peru. These plants grow all over the Andes and were fruit of the Incas. Cape Goose berry fruit is a berry which looks like a marble with a small round shape (1.25-2.5cm) in size. The fruit consists of numerous small yellow seeds. The fruit looks orange or yellowish in color and becomes bright yellow and sweet when ripe. The fruit is covered by wrinkled leaves, which form a Chinese lantern. The berry is encased in the inflated long, papery, tan husk (calyx) which is 3 cm to 3.5 cm in length.

Cape gooseberries are well known for its blood purifying capacity. They are also known for other medicinal qualities which are being a source of provitamin A, vitamin B and C, and are a rich source of carotene, phosphorus and iron, and also contains vitamin P. In India scientists have isolated physalolactone C from the leaves, a minor steroidal constituent removed from the paper-like husks. It may be eaten fresh, in salads or in cocktails. The attractive yellow marble-sized fruit makes an extremely tasty jam. It has a particularly delicious fruit with a tangy pineapple-like flavour. They make excellent pies and jellies and are very high in pectin. The fresh fruit may be served with husk pulled back for fondue. *Physalis peruviana* sauce is a nice accompaniment to a meat dish. While not well known by the retail consumer, the fruit has a strong following among chefs and the market is likely to grow for good quality, vine-ripened fruit. The husk is bitter and inedible. The fruits are covered or encased in a loose, papery husk shaded with purple, which is the persistent calyx and protects them from external injury. They resemble yellow cherry tomatoes. The fruit and husk will naturally dehisce (drop) with a good shake when they are fully ripe or nearly so.

Most of the research on Cape gooseberry deals with the development and improvement of growing techniques. There is a lack of breeding efforts; only some research related to selection among the different accessions or to the development of *in vitro* culture protocols to obtain soma clonal variation or as a first step for genetic transformation has been conducted. Genetic difference in yield and fruit quality characters found among accessions from different regions can be exploited for cape gooseberry breeding. A lot of variability in terms of vegetative growth, fruit size, quality, composition and colour among different genotypes/cultivars were reported by many workers. Evaluation and screening programme of Central Institute of Temperate Horticulture Srinagar J&K on production, fruit quality, colour characters sketch four genotypes (CITH CGB S20- CITH CGB S3- CITH CGB S12- CITH CGB S1- CITH CGB S6) of cape goose berry found suitable for temperate regions.

Under temperate climatic conditions *Physalis peruviana* propagates by seeds which are sown in flats of sandy soil or directly outside. In India, the seeds are mixed with wood ash or pulverized soil for uniform sowing. Sometimes propagation is done by means of 1-year-old stem cutting treated with hormones to promote rooting. *Physalis* shows an intermediate behaviour with increased respiration during ripening of climacteric fruits. The skin color of the cape gooseberry can be used as a maturity index. Harvesting should begin when the calyx begins to turn yellow, avoiding over maturity. With the fruit development fruit size, weight, soluble solids, acidity increases linearly up to full maturity stage. Visual harvest determination utilizes the synchronous color changing of both calyx and fruit, having nearly the same color. It should be done carefully to avoid the stem breaking and knocking off ripe fruits. Fruits presented an intense green color during the first 35 days after anthesis, starting to change slowly to yellow, which was an intense yellow-orange in the skin and pulp at 64 days when consumer maturity was reached. However, under temperate climate in Kashmir it has been reported that the cape gooseberry should be harvested 6-8 weeks after anthesis when the fruit are well formed and substantially filled the calyx. Fruit harvested at maturity grade 4 (Yellowish green colour) and those at grade 5 (yellow) dried at 18° C best conserved sucrose concentration. Sometimes premature fruit-fall is observed, probably related to over maturity, abruptly changing soil moisture and climate and varietal hormonal factors. Pre blossom spray with Ethrel (500 ppm) enhances fruit ripening by 10 days. At physiological maturity (56 days after anthesis) fruit measures 12.7°Brix, 3.52 pH and 1.215 g of citric acid per 100 g fruit fresh weight.

Harvest initiates, depending on site conditions (principally temperature), between 3-6 months after transplanting. Cape gooseberry fruits have a relatively low perishability, allowing for greater flexibility in harvesting. The fruit is harvested when it falls to the ground, but not all fallen fruits may be in the same stage of maturity and must be held until they ripen. It may take some experience to tell when the calyx-enclosed fruits are fully ripe. Properly matured and prepared fruits will keep for several months. It is recommended that the harvest is done two to three times per week during a harvest peak, in the early morning hours, avoiding picking during the rain or fruits with wet husks. Fruits should always be picked with their husk and with a peduncle with a maximum length of 25 mm. In ecotypes and varieties where the peduncle is strongly attached, picking has to be done with small scissors. Harvesting can be accomplished by allowing the fruit to fall on fabric or plastic placed under the plants. Collection is either done by hand picking, or by gathering up the plastic and pouring the fruit into containers.

In rainy or dewy weather, the fruit is not picked until the plants are dry. Berries that are already wet need to be lightly dried in the sun. Hand collection is preferable if the fruit is to be sold on the fresh market, to avoid bruising. At the peak of the season, a worker can pick 90 kg a day, but at the beginning and end of the season, when the crop is light, only 18 kg can be pick up in a day. The ripening of cape goose berry is associated with a conspicuous climacteric rise in CO_2 and ethylene production. Its respiration rate ethylene bio synthesis can be classified as extremely high. Ethylene yields between 7 and 24 nmol n^{-1} per gram in the ripe/over ripe stages thus compare favourably with production rates reported for tomato. As the fruit colour turns green (chlorophyll) to yellowish orange (carotenoids) and progressive softening occurs, several cell wall changes occur. The most noticeable change of cape gooseberry fruit during maturation and ripening is the change in skin color from green to yellow when physiological maturity occurs and synchronously the calyx changes from light green to yellow, which makes this characteristic adequate for use as a maturity index as discussed before. In cape gooseberry fruit, at the same time as the color change, the weight of the fruit increases, reaching a maximum around maturity. Total soluble solids (TSS), in green fruits with 9.3°Brix, peaked with between light orange and orange stage and then fell to about 13.7 at the overripe stage. Total titratable acidity (TTA), due to its demand in respiration decreases constantly from 39.5 mval/100 mL in green to 17.6 mval/100 mL in overripe fruits. Interestingly, β-carotene content, taking into account that the cape gooseberry is classified as a carotenoid fruit, peaks at the orange color stage, and after that its fall is interrupted because of a concentration effect on the decreasing fruit size at stage red orange stage. *Physalis peruviana* plants typically are heavy fruit producers. A single plant can produce up to 1.5 kg of fruits. According to other sources, a single plant may commonly yield from 130 to 300 fruits. The average yield from a plant covering 2.5 square meters was found to be 545 g. The fruits are usually dehusked before delivery to markets or processors. Manual workers can produce only 4.5 kg to 5.5 kg of husked fruit per hour.

Generally, the berries are not washed or disinfected, regardless of whether the husk is attached or not. A very high percentage of fruits are usually clean owing to protection from the calyx. Washing and disinfecting with low concentrations of chlorine (20 to 50 ppm) decrease the microbial contamination but eliminate the natural anti-feeding compounds of the fruit. In all fruits destined for transport with the calyx, the husk is carefully opened at the apical end and checked for defects (physical, physiological, pathological and entomological), size and color. Just like the fruit, the husk has to be free of defects. Generally, in the course of this process, fruit with discolorations (too green or too orange) or cracks are sorted out. Fruits have to be firm and fresh in appearance, with a smooth and shiny skin. If the calyx is present, the peduncle must not exceed 25 mm in length.Since cape gooseberry is mainly exported with an attached calyx, conserving the quality characteristics and protecting the fruit from damage, calyx drying is of great importance in the post-harvest life of this fruit. This is the reason that cape gooseberry importers require fruits with totally dried husks. The calyx, which is not edible, has a very similar carbohydrate pattern to that of the fruit, and can weigh up to 2 g, although dried ones have only one-tenth this weight.

Generally, calyx drying is done after fruit grading, before or after packaging (commonly done in different types of small plastic baskets). It is recommended that drying be done after packing due to the ease of packing fruit with a fresh calyx compared to fruit with dry husks. Colombian fruit export firms, generally, dry calices by convection, with forced air with RH lower than 50 percent, but the temperature can vary. Good results were found by drying fruit calices at an air temperature of 24°C for 6 hours without affecting physico-chemical and sensorial characteristics during or after posterior storage. Longer drying periods cause fruit weight losses and accelerate the ripening process. Ten hours of drying with temperatures up to 25°C can be used for fruits with green or wet calices. The best packaging for cape gooseberry fruit is the natural attached dried calyx: it protects the fruit not only from physical damage but also against fungus, and favors the modification of the atmosphere surrounding the fruit. For distribution to the wholesale markets, cape gooseberries are preferentially packed in plastic boxes or also those made of cardboard or wood (for 8 to 10 kg fruit weight). For consumers, fruits are packed in perforated small plastic fruit containers or in small plastic baskets with a total maximum fruit weight of 200 g (with calyx) and 500 g (without calyx).

The fruits of *Physalis peruviana* are long-lasting. The fresh fruits can be stored in a scalded container and kept in a dry atmosphere for several months without refrigeration. They will still be in good condition. If the fresh fruits are to be shipped, it is best to leave the husk on for protection. The unhusked fresh fruits of *Physalis ixocarpa* can be stored in single layers in a cool dry atmosphere for several months. The fruit can be kept more than 6 months in a ventilated spot by prolonging the ripening when it is being protected by its calyx. Optimum conditions for medium- and long-term storage of cape gooseberry fruit (<6 months) vary between 2°C and 4°C, and 80 to 90 percent RH. In fruits stored with calyx, a high RH can cause diseases, which highlights the importance of good calyx drying before storage.

The Cape gooseberry fruit which is acid sweet in taste and with a pleasant flavor is typically consumed fresh, whole or dried, without the calyx, but with the skin. Fruits of *Physalis peruviana* are juicy, widely astringent and sweet with a pleasant blend of acid. The overall quality of the fruit is good. In addition to being canned whole or preserve it d can be used in order to prepare desserts (pies, cakes, jellies, jams, puddings, chutneys, sauces, ice cream, yoghurt).

Its flavor has been defined as a pleasant, unique tomato/pineapple like blend. The osmo dehydrated products fruit can be used as a raisin and yeast substantiate, though it is not so sweet. These fruits are excellent when differ in character, drier and eaten. The fruits taste excellent when dipped in chocolates. Ripe good-quality fruits can be used as a dried food, similar to raisins and exported as such. Vitamin E-impregnated cape gooseberry fruits (using an isotonic dissolution of sucrose in the aqueous phase of the emulsion), were juicier, sweeter and less acidic than natural samples. Due to their 'exotic touch', it is common to use cape gooseberry fruits, half opened or without the husk, for decoration purposes on cold buffets in restaurants and hotels.

Chemical Composition of Berries

The chemical composition of berries is presented in Table 4.2. the total soluble solids content ranges from 10.2 to 19.7 per cent. The sugars constitute one of the major soluble components of berries. Reducing sugars from a major proportion of total sugars in the berries. In the majority of berries, citric acid is the predominant acid; with the exception of blackberry, wherein isocitric acid and its lactone predominate. The cranberry contains an appreciable amount of benzoix acid. The next important organic acid present in berries is malic acid. The most important vitamin in berries is vitamin C. Berries are also regarded as good source of β-carotene, thiamin, riboflavin and nicotinic acid.

Anthocyanin is a major pigment in berries. Raspberries contain mainly mono- and diglycosides of cyaniding. Strawberries contain mainly 3-glucosides of cyaniding and pelargonidin. The anthocyanins present in the skin of blueberries are 3-galactosides, 3-glucosides and 3-arabinosides if delphindin,petunidin, malvidin and cyaniding. Gooseberries and currants contain cyaniding and delphinidian

Table 4.2: Chemical Composition of Berries

Constituent	*Black-berries*	*Cran-berries*	*Goose-berries*	*Rasp-berries*	*Straw-berries*
TSS (per cent)	15.2	13.0	11.1	13.9	10.2
Total sugars (per cent)	4.3	3.5	4.6	1.57-5.34	5.0
Acidity (per cent)	0.68-1.84	2.90-3.17	1.21-2.91	0.74-3.62	0.52-2.26
Ascorbic acid (mg/100 g)	20	11-33	20-50	19-38	89
Carotene (mg/100 g)	0.1-0.59	0.02-0.58	0.18	0.05-0.08	0.15
Thiamine (mg/100 g)	0.03	0.03	-	0.02-0.03	0.03
Riboflavin (mg/100 g)	0.034	0.02	0.024-0.03	0.03-0.09	0.027
Nicotinic acid (mg/100 g)	0.40	0.10	0.30	0.40-0.9	0.6
Minerals (per cent)	0.5	0.20	0.40	0.5	0.5
Phosphorus (mg/100 g)	23.9	11.2	19.0	22	23
Potassium (mg/100 g)	208	119	170	168	161
Sodium (mg/100 g)	3.7	1.8	1.2	1.0	1.5
Calcium (mg/100 g)	63.3	14.7	18.5	22	22
Magnesium (mg/100 g)	29.5	8.4	8.6	-	11.7

Table 4.3: Ratio of Saturated to Polyunsaturated Fatty Acids

Berry	*Ratio of Saturated to Polyunsaturated Fatty Acids*
Blackberries	0.05
Strawberries	0.1
Blueberries	0.19

glycosides in their skin. The anthocyanins of cranberries (*V. macrocarpon*) are mainly the 3-galactosides and 3-glucosides. peonidin-3 glucoside (41.9 per cent) and cyanidin-3 glucoside (38.3 per cent) were the main anthocyanins isolated from fruits of *V. oxycoccus*. Smaller amounts of the 3-galactosides and 3-arabinosides of peondin and cyaniding were found in adiditon to 3-glucosides of delphinidin, petunidin and malvidin. Processing such as freezing and cooking effects the content of anthocyanin significantly. Other phenolic compounds contribute to the color of berry fruits mainly by copifmentation effects. There is considerable changes in the phenolic compounds as fruit matures and these are closely related to the oxidative enzyme system. Ellagic acid is a naturally occurring phenolic constituent present in fruit, especially in strawberries, and other berries. It has been shown to be effective as an antimutagen, anticarcinogen, and a potential inhibitor of chemically induced cancer.

Kiwi fruits are an excellent source of vitamin C, the average-sized fruit containing around 90 mg. Vitamin C content up to 300mg/100 g has been reported from Indian Samples. There is a small decrease in Vitamin C content in early stages if storage, but the concentration then becomes stable so that after 6-7 months in storage,

90 per cent of the Vitamin C present at harvest still remains. Actinidia eriantha and A. kolomikta can both have levels as high as 700-1000 mg/100 g.

The total chlorophyll (a+b) in ripe kiwi fruit is about 100mg/100 g, which decreses gradually during post-harvest ripening. *A proteolytic enzyme actinidin,* polyphenol oxidase *and peroxidase* are also present in kiwi fruit. The properties of actinidin are similar to those of papain. Actinidin concentration in the fruit is approximately 0.4g/100 g) fresh weight. Calcium oxalate has been identified as an irritant factor in the fruits. The high fiber content of the fruit and the special characteristics of the mucilage make kiwi fruit an excellent laxative.

More than 90 volatile flavor compounds have been identified from kiwi fruit. If fruit is ripened soon after harvest, there is a marked increases in the concentration of volatile esters, especially ethyl and methyl butanoate; but when fruit is ripened after storage, ethyl acetate becomes relatively more important. Hexanal and hex-2-enal are also major volatile components. The contributions that individual components make to the perceived aroma and flavor are still unclear. However, a tangy or acid flavor, typical of slightly under ripe fruit, is associated with low concentrations of the major volatile esters and high levels of citrate and soluble solids. Sweeter fruit contains higher levels of volatile esters.

Nutrients and Health Benefits of Berries

Not only are berries incredibly tasty, but they are also one of those foods universally recognized as healthy. They are certainly the healthiest types of fruit in the world. Whatever, dietary system people follow, meat-eater or vegan, almost everyone loves berries.In fact, it's hard to find anyone who doesn't like them. This article will look at the health benefits of more than 20 different types of berries.

1. Acai Berry

With a pronunciation of "ah-sah-ee," acai berries are native to South American rainforests. Over recent years, they have experienced an explosion in global popularity. Unfortunately, fresh acai berries are near impossible to source. As a result, people usually buy acai berries in powdered form which you can use to make a drink. Due to their purported antioxidant content, acai berry products are popular with mainstream health crowds.

Calories and Macronutrients (Per 3g Serving Acai Powder)

Calories: 20

Carbohydrate: 1g

Fat: 1.5g

Protein: 0g

Acai berries contain vitamin A, but in the specified serving sizes of acai powder, this is not a significant amount. The benefits of acai powder come from the antioxidant and polyphenol content. In fact, studies show freeze-dried acai powder has an exceptional polyphenol count – higher than any other fruit or vegetable.

2. Blackberry

Blackberries grow all around the world, and they also taste amazing.

Calories and Macronutrients Per Cup Blackberries

Calories: 62

Carbohydrate: 15g (Fiber: 8g, Sugar: 7g)

Fat: 1g

Protein: 2g

Blackberries are an excellent source of:

- ✰ Vitamin C: 50 per cent RDA
- ✰ Vitamin K1: 36 per cent RDA
- ✰ Manganese: 47 per cent RDA

As with all berries, blackberries also contain various health-protective polyphenols. Blackberries show a significant protective effect against LDL-oxidation, a prominent cardiovascular risk factor, in human intervention studies. Studies also show that wild blackberries are 3 to 5 times higher in polyphenols.

3. Black Raspberry

Despite looking similar to blackberries, black raspberries are an altogether different fruit.

Calories and Nutrients Per Cup Black Raspberries

Calories: 70

Carbohydrate: 16g (Fiber: 8.7g, Sugar: 5.92g)

Fat: 0.87g

Protein: 1.61g

Black raspberries are a significant source of Vitamin C: 58 per cent RDA

Black raspberries have also demonstrated strong anti-carcinogenic properties in clinical studies.

These studies show that black raspberries have anti-inflammatory, antiproliferative, and tumor-suppressive activity. Eating black raspberries with a bit of heavy cream is a great, tasty option.

Or can even make a healthy, low-carb black raspberry ice-cream:

- ✰ Blend the black raspberries with two egg yolks and some heavy cream
- ✰ Add some vanilla bean extract and a touch of salt (if you want, you can also add your choice of sweetener)
- ✰ Put it in the freezer
- ✰ Take the ice-cream out of the freezer about 30 mins before you want to eat it

4. Blueberry

As one of the most common types of berry, almost everyone knows about the health benefits of blueberries.

Calories and Macronutrients Per Cup Blueberries

Calories: 84

Carbohydrate: 21g (Fiber: 4g, Sugar:15g)

Fat: 0g

Protein: 1g

Compared to the previous berries, blueberries have less fat and more carbohydrate, and a sweeter taste.

Blueberries provide a reasonable source of:

- ☆ Vitamin C: 24 per cent RDA
- ☆ Vitamin K1: 36 per cent RDA
- ☆ Manganese: 25 per cent RDA

In the same fashion as blackberries, blueberries exert a protective effect on LDL and protect the particles from oxidation. Studies also show that blueberries have anti-carcinogenic properties, as well as improving insulin sensitivity.

Blueberries are easy to find, and you can buy them either fresh or frozen. Recently, blueberry wine is also becoming popular.

Regarding their nutritional content, there's no real difference between the two and frozen blueberries are just as good for you

5. Boysenberry

They are similar in appearance to raspberries and blackberries.

Calories and Macronutrients Per Cup Boysenberries

Calories: 66

Carbohydrate: 16g (Fiber: 7g, Sugar: 9g)

Fat: 0g

Protein: 1g

Boysenberries contain a good amount of:

- ☆ Folate: 21 per cent RDA
- ☆ Manganese: 36 per cent RDA
- ☆ Vitamin K: 13 per cent RDA

Boysenberries also have some impressive health benefits, specifically concerning the cardiovascular system. In one particular study, boysenberries reduced blood pressure and improved endothelial function. Notably, the impact of a single administration showed the same effects as chronic daily ingestion.

6. Chokeberry (Aronia Berries)

Also going by the name of Aronia, chokeberries are one of the most bitter tasting types of berries.Also, they have a very dry taste due to their high tannin content.

However, they are full of lots of beneficial nutrients.

Calories and Macronutrients Per Cup Chokeberries

Calories: 66

Carbohydrate: 13g (Fiber: 7g, Sugar: 6g)

Fat: 0g

Protein: 1g

Chokeberries contain a good amount of the following micronutrients:

- ✰ Vitamin C: 49 per cent RDA
- ✰ Iron: 11 per cent RDA
- ✰ Vitamin A: 10 per cent RDA

In fact, there are two different types of chokeberries: red and black. The black ones are quite prevalent, but red chokeberries are harder to find. Lots of Aronia powder supplements are also available, which claim to improve your health significantly. However, eating real fruit is better than some powdered product.

The chokeberry is probably one of the healthiest types of berries, but for me, it's far from the tastiest.

Various studies show the beneficial impacts of chokeberries, with some of the latest research showing that:

- ✰ Polyphenols in chokeberries help strengthen the immune system and have anti-inflammatory mechanisms
- ✰ In animal studies, polyphenols in chokeberries protect against oxidative damage during intense exposure to UV radiation

7. Cloudberry

Cloudberries are an amber-orange colored fruit that has a shape like a cloud.

Calories and Macronutrients Per Cup Cloudberries

Calories: 71.5

Carbohydrate: 10g

Fat: 0.2g

Protein: 3.5g

Cloudberries are a **major** source of vitamin C:

- ✰ Vitamin C: 368 per cent RDA

The best thing about cloudberries might be their unusual appearance, but they are also an impressive source of vitamin C without the large amounts of fructose some other fruits provide.

Cloudberries are also quite high in protein for a berry.

Regarding their benefits, like other types of berries, cloudberries have been shown to protect against LDL oxidation in a clinical setting.

8. Cranberry

Cranberries are one of the most famous types of berries in the world.

From juice to dried berries, alcohol, and jams, they're used to make all sorts of different things.

Calories and Macronutrients Per Cup Cranberries

Calories: 51

Carbohydrate: 13g

Fat: 5g

Protein: 4g

Cranberries contain a good source of vitamin C:

- ☆ Vitamin C: 24 per cent RDA

First of all, cranberries have a slightly sour taste.

As a result of this, many cranberry products are loaded with sugar as well as vegetable oil.

Cranberry juice is often recommended as a treatment for urinary tract infections (UTIs), based on a wealth of studies suggesting they have a beneficial effect.

9. Elderberry

Elderberries are a tiny variety of berry that people often use to make tea.

Calories and Macronutrients Per Cup Elderberries

Calories: 106

Carbohydrate: 26.7g

Fat: 0.7g

Protein: 1g

Nutrients and Health Benefits

Elderberries provide an excellent source of the following nutrients:

- ☆ Vitamin C: 87 per cent RDA
- ☆ Vitamin A: 17 per cent RDA
- ☆ Vitamin B6: 17 per cent RDA

- ☆ Iron: 13 per cent RDA
- ☆ Potassium: 12 per cent RDA

Generally, most elderberries have a tart and bitter taste.

Because of this, it's easy to find many sweetened elderberry products such as elderberry tea and jam. Similar to many other dark berries, elderberries contain flavonoids called anthocyanins which have many health benefits. Elderberries have a history as a traditional anti-viral treatment, and in clinical studies, elderberry flavonoids compare well to anti-influenza meds such as Tamiflu

10. Goji Berry (Wolfberry)

Appearing over the last decade or so, goji berries are now available almost everywhere, usually as a dried berry. Also known as 'wolfberry,' goji berries are native to East Asia and are traditionally made into a tea in China and Korea.

Calories and Macronutrients Per Cup Goji Berries

Calories: 92

Carbohydrate: 24g (Fiber: 8g, Sugar: 16g)

Fat: 0g

Protein: 3g

Nutrients and Health Benefits

Goji berries provide an excellent source of the following nutrients:

- ☆ Vitamin A: 50 per cent RDA
- ☆ Copper: 28 per cent RDA
- ☆ Selenium: 25 per cent RDA
- ☆ Riboflavin: 21 per cent RDA
- ☆ Iron: 14 per cent RDA

As shown above, goji berries are fairly high in nutrients but with a higher sugar content than other types of berries. The main reason for this is that they are dried, so they have a higher sugar concentration. Goji berries contain the antioxidant zeaxanthin, which has proven benefits for our eyesight. A recent study shows that 90-day supplementation significantly increases plasma zeaxanthin levels.

On the positive side, goji berries taste amazing.It's hard to describe the taste because they're unique.Chewy, flavorful, tough, but soft inside... a little sweet and slightly bitter. And that probably makes no sense, but yes – goji berries taste pretty good.

11. Gooseberry

Gooseberries are a very sour, tart berry that grow all over the world.

Calories and Macronutrients Per Cup Gooseberries

Calories: 66

Carbohydrate: 15g (Fiber: 6g, Sugar: 9g)

Fat: 1g

Protein: 1g

Gooseberries provide the following nutrients:

- Manganese: 11 per cent RDA
- Vitamin C: 69 per cent RDA

Not only is it possible to see fresh gooseberries, but also frozen, canned, and dried ones.

As previously mentioned, they have a very sour taste – something akin to a sour grape.

However, they are also a little sweet which balances the taste more than a lemon, for instance.

The berry contains a good amount of antioxidants, and it's quite nutritious.

12. Huckleberry

Calories and Macronutrients Per Cup Huckleberries

Calories: 55

Carbohydrate: 9g

Fat: 0g

Protein: 0g

Huckleberries provide some vitamins and minerals, but in non-significant amounts.

Like other dark purple fruit, huckleberries are also high in anthocyanin flavonoids.

Despite looking the same, huckleberries and blueberries have quite a few differences.

For one thing, blueberries contain a lot more carbohydrate (mainly sugar) – and thus a higher amount of calories too. Therefore, the taste is also a little different – with blueberries having an understandably sweeter taste. While blueberries are usually commercially cultivated, huckleberries are mainly found in the wild. As growers often breed commercial fruit for sweetness, this is likely one of the reasons for the difference in taste.

13. Lingonberry

Lingon berries are another highly touted berry full of healthy flavonoids, with a range of commercial powders and drinks springing up around them.

But is there anything unique about them?

Calories and Macronutrients Per Cup Lingonberries

Calories: 71

Carbohydrate: 16.3g (Fiber: 3.7g, Sugar: 8.3g)

Fat: 0.5g

Protein: 1g

Lingonberries provide a good source of the following nutrients:

- ✰ Vitamin C: 72 per cent RDA

In healthy human volunteers, consuming lingon berries along with 50g of glucose lessened the glycemic response. However, consuming a lingon berry powder drink had zero effect, likely because the lingon berry compounds are lacking their natural fibrous structure.

Lingonberries are also associated with a healthier gut microbiota and reduced plasma markers of inflammation.

14. Loganberry

While sounding quite similar to lingonberry, loganberries are an altogether different fruit.

Calories and Macronutrients Per Cup Loganberries

Calories: 91

Carbohydrate: 19g (Fiber: 8g, Sugar: 11g)

Fat: 0g

Protein: 2g

Lingonberries provide the following nutrients:

- ✰ Manganese: 92 per cent RDA
- ✰ Vitamin C: 72 per cent RDA
- ✰ Vitamin K: 14 per cent RDA
- ✰ Folate: 10 per cent RDA

The loganberry is also a hybrid cross between a raspberry and a blackberry, but it measures slightly longer in length. Regarding taste, it is also somewhere in between the two.

Apparently, it was accidentally created by a horticulturist in the late 19th century.

Loganberries can be eaten fresh or used to make various condiments.Similar to other types of berries, loganberries are high in vitamin C and contain beneficial flavonoids.

15. Raspberry

As one of the most mainstream berries, raspberries are a popular fruit around the world.

Calories and Macronutrients Per Cup Raspberries

Calories: 64

Carbohydrate: 15g (Fiber: 8g, Sugar: 5g)

Fat: 1g

Protein: 1g

Raspberries provide the following nutrients:

- ☆ Vitamin C: 54 per cent RDA
- ☆ Manganese: 41 per cent RDA
- ☆ Vitamin K: 12 per cent RDA

As shown above, raspberries have more fiber than they do sugar, as well as a good amount of vitamin C.

Raspberries contain a wide variety of polyphenols, and a growing body of evidence suggests they help reduce the risk of metabolic disease.

Taste wise, raspberries are also a delicious choice and make a particularly good match with some fresh cream.

16. Red Mulberry

First, there are two different varieties of mulberries: red and white.

As can be seen, the red kind looks slightly similar to a raspberry but longer and thinner.

Calories and Macronutrients Per Cup Red Mulberries

Calories: 66

Carbohydrate: 13.7g (Fiber: 2.4g, Sugar: 11.3g)

Fat: 0.5g

Protein: 2g

Red mulberries provide the following nutrients:

- ☆ Vitamin C: 85 per cent RDA
- ☆ Vitamin K: 14 per cent RDA
- ☆ Iron: 14 per cent RDA

17. Salmonberries

There are many interesting types of berries, but salmonberries have a unique orange color which looks impressive.

Calories and Macronutrients Per Cup Salmonberries

Calories: 66

Carbohydrate: 14g (Fiber: 2.5g, Sugar: 5g)

Fat: 0.5g

Protein: 1g

Salmonberries provide a good source of the following nutrients:

- ✰ Manganese: 75 per cent RDA
- ✰ Vitamin K: 25 per cent RDA
- ✰ Vitamin C: 20 per cent RDA
- ✰ Vitamin A: 15 per cent RDA
- ✰ Vitamin E: 10 per cent RDA

The color of salmonberries is a bright orange, and the taste is slightly sweet, a tiny bit sour, and very juicy. Regarding their health benefits, they are a very good source of manganese, and vitamins A, C, and K. Additionally, studies suggest that wild salmonberries are an exceptional source of antioxidants.

18. Strawberries

As one of the most popular types of berries in the world, not much needs to be said about strawberries.

Calories and Macronutrients Per Cup Strawberries

Calories: 48.6

Carbohydrate: 11.7g (Fiber: 3g, Sugar: 7.4g)

Fat: 0.5g

Protein: 1g

Strawberries provide a good source of vitamin C and manganese:

- ✰ Vitamin C: 149 per cent RDA
- ✰ Manganese: 29 per cent RDA

With attention to their size, strawberries are one of the biggest berry varieties around.

Fresh strawberries are available almost everywhere, but frozen strawberries are a great option too. A randomized controlled trial examining the benefits of strawberries found that in metabolic syndrome patients, daily supplementation decreases cardiovascular risk factors.

The results showed that strawberries have both anti-hypertensive and HDL-raising properties.

Strawberry Desserts and Drinks

Strawberries are also used in so many dessert recipes. Perhaps the easiest and simplest is berries and cream, which takes no time to make and tastes amazing. And that's not all; strawberries are also perfect for making wine and cocktails.

19. Tayberry

Tayberries are another species of berry closely related to raspberries. In fact, they are a cross between raspberries and blackberries.Although they have many similarities, the **difference between loganberries and tayberries is** the size and sweetness.

Originally loganberries were an unintentional cross-breed, whereas tayberries are specially cultivated for size and a sweet taste.

Calories and Macronutrients Per Cup Tayberries

Calories: 81.77

Carbohydrate: 16.7g (Fiber: 9.1g, Sugar: 6.18g)

Fat: 0.91g

Protein: 1.68g

Like other berries related to raspberries, tayberries are rich in vitamin C and flavonoids.

20. White Mulberry

Unlike the red mulberry which is native to the United States, the white mulberry is a Chinese native berry.However, it is now widespread in America and slowly displacing the native red mulberry.

White mulberries are also on sale in a dried form.

Calories and Macronutrients Per Cup White Mulberries

Calories: 66

Carbohydrate: 13.7g (Fiber: 2.4g, Sugar: 11.3g)

Fat: 0.5g

Protein: 2g

White mulberries provide the following nutrients:

- ✰ Vitamin C: 85 per cent RDA
- ✰ Vitamin K: 14 per cent RDA
- ✰ Iron: 14 per cent RDA

White mulberries are high in vitamin C and flavonoids and have an all-around healthy profile.

Similar to other types of berries, people say they help prevent diabetes. While they may assist in some small way, a healthy diet low in digestible carbohydrate is significantly more effective.

Storage

Blackberries

Due to their high perishable nature, prompt precooling of blackberries to -0.6 to 0°C and 90-95 per cent relative humidity is essential, especially when the berries are destined to be shipped to distant markets. These berries cannot be stored satisfactorily for more than about 2-3 days, because longer storage results in loss of good marketing quality. Storage of machine-harvested blackberries in 20 and 40 per cent CO_2 at 20°C for up to 48 h maintained their raw and processing quality. These investigators further stated that the use of high-CO_2 storage atmosphere with blackberries held at 20°C partially offset the need for refrigeration to reduce their post-harvest quality loss. Handling of blueberries should be minimized to avoid damaging the bloom. Blueberries stores at temperatures of 4.5°C and above gradually develop an undesirable, tough-textured skin. The use of controlled atmosphere to extend the self life of blueberries has not been successful.

Cranberries

Cranberries can safely be held at 0°C only for about 2 weeks. At low temperatures (less than 2.2°C), however, they usually develop flesh breakdown, characterized by a rubbery-textured red flesh. Storage environment of 2.2-4.4°C with 90-95 per cent relative humidity is ideal to store sound berries for up to 2-4 months. Refrigerated storage extended the shelf life of berries up to 10 weeks at 4.4°C with less water loss than in those at simulated common storage (15.6°C).

Most of the cranberry crop is now processed into sauce or juice. Fruit intended for processing is screened as rapidly as possible and utilized or frozen shortly after harvest. Such fruit can be frozen at 18°C until needed without quality loss. Freezing berries enhances development of the desired color in the juice.

Gooseberries and Currants

Gooseberries and currants are rarely stored, owing to their high perishability. Storage conditions of 0°C with 90-95 per cent relative humidity holds the fruit in a marketable condition for about 2 weeks. As soon as they are picked, red currants, black currants and gooseberries should be precooled to a temperature lower than 4°C to reduce damage during transport to market.

Storage of hard green gooseberries forlonger period at 0°C in perforated polyethylene bags is possible if some CO_2 is allowed to accumulate.

Kiwi Fruits

The best protection of kiwi fruit after harvest requires cooling to near storage temperature within 6 h of harvest, avoiding any ethylene exposure, and storing at 0°C. The cooling can best be achieved by using forced-air cooling. Kiwi fruit should be stored at a flesh temperature of 0°C. At this temperature, flesh softening is slowed substantially when the ethylene level is below 100 ppb. Control pf water loss and shriveling of kiwi fruit requires a relative humidity in the storage room at 90-95 per cent and air flow during storage be no higher than is needed to maintain fruit temperature. Freshly harvested kiwi fruit can be effectively ripened for early-season marketing by exposure to 10 ppm ethylene for 24 h at 20°C followed by maintenance at 20°C until soft.

Several studies have been conducted on the potential benefit of controlled atmosphere (CA) storage for kiwi fruit. The major benefits are a delay in flesh softening during storage and a reduction in Botrytis rot problem. Best results have been obtained in a controlled atmosphere of about 5 per cent CO2 and 2 per cent O2 at 0°C. The fruit must be promptly cooled and placed under CA conditions as soon after harvest benefits and enhance flesh softening. Carbon dioxide levels of 10 per cent in storage may be injurious to kiwi fruit. The kiwi fruit is very sensitive to ethylene. During packing and storage, fruit must be protected from exposure to even low levels of ethylene gas; concentrations above 6.03 ppm are considered unacceptable. For this reason, kiwi fruit should never be packed or stored together with other fruit crops.

Botrytis stem end rot (*Botrytis cinerea*). Alternaria rot (*Alternaria alternata*). Dothiorella soft rot (*Dothiorella gregaria*). Phoma rot (*Phoma* sp.) Phomopsis stem-end rot (*Typhula sp.*) are some of the important post-harvest diseases of kiwi fruits. It is important to maintain cleanliness in the vineyard to avoid fruit injuries during handling, to brush the fruit to remove dead dloral parts and other material on the fruit surface, to avoid contamination, to control the fruit rapidly and to maintain a constant 0°C storage temperature.

Raspberries

Prompt and continuous cooling of harvested raspberries to 0°C is essential to preserve their marketable life for 2-3 days. Holding conditions of 0°C and 90-95 per cent relative humidity through out postharbvest handling have been recommended. This temperature retards ripening and development of gray mold rot. Rhizopus rot. And Cladosporium rot and allows 2 or 3 days of storage prior to marketing. Smith reported that an atmosphere containing 20-25 per cent CO_2, slowed ripening and reduced decay, especially during refrigerated transport.

Strawberries

All types of berries, including strawberries are precooled rooms or forced-air coolers. Wetting the fruit by hydrocooling is generally not advocated. Goble and Cooler, however, reported that hydrocooling of strawberries to 4.4°C increased consumer acceptance due to the increased freshness and reduction in weight loss.

Owing to their high perishability, strawberries are rarely stores. Storage at a temperature of -0.6 to 0°C and 90-95 per cent relative humidity can extend the shelf life of strawberries for about a week. Prompt removal of field heat extends the marketable life of the fruit by about a week if low temperature is provided. It has been found that the rate of respiration of fresh Strawberries soon after harvest was about 7.2 times as high at 20°C as at 0°C. Although strawberries are quite tolerant to higher CO_2 levels in the storage atmosphere, low O_2 concentrations (below 2 per cent) have been known to cause fermented flavor. It has been found that with increased CO_2 the useful life of strawberries could be extended by about 50 per cent over that in air at the same temperature. Since, strawberries can tolerate high levels of CO_2 (up to 20 per cent), such high CO_2 concentration can be used to inhabit decay and retard the softening of strawberries without impairing the delicate flavor of the fruit.

Processing of Berries for Value Added Products

Jam

Strawberry jam is popular among consumers, because of its natural color and flavor. The maturity of the fruit affects the quality and storage stability of the jam. Low-sugar jam with replacement of sucrose by non caloric sweeteners is becoming popular.

Juice

Juices from berries are widely used for preparation of fruit jellies, beverages and ice cream. A significant quantity of juice is utilized for preparation of wine. In preparation of juice, frozen berries are thawed over right at room temperature, mashed and heated to 50°C to extract the pulp. Pectolytic enzymes are added to the pulp (0.1 g/kg of fruits). The enzymes are allowed to act for 2 h at 50°C. The juice is recovered by centrifugation of the pulp followed by filtration. The juice is preserved by the addition of antimicrobials such as potassium benzoate or sodium sorbate and pasteurization at 85-90°C for 1 min. The juice yield and quality are influenced by fruit maturity and processing parameters such as freezing-thawing, heat, pectinase treatment and fining.

Pulp or puree is one of the processed kiwi fruit products finding use in the international market. Although heat can be used to destroy microorganisms in production of kiwi fruit pulp, undesirable changes take place in the color and flavor of the product. Consequently, it is recommended that the pulp of kiwi fruit be frozen immediately after manufacture and stored at -18°C.

Concentrates

Juice concentrate is widely produced as natural flavoring. However, its use is hindered by its susceptibility to browning and undesirable flavor changes when held at room temperature for a short time or for longer periods at refrigerated temperatures. In spite of the general popularity of natural strawberry flavor, this problem has hindered the application of strawberry concentrates to foods and beverages.

Wine

Berries can be used for preparation of wine. Frozen strawberries are thawed at room temperature and juice is prepared. The juice is ameliorated to 22° Brix by adding sucrose. One percent ammonium phosphate and 1 per cent active yeast culture are added and mixed. The juice is transferred to jars and allowed to ferment at 16°C until it reaches 0.1-0.2 per cent reducing sugars. The wines are bottled, racked and stored in the dark. Strawberry wine has appealing color. The composition, maturity and mold contamination affects the quality of the wine. In case of raspberry wine it has been observed that processing conditions and storage influenced the degradation of anthocyanins, color and appearance of wine.

A good-quality wine can be prepared from kiwi fruit pulp. Earlier attempts to make using conventional practices were unsuccessful because the product was described as grassy, green salty in aroma and taste and as being of unacceptable bitterness and astringency. White wine of outstanding character can be prepared from kwi fruits. Juice can be extracted from crushed fruit with a rack and cloth press. The yield of juice (55.60 per cent) was increased to 84 per cent by pectolytic enzyme or press-aid treatment of pulp before pressing. The juice had high acidity (2 per cent) and low sugar (10 per cent), which necessitated juice amelioration for production of balanced wines. Wine made by fermenting juice and clarifying to brilliance with commercial pectolytic enzymes developed an intense, fruity, Riesling Sylvaner-type aroma during fermentation and retained no undesirable astringency or bitterness.

Medicinal Properties

In addition to their uses in food products, many of these berries such as cranberries, blueberries and blackberries are popular for their medicinal properties. The extracts of berries have been used as herbal remedies or as well-defined pharmaceutical products for a long time. Some of their applications are as antiseptics, diuretics, anti-inflammatory, antihyperglycemic and anticarcinogenic agents. However, very little is known about the active compounds or the mechanism of action. In recent years, there has been a renewed interest in the elicidation of medicinal properties and nature of active phyto chemicals in these berries.

Antimicrobial Properties

The antimicrobial properties of various berries are well known. Studies show that strawberry extract inactivated several enteric viruses and herpes simplex virus. Inactivation of poliovirus type 1 with strawberry extract was dependent on time, pH and concentration. Strawberry extract also reduced subsequent infection with the virus. The inactivation of poliovirus type 1 by raspberry extract has been reported. Fruits of vaccinium species have been antifungal, antibacterial and antiviral properties. The fruits contain a wide array of potentially antimicrobial phenolic compounds and high levels of organic acids. Extracts of cranberry and high bush blueberry were effective against various fungi such as *Penicillium* spp. and *Aspergillus niger*. Fractionation studies have demonstrated that the primary sources of antifungal activity are water-soluble, nonalkaloidal chemicals such as phenolics and acids. It is also reported that the antifungal activity of the fall-ripening species such as cranberries was much higher than the summer-ripening species such as high bush blueberries. The antiviral properties of cranberry and cowberry extracts have also been reported. Cranberry juice is found effective against poliovirus type 1 and the response was greater at pH 7.0 than at pH 2.6, similar to strawberry extract. Cowberry extracts have been reported to be active against influenza virus A2 and herpes virus A2.

The antibacterial properties of cranberry juice have been known for a long time. For decades cranberry juice has been prescribed in the treatment of bacteriuria and pyuria. Earlier studies suggested that the acidification of the urine was the mechanism by which cranberry juice produced bacteriostatic effect. However, recent studies have indicated that the effect could be due to an inhibition of bacterial adherence to mucosal surface by cranberry.

Antioxidant, Anticarcinogenic and Cardioprotective Properties

The extract of many of these berries rich in various flavonoids and related compounds have antioxidant properties. The properties of anthocyanins, other flavonoids and phenolic acids as free radical scavengers and inhibitors of lipid peroxidation have been described by many investigators. Several researchers in recent study reported the antiradical activity of several cultivars of red currant, red raspberry, blackberry and high-bush blueberry. All the crude extracts also showed a remarkably high activity towards chemically-generated superoxide radicals. The extracts also showed inhibitory activity towards xanthine, oxadise an enzyme which initiates free radicals formation at the cellular level. Black currant extracts exhibited the highest activity, being the richest in both anthocyanins and polyphenols. On the other hand, red currant extract seemed to contain more active substances than the other crude extracts.

The antiradical activity of these berries may have significant beneficial effects against various chronic diseases associated with free radicals such as cancer, cardiovascular diseases, arthritis and autoimmune diseases. Crude anthocyanin extract of bilberry (V. myrtillus), which has a very similar anthocyanin compositions to high-bush and low-bush blueberries, is highly effective in the treatment of various microcirculation diseases. The extract inhibits platelet aggregation and reduces adhesiveness in vitro and in vivo. It also functions as a vasodilatory agent and is effective in promoting and enhancing arterilar rhythemic diameter changes. In Europe, the extract (myrtocyan) is marketed (Indena, Milan) as a pharmaceutical product. Similarly, cowberry fruit extract has been reported to inhibit transplanted sarcomas (M-1 and 45, Piss lymphosarcoma and cholangioma RS-1 in rats.

REFERENCES

Arpaia, M. L., F. G. Mithchell, A.A. Kader 1984. Effects of delays in establishing controlled atmosphere in kiwi fruit softening during and following storage. *J. Am. Soc. Hort. Sci.* 109: 768.

Arpaia, M. L., F. G. Mithchell, and G. Mayer. 1980. The effects of ethylene and high CO_2 on the storage of kiwi fruit. *Hort Science* 15: 423.

Arpaia, M. L., F.G. Mitchell, A.A. Kader and G. Mayer. 1985. Effects of 2 per cent O_2 and varying concentration of CO_2 with and without $C_2 H_4$ on storage of kiwi fruit. *J. Am. Soc. Hort. Sci.* 110: 200.

Astridge, S. J. 1975. Cultivars of Chinese gooseberry (*Actinidia chinensis*) in New Zealand, *Econ. Bot.* 29: 357.

Avorn. J., M. Monane, J. H. Gurwaitz, R. J. Giynn, I Choodnovasky, and L. A. Lipstiz. 1994. Reduction of bacteriuria and pyuria after ingestion of cranberry juice. *J. Am. Med. Ass.* 271: 751.

Baj, A., B. Bombardelli, B. Gabeeta, and E. M. Martienelli. 1983. Qualitative and quantitative evaluation of Vaccinium myritillus anthocyanins by high – resolution gas chromatography and high performance liquid chromatography, *J.Chromato.* 279.

Ballinger, W. E., E, P. Maness, and W. F. MeClure. 1978. Relationship of stages of ripeness and holding temperature to decay development of blueberries. *J. Am. Soc. Hort. Sci.* 103: 130.

Beever, D.J. 1993. Kiwi Fruit, Encyclopaedia of Food Science. Food Technology Nutrition (R. Macree, R.K. Robinson, and S.J. Sadler, eds.) Academic Press, New York, p. 2626.

Bettani. V., R. Aragon, M.B. Bettini, G. Braggion, L. Calore, M. T. Concolato, P. Favorao, G. Penada and S. Montin 1991. Vasodilator and inhibitory effects of Vaccinium myrtillus anthocyanosides on the contractile responses of coronary artery segments to acetycholine: Role of the prostacylins and of the endothelium –derived relaxing factor, *Fitoterapia* 62: 159.

Brooks, R.M., and H.P Olmo 1982. Register of new fruits and nut varieties List 32: *Horticulture* 17(70): 17.

Cappelini, R.A., A.W. Stretch, and J.M. Maiello. 1972. Fungi associated with black berries held at various storage times and temperature, *Phtopathology,* 62: 68.

Cipoilini, M. L., and E.M. Stiles, 1992. Antifungal activity of ripe Ericaceous fruits: Phenolic-acid interactions and palatability for dispersers, *Biochem. Syst. Ecol.* 20: 501.

CISR, 1980. Wealth of India; Raw material, Council of scientific and industrial Research (CISR), New Delhi India, p. 69.

Colatuoni, A.., Betuglia, M. J. Magistretti, and L. Donato 1991. Effects of Vaccinium myrtillus anthocyanosides on arterial vasomotion, *Arzneim Forsch.* 41: 905.

Cossa, G., C. Trova and G. Gandolfo, 1988. Extraction of identification of kiwi fruit volatile constituents. Industries Alimentari 27: 531.

Costantino, L., A. Albasini, G. Rastelli, and S. Benvenuti, 1992. Activity of polyphenolic crude extract as scavengers of superoxide radicals and inhabitors of xanthine oxidase, *Planta Med.* 58: 342.

Croteau, R. J., and I. S. Furgerson, 1968. Volatile compounds in cranberries (*V. macrocarpan*). *J. Food Sci.* 33: 386.

FAO, 1990. Production Yearbook. Vol. 44 Food and Agriculture Organization, Rome, p.171.

Green, A., 1971. Soft fruits, The Biochemistry of Fruits and their Products. Vol. 2 (A. C. Hulme, ed.), Academic Press, New York, p. 375.

Hang, Y.D., B.S. Luh, and E.E. Woodams, 1986. Microbial Production of citric acid by solid state fermentation of kiwi fruit peel. *J. Food Sci.* 52(1): 226.

Harris, C. M., and J.M. Harvey 1973. Quality and decay of California strawberries stored in CO_2 enriched atmosphere. Plant Dis. Rep. 57: 44.

Harris, S., 1981. Ethylene and kiwi fruit. *The Orchardists (N.Z)* 54: 105.

Honkaren, E., P. Tapani, and T. Hirvi, 1980. The aroma of Finish wild raspberries (*Rubus idaeus* L.) Z. Lebensin Unters, Forsh. 171: 180.

Hruschka, H.W. 1970. Physiological breakdown in cranberries inhibition by intermitted warming during cold storage. *Plant Dis. Rep*. 54: 219.

Hudson, D.E, and W.H. Tietsen, 1981. Effects of cooling rate on shelf life and decay of high brush blueberries. *Hort Science* 11: 5: 656.

Hyvonen, L., and R. Torma, 1983. Examination of sugars, sugar, alcohols and artificial sweeteners as substitute for sucrose in strawberry jams: products development. *J. Food Sci*. 46: 183.

Kenny, A., 1971. Storage of gooseberries in plastic bags. *J. Food Technol.* 61: 403.

Kiepert, K., 1978. Pre-cooling and storage of currants, gooseberries and black currants, Landwristschafsk Rheinland 12: 459.

Kinney, A. B. and M. Blount 1979. Effects of cranberry juice on urinary Ph. *Nursing Res.* 28: 287.

Konowalchuk, J., and J. I. Speirs. 1976. Antiviral activity of fruit extracts. *J. Food Sci.* 41: 1013.

Konowalchuk, J., and J. I. Speirs. 1978. Antiviral effects of commercial juice and beverages. *Appl. Environ. Microbial,* 35: 1219.

Lewis, D.A., and B.S. Luh, 1988. Development and distribution and actinidin in kiwi fruit and its partial characterization. *J. Food Biochem.* 12: 409.

Lock Hart, C. L., F. R. Forsyth, R. Stack, and I. B. Hall. 1971. Nitrogen gas suppresses microorganisms on cranberries in short term storage. *Phytopathology* 61: 335.

Lodge, N., T.T. Ngyyen, and D. Mclntyre. 1987. Characterization of crude kiwi pectin extract. *J. Food Sci.* 52: 1095.

Lokar, L.C., and I. Poldini, 1988. Herbal remedies in the traditional medicine of the Venezia Giulia region (northeast Italy). *J. Ethnopharmacol.,* 22: 231.

Lundahl, D.S., M.R. McDaniel, and R.E. Wrolstad, 1988. Flavour aroma and compositional changes in strawberry juice concentrated stored at 20°C. *J. Food Sci.* 54: 396.

Mass, J.L., G.J. Galeeta, and G. D. Stoner, 1991. Elliagic acid, a anticarcinogen in fruit, especially in strawberries: A review. *Hort. Science* 26: 10.

Mass, J.L., S.Y. Wang, and G.L. Galleta, 1991. Evaluation of strawberry cultivars for ellagic acid content. *Hort. Science* 26: 266.

May, G., and G. Willuhn, 1978. Antiviral activity of aqueous extracts from medicinal plants in tissue culture. *Arzneim, Forsh.,* 28: 1.

Meunier, M.T., E. Duroux, and P. Bastide, 1990. Antiradical activity of procyanidolic oliromers and anthocyanosides towards superoxide anion and lipid peroxidation. *Plantes Med. Phytother*. 23: 267.

Mitchell, F.G., 1992. Post-harvest handling systems: Small fruits (table grapes, strawberries, kiwi fruit), Post-harvest Technology of Horticultural Crops (A.A. Kader, ed.), University of California, Davis, CA, Pub. 3311, p.223.

Mitchell, F.G., Arpia, M.L., and G. Mayer, 1982. Modified atmosphere storage of kiwi fruit. Proc. 3rd Natl. Controlled Atmosphere Research of Conf. (D. G. Richardson and M. Mehenuk, eds.), Timber Press, Beaverton, OR, p. 235.

Mithchell, F.G., 1990. Post-harvest physiology and technology of kiwi fruit. *Acta Hort.* 282: 291.

Miyazaki, M., S. Miya, and Yabbuchi, 1980. Factor causing seasonal changes in the number of hands calyx free strawberry fruit processing. *J. Jpn. Soc. Technol* 27: 564 (in Japanese).

Morazzoni, P. and M.J. Megistretti, 1990. Activity of Myrtocyan, AN anthocyanosiides complex from Vaccinium myrtillus on platelets aggregation and adhesiveness. *Fitoterapia* 61: 13.

Morris, J, R., S. E. Spayd, J.G. Brooks. and D. L. Cawthon, 1981. Influence of post-harvesting holding on raw and process quality of machine harvesting blackberry. *J. Am. Soc. Hort. Sci.,* 10: 69.

Namiki, M., 1990. Antioxidant/antimutagens in food. *Crit. Rev. Food Sci. Nutr.* 29: 273.

Naumann, W.D. and U. Wittenberg, 1980. Anthocynins, soluble solids, and titrable acidity in blackberry as influenced by preharvest temperature, Proc. Symp. on "Breeding and Machine Harvesting of Rubas". *Acta Hort.* 112: 304.

Ogura, N., K. Kumakawa, M. Fukishima, I. Soda, T. Sato, and Nakagawa, 1988. Studies on peroxidase of kiwi fruit. *J. Jpn. Soc. Food Sci. Technol.* 35(1): 1.

Park, E. Y., and B. S. Luh, 1985. Polyphenol oxidase of kiwi fruit. *J. Food. Sci.* 50(3): 678.

Perera, C.O., I. C. Hallett, T. T. Naguyen, and J.C. Charles, 1990. Calcium oxalate crystals: the irritant factor in kiwi fruit. *J. Food Sci.* 55: 10: 666.

Pilando, L., R. E. Wrolstad, and D. A. Heatherbell, 1985. Influence of fruit composition, maturity and mold contamination on color and appearance of strawberry wine. *J. Food. Sci.* 50: 1121.

Prestamo, G., 1989. Peroxidase of kiwi fruit. *J. Food Sci.* 54: 760.

Rommel, A., 1988. Red Raspberry and blackberry juice and wine: The effects of processing and storage on color and appearance. *M.Sc. Thesis,* Oregon State University, Corvallis, OR.

Rommel, A.D., A Heatherbell, and R. E. Wrolstad, 1990. Red raspberry juice and wine. Effect of processing and storage on anthocyanins pigment composition color appearances. *J. Food Sci.* 55: 1011.

Ryall, A.L and Pentzer, 1974. Handling and transportation and storage of fruits and vegetables, Vol. 2 Fruit and tree nuts, *AVI, Westport ct.*

Sabota, A.E., 1984. Inhibition of bacterial adherence by cranberry juice: Potential use for the treatment of urinary tract infections. *J. Urol.* 131: 1013.

Sale, P.R., 1985. Kiwi Fruit Culture, V. R. Ward Govt. Printers, Wellington, New Zealand, p. 96.

Sapers, G. M., S.B. Jones, and G.T. Maher, 1983. Factors affecting the recovery of juice and anthocyanin from cranberries. *J. Am. Soc.* 108: 245.

Sapers, G.M., S. B., Jones and J.G. Philips, 1985. Leakage of anthocyanins from skin of thawed frozen high-bush blueberries (*V. corymbosum* L.). *J. Food Sci.,* 50: 43.

Schmidt, D.R. and A.E. Sabota, 1988. An examination of the anti-adherence activity of cranberry juice on urinary and non-urinary bacterial isolates. *Microbes.* 55: 173.

Scott, K.J., S.A. Spraggon, and R.L. McBride, 1986. Two new maturity tests for kiwi fruit. *CSIRO Food Res. Quart.* 46: 25.

Shamaila, M., W.D. Powrie, and B.J. Shura, 1992. Analysis of volatile compound from strawberry fruit stored under modified atmosphere packaging. *J. Food Sci.* 57: 1173.

Soda, L, M. Kaneko, T. Sato, H. Nakagawa, and N. Ogura, 1987. Studies on utilization of kiwi fruit protease. *J. Jpn. Soc. Food Sci. Technol.* 34: 36.

Spayed, S.E. and J.R. Morrris, 1981. Influence of immature fruit on strawberry jam quality and storage stability, J. Food Sci. 46 414.

Stang, E.J., and E. L. Denison, 1981. Inflorescence and fruit development in concentrated and non-concentrated ripening strawberries. *J. Am. Soc. Hort. Sci.* 106: 101.

Valavichyusk-Yu, M., K.K. Yankyavichyus, A. Vaichyunene-Ya, B.Virbitskas-Yu, S. F. Budrene, and A. A. Pyasyatskene, 1989. Antitumor activity of medicinal plant in the Lithuanian SSR: 9. Garden angelica, common comfrey, mars violet, wonder violet and cowberry. *Liet. Tsr. Mokslu Akademijos Darbai Serija C Biol. Mokslai* 3: 146.

Wang, S. Y., J.L. Mass, E.M. Dineal, G. J. Gallettta, 1990. Improved HPLC resolution and quantification of ellagic acid from strawberry, black berry and cranberry. *Hort. Science* 25: 1078.

Warrington, I. J. and G. C. Westech (eds) 1990. Kiwi Fruit: Science and management, Ray Richard, in Association with the New Zealand Society of Horticulture of Science, Aukland.

Wesche-Ebeling, P., and M.W. Montgomery, 1990. Strawberry polyphenol oxidase: Extration and partial characterization. *J. Food Sci.* 55: 1315.

Wilkinson, A. E. 1987. The Encyclopaedia of fruit and berries and nuts and how to grow them, Shree publishers, New Delhi, India, p. 142.

Withy, L. M., and N. Lodge. 1982. Kiwi fruit wine. Production and evolution. *Am. J. Enol. Vitic.* 33(4): 191.

Wrolstad, R.E., G. Skade, P. Lea, and G. Enersen, 1990. Influence of sugar on anthocyanin pigment stability in frozen strawberries. *J. Food Sci.,* 55: 1064.

Chapter 5

Cherry

Introduction

Prunus avium, commonly called wild cherry, sweet cherry is a species of cherry, a flowering plant in the rose family Rosaceae. Cherry is the fruit of many plants of the genus *Prunus*, and is a fleshy drupe (stone fruit) and packed with healthy nutrients and excellent antioxidants. Cherries are mostly eaten as desert fruit or more conveniently in the brined, frozen or canned state. The cherry fruits of commerce usually are obtained from a limited number of species such as cultivars of the sweet cherry, *Prunus avium*. The name 'cherry' also refers to the cherry tree, and is sometimes applied to almonds and visually similar flowering trees in the genus *Prunus*, as in "ornamental cherry", "cherry blossom", *etc.* Wild cherry may refer to any of the cherry species growing outside of cultivation, although *Prunus avium* is often referred to specifically by the name "wild cherry" in the British Isles. Basically, cherries are native to Europe and Asia regions. Deandolle reported that the cherry first grew wild in northern Persia and the Russia provinces south of the Caucasus. From there, it spread rapidly because they are attractive to birds; hence the name *Prunus avium* L. or bird cherry.

Cherry trees offer products other than the fruit itself. The lovely, fragrant cherry blossoms are a rite of spring and are actually a tourist draw in places such as Washington, DC, and Door County, Wisconsin. In addition, parts of the tree itself have long been used for medicinal purposes. The bark, leaves, and seeds of the cherry trees contain cyanogenic glycosides—poisons that are lethal if ingested by children or animals. Native Americans and others use the leaves and carefully prepare teas with them for the treatment of colds or coughs. Others have experimented with cherry stalk tea in the treatment of kidney diseases. The cherry has also been associated with virginity from ancient times to the present day. The association may be derived from the fact that the red colored fruit that encircles a small seed symbolizes the uterus of Maya, the virgin mother of Buddha, who was offered fruit and succor by a holy cherry tree while she was pregnant.

Cherries are cultivated all over the world and top 3 producers of cherries are Turkey, USA and Iran. In India its cultivation is confined to Jammu and Kashmir Himachal Pradesh and hills of Uttar Pradesh. A delicious fruit, cherry is rich in protein, sugars and minerals, has more calorific value than apple. Due to higher return cherry is gaining popularity in temperate regions of the country.

Origin, History and Distribution

The sweet cherry originated in the area between the Black and Caspian Seas in Asia Minor. It is likely that bird feces carried it to Europe prior to human civilization. Greeks probably cultivated the fruit first. Romans cultivated the fruit as it was essential to the diet of the Roman Legionnaires (their use likely spread the fruit throughout Western Europe). It is believed that English Colonists brought the fruit to the New World prior to 1630, but they do not seem to have flourished in the eastern United States. Spanish Missionaries brought sweet cherries to California, and varieties were brought west by pioneers and fur traders as well. Sour cherries also are native to Asia Minor, and were brought over to the New World by settlers rather early as well. Basically, cherries are native to Europe and Asia regions. Deandolle reported that the cherry first grew wild in northern Persia and the Russia provinces south of the Caucasus. From there, it spread rapidly because they are attractive to birds; hence the name *Prunus avium* L. or bird cherry.

Botany

Many cherries are members of the subgenus *Cerasus*, which is distinguished by having the flowers in small corymbs of several together (not single or in racemes), and by having smooth fruit with only a weak groove along with one side, or no grooves. Most of eating cherries are derived from either *Prunus avium*, the sweet cherry (also called the wild cherry), or from *Prunus cerasus*, the sour cherry. The sweet cherry (*Prunus avium*) and the sour (red) cherry (*P. cerasus*) have their commercial origin in Europe. Sweet cherries were once known as "bird cherries" as their name (avium) indicate. Duke cherries are probably hybrids between sweet and red tart cherries. The most common native cherry of the United States is the choke-cherry (*P. virginiana*), grown in eastern parts and *P. dimissa*, cultivated largely in western regions of the United States. Sweet cherries can be divided into two major types based on fruit characteristics. Heart-type cherries are void or heart-shaped, with relatively soft flesh and often with early ripening. Most of the commercially important cultivars, however, are of the bigarreau type, with firmer, crisp-fleshed fruit and ripens during mid to late season. Fruit flesh skin may be dark (red nearly black) or light (yellow-red to yellow-white).

Cultural History

Pliny distinguishes between *Prunus*, the plum fruit and *Cerasus*, the cherry fruit. Already in Pliny quite a number of cultivars are cited, some possibly species or varieties, Aproniana, Lutatia, Caeciliana, and so on. Pliny grades them by flavour, including dulcis ("sweet") and acer ("sharp") and goes so far as to say that before the Roman consul Lucullus defeated Mithridates in 74 BC, *Cerasia ... non fuere in Italia*, "There were no cherry trees in Italy". According to him, Lucullus brought them in from Pontusand in the 120 years since that time they had spread across Europe to Britain. Although cultivated/domesticated varieties of *Prunus avium* (sweet cherry) didn't exist in Britain or much of Europe, the tree in its wild state is native to most of Europe, including Britain. Evidence of consumption of the wild fruits has been found as far back as the Bronze Age at a Crannog in County Offaly, in Ireland· Seeds of a number of cherry species have however been found in Bronze Age and Roman archaeological sites throughout Europe. The reference to "sweet" and "sour" supports the modern view that "sweet" was *Prunus avium*; there are no other candidates among the cherries found. In 1882 Alphonse de Candolle pointed out that seeds of *Prunus avium* were found in the Terramare culture of north Italy (1500–1100 BC) and over the layers of the Swiss pile dwellings. Of Pliny's statement he says, since this error is perpetuated by its incessant repetition in classical schools, it must once more be said that cherry trees (at least the bird cherry) existed in Italy before Lucullus, and that the famous gourmet did not need to go far to seek the species with the sour or bitter fruit. De Candolle suggests that what Lucullus brought back was a particular cultivar of *Prunus avium* from the Caucasus. The origin of cultivars of *P. avium* is still an open question. Modern cultivated cherries differ from wild ones in having larger fruit, 2–3 cm diameter. The trees are often grown on dwarfing rootstocks to keep them smaller for easier harvesting. Folkard (1892) similarly identifies Lucullus's cherry as a cultivated variety. He states that it was planted in Britain a century after its introduction into Italy, but "disappeared during the Saxon period. He notes that in the fifteenth century "Cherries on the ryse" (*i.e.* on the twigs) was one of the street cries of London, but conjectures that these were the fruit of "the native wild Cherry, or Gean-tree. The cultivated variety was reintroduced into Britain by the fruiterer of Henry VIII, who brought it from Flanders and planted a cherry orchard at Teynham.

Nutritional Composition and Uses

Cherries are rich in sugars, which contribute 50-60 per cent of total dry matter in red tart cherries. Fructose is a major sugar in desert cherries. Sorbital and mannitol have also been detected in cherry pulp. Maleic acid has been found to represent 75-95 per cent of total nonvolatile acid in red tart cherries. The presence of cyanidin-3 monoglucoside is reported in tart cherries. Carotenoides are present in some cherries and range in concentration from 1.2 to 9.5 μg/g tissue. The vitamin C content of cherries was found to range from 13 to 56 mg/100 g of edible fruit, depending on variety. Volatile compounds which have been reported in cherry include methanol, ethanol, butanol, pentanol, octanol, geraniol, ethyle acetate and benzyldehyde.

Medicinal Properties of Cherry Tree

Diuretic

It stimulates urine elimination, very useful in those cases in which it is necessary to stimulate the kidneys to increase maturation, in illnesses like: obesity, cellulite,

Table 5.1: Nutritional Composition of Cherries (per 100 gr)

Water	82gr
Carbohydrates	35gr
Proteins	0.9gr
Fiber	1.90gr
Fat	0.17gr
Potassium	230mg.
Copper	94mg
Manganese	63mg
Phosphorus	20mg
Calcium	17mg
Vitamin C	15mg
Vitamin B1	0.02mg
Vitamin B2	0.04mg
Vitamin B6	0.04mg
Vitamin E	0.1 mg

dropsy, (accumulation of liquids in the body with swelling of body tissues). Edemas, Kidney pain, nephritis, renal insufficiency, swollen eyes. *etc.* There are fundamentally two treatment types.

a. **Decoction of the floral peduncles or dry fruit decoction:** Boil during 10 or 12 minutes 40 g of peduncles after having allowed them to macerate in water for 6 or 7 hours. Filter the preparation. Take three cups a day, after the main meals. It is a treatment that has a great diuretic power and it has to be used with wisdom in those people that have hypotension problems (with low blood pressure, using this treatment can lower blood pressure too much).

b. **Infusion of peduncles:** Carry out an infusion of a spoonful of peduncles for each cup of water, for 5 minutes. Cool and take three cups a day after the main meals.

Antirheumatic: For its diuretic capacity it is used to help in the treatment of rheumatic illness: gout, arthritis, rheumatism, *etc.*

Circulator system: It fluidiflies the blood and improves blood circulation, being very appropriate for the treatment of illness related with a faulty circulation: varixes, hemorrhoids, ocular pressure, *etc.*

Cardiotonic: It has got cardiotonic properties, making the heart muscle to contract more powerfully, so it has been used in cases of light heart weakness that don't require the use of foxgloves.

External Use Preparations with Cherry Tree

Skin disease: The preparation of slighter decoctions that the one previously seen exercises a healing power on the skin, to get rid of pimples, acne and other skin disorders and favoring the wound scaring. (Boil about 80 g of dry peduncles in a liter of water. Cool and apply with a gauze on the affected surface)

Emmenagogue: Decoctions of floral summits are very adequate to favour menstruation, in cases such as dismenorrhea. It also alleviates women from premenstrual syndrome (Boil a teaspoon of floral summits for every glass of water during 10 minutes. Leave it cool and drink a couple of cups a day)

Edible Properties of Cherries

The fruit of the cherry is a very convenient food that should be taken in abundance in the short time in which we have it at hand, during spring time. Its little caloric power (52 calories each 100 gr.) and its richness in fibers is very useful to favor the intestinal evacuation, avoiding the constipation. In the same way, as it is very rich in vitamins and minerals it is very appropriate to strengthen the organism and to avoid the spring asthenia.

Its high potassium content makes it especially interesting in the control of the heart activity, the good state of the muscles and of the nervous system, as well as in the calcification of the bones- very useful in the treatment of osteoporosis, too. It is very advisable to undertake a cure of cherries in spring. This cure consists on eating during three days 1500 gr of cherries and three yoghurts every day.

Industrial Uses of Cherries

The wood of the cherry tree is very appropriate in turnery and joinery. If it is used for the fire, its smoke produces a very characteristic aroma.

Other Uses

- ☆ The gum from bark wounds is aromatic and can be chewed as a substitute for chewing gum.
- ☆ Medicine can be prepared from the stalks (peduncles) of the drupes that is astringent, antitussive, and diuretic.
- ☆ A green dye can also be prepared from the plant.
- ☆ Wild cherry is used extensively in Europe for the afforestation of agricultural land and it is also valued for wildlife and amenity plantings. Many European countries have gene conservation and/or breeding programmes for wild cherry.

Toxicity of Cherries

The cherry tree as the rest of trees belonging to the Prunus genus, such as the plum tree (*Prunus domestica*); the apricot tree (*Prunus armeniaca*); the peach tree (*Prunus persica*); or the almond tree (*Prunus dulcis*) contains in its seeds, flowers and leaves the cyanogenetic glycoside amygdalin. This compound that, in the case of the almond tree, appears with a very big proportion in the seeds of the bitter almonds it is responsible for much intoxication Amygdalin – because of the action of the ferment emulsion, in contact with the saliva, becomes cyanhydric acid, -cyanide –

a very potent poison. In a same way cases of intoxications have been detected in animals, especially pigs that have eaten seeds, flowers or leaves of these plants. The intoxication symptoms are as: Suffocation, bad breath, vomiting, sickness, increase of the heart rhythm, respiratory failure and death.

Harvesting and Yield

The cherries are harvested at full maturity when surface color changes from green to red. The fruits are packed in boxes lined with paper. Generally 5 kg boxes are used for packaging. For fresh market the cherries are harvested manually. Practically all of the red tart cherry which is canned or frozen is harvested mechanically. Limb shakers or trunk shakers mounted either on self-propelled catching frames or on tractors which move the frames are commonly employed for this purpose. Special care is needed to handle the fruit gently during and following harvest to avoid physical injury to the fruit. Cherries from the catching frames are transferred into pallet tanks containing water to avoid bruising. Internal darkening of bruised fruit can be minimized by prompt and continuous cooling of the fruit from tree to the processing line. Except for a small proportion of cherries required for brining, sweet cherries are harvested after full maturity. For the fresh market they are rarely harvested mechanically. The harvest maturity of red sweet cherries can be judged by a change in color from light red to black. The soluble solids content of sweet cherries varies from 15to 22 per cent. Maturity indices, such as color, specific gravity, flesh firmness, light reflectance of surface color and soluble solids by refractometer or balling hydrometer, have been employed successfully to judge the packing maturity of cherries. The fruit moisture status influences the textural and mechanical properties of harvested sweet cherries.

Yield depends on the fertility of the soil, rootstock used, plant density and finally on farm management practices. An average of 20 kg cherries per tree can be obtained.

Recommended and Popular Cultivars

Cultivars

Sweet cherry cultivars grown throughout the world include Napoleon (Royal Ann), Black Tartarian, Eagle, early Purple, Early Rivers, Elkhon, Hedelfingen, Knight's Early Black, Lyon Schmidt, Windson, Van, Sam, Vista, Victor Sue, Vega, Summit, Stella, Chinook, Rainer, Bing, Lambert, Black Republican, corum, Hoskins, Chapman, Burbank, Bush, Tartaraian, Mona, Larian, Jubillee, Berryessa and Bada, Ulster, Hudson and Angela. The most important sweet cherry cultivars in the western United States, where 80 per cent of the U.S. crop is grown, are being (the leading cultivar in North America), Van and Lambert, although others may be available because of their use as pollinizers. Bing is mainly a fresh market cultivar and Lambert is used both for canning and fresh market, Black Republican and other very dark cherries are good for freezing.

Sour cherry fruits are generally soft, juicy and depressed globose in shape, buit colors may range from Morello types, with red to dark red flesh and juice, to the

Amerelle types, with nearly colorless juice and flesh. There are several sour cherry cultivars grown in North America, ranging from the light red, early Richmond, to the medium red-skinned Montmorency, to the late dark red English Morello. Montmorency is the standard. In Western Europe, Schattenmorelle and Sternsbaer are common, but many others are grown in Russia, Yugoslavia, Romania and Hungary. Unlike sweet cherries, most are more or less self-fertile and generally do not require pollinizers.

Species

The list below contains many *Prunus* species that bear the common name cherry, but they are not necessarily members of the subgenus *Cerasus*, or bear edible fruit. Some common names listed here have historically been used for more than one species, *e.g.* "rock cherry" is used as an alternative common name for both *P. prostrata* and *P. mahaleb* and "wild cherry" is used for several species.

- ✰ *Prunus apetala* (Siebold and Zucc.) Franch. and Sav. - clove cherry
- ✰ *Prunus avium* (L.) L. - sweet cherry, wild cherry, mazzard or gean
- ✰ *Prunus campanulata* Maxim. - Taiwan cherry, Formosan cherry or bell-flowered cherry
- ✰ *Prunus canescens* Bois. - grey-leaf cherry
- ✰ *Prunus caroliniana* Aiton - Carolina laurel cherry or laurel cherry
- ✰ *Prunus cerasoides* D. Don. - wild Himalayan cherry
- ✰ *Prunus cerasus* L. - sour cherry
- ✰ *Prunus cistena* Koehne - purple-leaf sand cherry
- ✰ *Prunus cornuta* (Wall. ex Royle) Steud. - Himalayan bird cherry
- ✰ *Prunus cuthbertii* Small - Cuthbert cherry
- ✰ *Prunus cyclamina* Koehne - cyclamen cherry or Chinese flowering cherry
- ✰ *Prunus dawyckensis* Sealy - Dawyck cherry
- ✰ *Prunus dielsiana* C.K. Schneid. - tailed-leaf cherry
- ✰ *Prunus emarginata* (Douglas ex Hook.) Walp. - Oregon cherry or bitter cherry
- ✰ *Prunus eminens* Beck - German: *mittlere Weichsel* (semisour cherry)
- ✰ *Prunus fruticosa* Pall. - European dwarf cherry, dwarf cherry, Mongolian cherry or steppe cherry
- ✰ *Prunus gondouinii* (Poit. and Turpin) Rehder - duke cherry
- ✰ *Prunus grayana* Maxim. - Japanese bird cherry or Gray's bird cherry
- ✰ *Prunus humilis* Bunge - Chinese plum-cherry or humble bush cherry
- ✰ *Prunus ilicifolia* (Nutt. ex Hook. and Arn.) Walp. - hollyleaf cherry, evergreen cherry, holly-leaved cherry or islay
- ✰ *Prunus incisa* Thunb. - Fuji cherry
- ✰ *Prunus jamasakura* Siebold ex Koidz. - Japanese mountain cherry or Japanese hill cherry

- ☆ *Prunus japonica* Thunb. - Korean cherry
- ☆ *Prunus laurocerasus* L. - cherry laurel
- ☆ *Prunus lyonii* (Eastw.) Sarg. - Catalina Island cherry
- ☆ *Prunus maackii* Rupr. - Manchurian cherry or Amur chokecherry
- ☆ *Prunus mahaleb* L. - Saint Lucie cherry, rock cherry, perfumed cherry or mahaleb cherry
- ☆ *Prunus maximowiczii* Rupr. - Miyama cherry or Korean cherry
- ☆ *Prunus mume* (Siebold and Zucc.) - Chinese plum or Japanese apricot
- ☆ *Prunus myrtifolia* (L.) Urb. - West Indian cherry
- ☆ *Prunus nepaulensis* (Ser.) Steud. - Nepal bird cherry
- ☆ *Prunus nipponica* Matsum. - Takane cherry, peak cherry or Japanese alpine cherry
- ☆ *Prunus occidentalis* Sw. - western cherry laurel
- ☆ *Prunus padus* L. - bird cherry or European bird cherry
- ☆ *Prunus pensylvanica* L.f. - pin cherry, fire cherry, or wild red cherry
- ☆ *Prunus pleuradenia* Griseb. - Antilles cherry
- ☆ *Prunus prostrata* Labill. - mountain cherry, rock cherry, spreading cherry or prostrate cherry
- ☆ *Prunus pseudocerasus* Lindl. - Chinese sour cherry or false cherry
- ☆ *Prunus pumila* L. - sand cherry
- ☆ *Prunus rufa* Wall ex Hook.f. - Himalayan cherry
- ☆ *Prunus salicifolia* Kunth. (=*P. serotina*) - capulin, Singapore cherry or tropic cherry
- ☆ *Prunus sargentii* Rehder - Sargent's cherry
- ☆ *Prunus serotina* Ehrh. - black cherry, wild cherry
- ☆ *Prunus serrula* Franch. - paperbark cherry, birch bark cherry or Tibetan cherry
- ☆ *Prunus serrulata* Lindl. - Japanese cherry, hill cherry, Oriental cherry or East Asian cherry
- ☆ *Prunus speciosa* (Koidz.) Ingram - Oshima cherry
- ☆ *Prunus ssiori* Schmidt- Hokkaido bird cherry
- ☆ *Prunus stipulacea* Maxim.
- ☆ *Prunus subhirtella* Miq. - Higan cherry or spring cherry
- ☆ *Prunus takesimensis* Nakai - Takeshima flowering cherry
- ☆ *Prunus tomentosa* Thunb. - Nanking cherry, Manchu cherry, downy cherry, Shanghai cherry, Ando cherry, mountain cherry, Chinese dwarf cherry, Chinese bush cherry

- ✰ *Prunus verecunda* (Koidz.) Koehne - Korean mountain cherry
- ✰ *Prunus virginiana* L. - chokecherry
- ✰ *Prunus x yedoensis* Matsum. - Yoshino cherry or Tokyo cherry

Major Types/Varieties of Cherry Fruit in India

There are more than 100 varieties of sweet cherries that are divided into Bigarreau and heart groups.

Bigarreau Group: Under this group, cherries are round in shape and generally the fruit colour varies from dark to light red. The promising hybrid varieties in this group are lapins, summit, sue, sam, sunburst and stella.

Heart Group: Under this group, cherry fruits are in heart shape with tender flesh. Fruit color varied from darkish red to light coloured reddish.

Recommended Varieties in India

Himachal Pradesh: white Heart, Stella Lambert, Pink Early, Napoleon White, Black Tartarian, Van, Early Rivers and Black Republican.

Jammu and Kashmir: Bigarreau Noir Gross, Guigne Noir Hative, Early Purple Black Heart, Guigne Pour oa Precece, Black Heart, Bigarreau Napoleon and Guigne Pour ova Precece.

Uttar Pradesh: Governor's Wood, Bedford Prolific and Black Heart.

Maturity Indices

Skin color and soluble solids content (SSC) are the main criteria used to judge fruit maturity. Minimum maturity is that when the entire cherry surface have a minimum of light red color and 14 to 16 per cent SSC, depending on the variety. The red mahogany stage is recommended for harvest for many varieties as maturity and ripeness stages of cherries.

Quality Indices

Taste is related to SSC, titratable acidity (TA) and the ratio of SSC/TA. Freedom from cracks, ird pecks, shriveling, decay or misshapen fruit (doubles, spurs). Green fleshy stems are often associated with freshness and quality.

Pests

Generally, the cherry can be a difficult fruit tree to grow and keep alive.The first visible pest in the growing season soon after blossom usually is the black cherry aphid ("cherry blackfly", *Myzus cerasi*), which causes leaves at the tips of branches to curl, with the blackfly colonies exuding a sticky secretion which promotes fungal growth on the leaves and fruit. At the fruiting stage, the cherry fruit fly (*Rhagoletis cingulata* and *Rhagoletis cerasi*) lays its eggs in the immature fruit, where after its larvae feed on the cherry flesh and exit through a small hole (about 1 mm diameter), which in turn is the entry point for fungal infection of the cherry fruit after rainfall.

In addition, cherry trees are susceptible to bacterial canker, cytospora canker, brown rot of the fruit, and root rot from overly wet soil, crown rot, and several viruses.

Diseases and Disorders

The main cherry diseases are blossom and ripe fruit rots (caused by *Monilinia fructicola* and *Monilinia laxa*), bacterial canker (caused by *Pseudomonas syringae*), X-disease (caused by a phtyoplasma organism) and Phytophthora root and crown rot (caused by *Phytophthora* spp.).

Leaf Spot (fungus – *Wflsonia hiemalis*): This is the most common and most damaging leaf disease of cherries. It is most prevalent during rainy springs. The reddish spots drop out leaving circular holes in the leaves. With severe infection, defoliation may follow the appearance of the "shot hole" symptom. This disease can be controlled with foliar fungicides.

Shot-hole or Black Leaf spot (bacterium – *Xanthomonas pruni*): This bacterium also attacks peach and plum. The infected tissue dries up and falls out leaving a one-eighth inch diameter hole. It also causes a canker and gummosis of the stem.

Rust (fungus – *Tranzschelia discolor*): Rust also occurs on peach and plum. Yellow spots appear on the upper leaf surface with reddish pustules filled with spores on the lower surface. It can cause premature leaf drop lowering tree vigor. Apply fungicides in late summer and fall.

Bacterial Canker (bacterium – *Psuedomonas syrinigae pv. syringae*): This bacterium causes cankers, gummosis, and shoot blight. Sunken lesions appear in the spring with gummosis. Cracks and cankering follow, resulting in a girdled stem. The foliage becomes yellow, curled, and withered. Wet spring weather can result in shoot blight and a small angular leaf spot. This is usually only a problem in young orchards. Use resistant F12-1 Mazzard rootstock. Remove girdled branches. Spray with copper hydroxide in October and January. This bacterium also attacks peaches.

Crown Gall (bacterium – *Agrobacterium tumefaciens*): Warty tissue masses or galls may form on the roots, at the root collar, or anywhere on the stem. No practical controls are available.

Black Knot (fungus – *Apiosporina morbosum*): Black knot is a very conspicuous and common disease that is also found on apricots and plums. Large, rough black galls form on the twigs. Twigs may die beyond the galls. No chemical control is recommended. Infected limbs should be removed.

Brown Rot (fungus – *Monilinia fructicola* and *Monilitia laxa*): These fungi cause blossom blight, twig blight, and fruit rot. Control by fungicide application and sanitation.

Witches' Broom (fungus – *Taphrina cerasi*): This fungus causes large numbers of irregular small branches to develop on twigs and larger branches forming a witches' broom. Blossoms develop and the leaves come out on the brooms earlier than on the normal branches. Large numbers of brooms may eventually kill a tree. Cut and burn brooms.

Littleleaf: This is a condition caused by lack of zinc.

Little Cherry (virus): Fruit are half size. Flowering cherry acts as a reservoir for the virus. Buy only virus-free trees from reputable nurserymen and remove infected trees when they are discovered.

Necrotic Rusty Mottle: Foliage and blossoms are delayed in the spring. Brown necrotic or rusty chlorotic spots develop on the leaves, often falling out to leave shot holes. Defoliation may occur. Buy virus-free trees from reputable nurserymen. Virus infected trees should be removed from the orchard.

Rugose Mosaic (virus): General chlorosis of the leaf occurs between mid vein and margin with distortion. Fruit yield is reduced. The fruits become flattened and angular. Remove infected trees.

Rusty Mottle (virus): Many leaves turn bright yellow to red with green islands and drop before harvest. Those that remain have yellow-brown spots with a rusty appearance. The fruit is small, late, and tasteless. Remove diseased trees and select grafting material from virus-free trees.

Vein Clearing or Sweet Cherry Crinkle (probably genetic): A non-transmissible virus-like disease. Veins clear in localized areas or the entire leaf Margins of the leaves are irregular. Some are elongated with slot-like perforations. Small blisters appear on lower sides of veins, and the upper sides are silvery. Leaves fold along midrib, wilt, and drop in midsummer. Some branches will rosette. Flowers are plentiful, but few will develop into fruit. The fruit will be pointed and flattened on one side with a swollen ridge.

Ripe fruit rot, brown rot blossom, and **twig blight** (*Monilinia laxa* and *M. fruticola*) are most prevalent with rain or dew during bloom and moderate temperatures. Sweet cherry is the least susceptible of stone fruits to blossom rot and twig blight. Control is relatively simple with 1-2 fungicide sprays during bloom development; repeated applications may be necessary during frequent or heavy rain intervals, depending on the bloom stage.

Fruit rot and **twig blight**, also caused by *Monilinia laxa* and *M. fruticola*, occurs in both wounded and unwounded fruit. Wounding, fruit cracking and exposure to moisture increase the likelihood of infection. Infection begins in fruit and can spread to twigs, leading to dieback and branch cankers. Prompt cooling and storing fruit in refrigerated conditions can help reduce post-harvest losses from these diseases. Preventative sprays applied before harvest can sometimes help reduce post-harvest decay. Post-harvest fungicide applications applied during fruit packaging also provide good control.

Bacterial canker (*Pseudomonas syringae*) occurs under cold, wet conditions in trees that are stressed due to low nutrient availability, nematode feeding, poor soil, or mismanaged irrigation. Bacterial canker can affect trees of any age, but is usually most severe in young orchards from 1 to 6 years old. Cankers produced by the bacteria girdle and kill individual branches and sometimes entire trees.

X-disease is spread through vector pests (*e.g.* Cherry and Mountain leafhoppers) leading to tree decline and death. California. Symptoms differ among rootstocks.

Prompt removal of infected trees, combined with regular post-harvest orchard treatments using effective insecticides, can help to prevent the spread of X-disease within and among orchards.

Phytophthora root and **crown rot** typically occur in over irrigated soils. To prevent Phythophthora infection, maintain good soil drainage and irrigate so as to avoid prolonged periods of saturated soil conditions. Improved drainage around the crown of the tree (below-ground part of the trunk) and reduce disease incidence.

Cherry Cracking

Rain induced cracking injuries may vary from minor breaks, which are only skin deep to ruptures extending the length of the fruit and deep into the flesh. Three categories of cracking are generally noted. These include: 1) the small star or crescent-shaped or concentric rings at the apical end, 2) the circular or semi-circular cracks at the stem end, and 3) the long irregular cracks along the sides. These first two areas are commonly cracked since water accumulates at the tip of the fruit, which is the area of highest sugar content and in the cup surrounding the stem.

The severity of cracking depends upon a number of factors, which include the variety, the stage of maturity, the length of the rainy period, and the air temperature. The varieties most susceptible to cracking include Bing, Brooks, Chinook, Lambert, Royal Ann, Early Burlat, Macmar, Knight's Early Black, Sue, and Early Republican. Varieties considered of moderate susceptibility include Rainier, Van, Corum, Stella, Vega, Lamida, and Spalding. Low susceptibility varieties include Black Tartarian, Jubilee, Sam, Vista, Seneca, Viva, Schmidt, Vernon, Vic, Emperor Francis, Windsor, and Hudson.

As the sugar content increases within the fruit, the greater the cracking potential within each of these categories. As the fruit becomes more mature its potential for cracking increases and reaches its maximum cracking at tree ripeness. Cracking is not a problem in the sour cherry industry.

Causes of Cracking

For many years it was thought that the rainwater was taken up through the root system and moved into the fruit causing the cracking; however, more recent studies have proven that rain cracking is caused by absorption of water through the skin of the fruit. The firmer varieties are most susceptible to cracking and fruit which is turgid prior to a rain will crack more severely than fruit which is somewhat water stressed prior to a rain event.

The potential for skin cracking increases with warmer air temperatures. Oregon research noted that little splitting occurred at 40 degrees while the splitting potential increased by 50 percent at 77 degrees over that noted at 59 degrees.

The osmotic potential of the fruit dictates how much and how quickly water is absorbed through the skin of the fruit. As the sugar contents of the fruit increases so does the movement of water through the skin and into the flesh of the fruit.

The architecture of the wall or skin of the cherry is rather complex. It is made up of three parts that include the cuticle, epidermis, and hypodermis. The cuticle consists of a waxy outer layer, which is imbedded into a spongy carbohydrate matrix that provides some body to the waxy layer. This is glued to the outer layer of cells of the cherry by a pectin substance. This outer layer is called the epidermis, which is only one cell thick. In this surface there are pores, which develop in the fruit wall, which become nonfunctioning lenticels, but do allow gas exchange and moisture loss through transpiration. Below the epidermis is the hypodermis, which consists of smaller rounded cells from three to seven cells deep and below this is the flesh of the fruit. As the sugar contents increases inside the fruit so does its ability to absorb water through the skin; thus, causing the skin to crack.

Minimizing Losses

Harvesting fruit as soon as possible following a rain is the best option to reduce cracking. Lightly cropped trees or trees in weaker ground should be harvested first. The longer the fruit remains on the tree and the higher the humidity the greater the cracking potential. The larger fruit will crack before the smaller fruit.

Remove as much water as possible from the surface of the fruit and leaves as soon as you can enter the field. The use of speed sprayers, low flying helicopters, wind machines, orchard heaters, or limb shaking can be helpful depending upon how much water is removed.

There have been many methods attempted to reduce rain cracking including nutrient and mineral sprays, the use of plant growth regulators, plant growth retardants, and anti-transpirants, limb shaking, and blowing water off trees with speed sprayers. All of these methods have produced variable results.

The application of chemicals prior to the rain such as gibberellic acid can help toughen the skin somewhat, but delays maturity a few days.

A few growers who use gibberellic acid also use the metalized reflective film to enhance the color, which is delayed by the gibberellic acid treatment. Materials that provide a water penetration barrier such as the anti-transpirants may provide some protection provided good coverage of the fruit surface is provided.

Boron is responsible for increasing the elasticity of cell membranes and preventing the breakdown of vegetative tissues. Experimental applications of soil-applied boron have reduced cracking from 25 percent to 50 percent. Spraying Solubor on the soil surface two weeks before harvest may reduce cracking.

Calcium containing materials such as Bordeaux mix were first noted to reduce cracking in the 1930s. Several research trials from around the world have demonstrated that calcium in various forms (calcium chloride, calcium caseinate, calcium hydrate, calcium sulfate, and others) can reduce cracking in cherry by reducing but not stopping water absorption. Field-tested calcium applications with speed sprayers and overhead sprinkler applications of calcium chloride have successfully reduced cracking provided they are applied during the rain event and are repeated as each rain event occurs with good spray coverage. Multiple applications are generally necessary if the rain is persistent.

The calcium solution must be on the surface of the fruit when there is rainwater present on the fruit surface to prevent water uptake. Field tests have demonstrated that there is very little difference in reduced fruit cracking between applying calcium chloride with a speed sprayer or by applying it through an automated over tree micro-sprinkler system.

The automatic over tree rain activated systems are very expensive; however, they do offer other potential benefits. These additional benefits might include temperature activated frost protection or evaporative cooling during hot periods or during the time of fruit bud differentiation to prevent spurs or doubled fruit, to apply foliar nutrient sprays, or increase chilling hours during the winter months.

Rain covers are another option, but they are very expensive and delay maturity from three to five days. These covers must be well ventilated on the sides to prevent condensation and fruit rot from developing.

Storage

Prompt cooling of sweet cherries to 0°C and holding at -0.6 to 1.1°C at 90-95 per cent relative humidity have been recommended. However the post-harvest life of cherries, even under low-temperature conditions, is limited to about 2-3 weeks due to their highly perishable nature. Sweet cherries lose flavor and color if stored longer than 3 weeks. Red cherries can be stored for about 1 month if healthy but not too mature. Cherry fruits are vacuum-precooled rather than water- cooled and stored at 1°C in sealed bags in a controlled atmosphere (5 per cent O_2 and 20 per cent CO_2).

(a) Optimum temperature

-0.5 ± 0.5°C

(b) Optimum relative humidity

90-95 percent; high humidity is particularly important to maintain green stem color.

(c) Rates of respiration/CO_2 Production

Temperature	0°C	5°C	10°C	20°C
ml CO_2/kg –hr	3-5	5-9	15-17	22-28

to calculate heat production multiply ml CO2/kg-hr by 122 to get kcal/ metric ton/day.

(d) Rates of Ethylene

<1μl/kg-hr at 20°C

(e) Responses to Ethylene Production

Cherry responses to ethylene are minimal. Ethylene is minimal. Ethylene does not accelerate cherry ripening.

(f) Response to controlled atmosphere

CA reduces respiration rate and thereby increases post-harvest life. Elevated CO_2 suppresses decay development. Modified atmosphere packaging within boxes has been very successful. Successful atmospheres are generally within the following ranges:

- ☆ 3 to 10 per cent O_2

- ☆ 10 to 15 per cent CO_2
- ☆ < 1 per cent O_2 can result in skin pitting off- flavors
- ☆ > 30 per cent CO_2 can result in brown skin discoloration and off- falvours.

Flavor volaties may be reduced following several weeks of CA storage resulting in fruit of good visual quality but poor sensory quality.

Processing

Cherries are processed into various canned, frozen and dehydrated products Red tart pitted cherries can be freeze-dehydrated and compressed. Maraschino cherries are prepared by bleaching with 2 per cent sodium chloride solution containing about 0.5 per cent each of sulfur dioxide and calcium carbonate. The washed fruit is packed in jars with about 12-20 per cent sugar solution and pasteurized at about 74°C for 3 min. alternately, cherries can also be preserved with 15-20 per cent sugar solution containing about 0.1 per cent sodium benzoate or 0.5 per cent potassium metabisulfite. Cherries can be frozen after soaking the fruit in water containing calcium chloride and packing in slipcover cans with sugar in 3:1 (fruit: sugar) proportion.

Dehydrated cherries (raisins) are prepared by dipping for 4 min in 2 per cent bisulfite-citric acid solution and drying in a dehydrator. Some cherries are subjected to sulfur fumigation and then dried in a dehydrator. Some cherries are soaked in sugar solution for 2 h before sulfuring and drying. To make picked sweet cherries, cherries are graded, washed, and preserved in quart jars using a pickling solution containing vinegar, water, sugar and sodium chloride or calcium chloride. The solution is heated to 71°C before being poured over the cherries. The jars are sealed and stored at ambient temperature.

Post-harvest Disorders, Disease and their Control

Physiological and Physical Disorders

Pitting

An indentation in the surface of the fruit caused by the collapse of cells under the skin. Thought to be result of impact injury.

Bruising

Results from compression and impact of fruit.

Post-harvest life is closely related to respiration rate. Respiration rate increases as a result of increased temperature and physical injury.

Pathological Disorders

Brown Rot

Caused by *Monilinia fruticola*, disease can begin in the orchard or post-harvest. Pre and post-harvest control measures are necessary.

Grey Mold

Caused by Botrytis cinerea, a fungus that continues to grow slowly at 0°C.

Rhizopus Rot

Caused by Rhizopus stolonifer, a fungus that is found in fruit exposex to temperatures of 5°C or greater.

Cherry Pitting Damage

Proper temperature management (rapid cooling to optimum storage temperature) can completely control Rhizopus Rot and significantly reduce Brown Rot and Grey Mould. Eliminating injured and diseased fruit from the packed box is important. Fungicide treatments, pre and post-harvest are often beneficial.

REFERENCES

Arus P, Yamamoto T, Dirlewanger E, Abbott AG 2006. Synteny in the Rosaceae. *Plant Breeding Reviews* 25, 175-211.

Bargioni G. 1996. Sweet cherry scions: characteristics of the principal commercial cultivars, breeding objectives and methods. In: Webster WD, Looney NE (Eds) *Cherries: Crop Physiology, Production and Uses*, CAB International, Wallingford, UK, pp 73-112.

Bhagwat B, Lane WD. 2004. *In vitro* shoot regeneration from leaves of sweet cherry (*Prunus avium* L.) 'Lapins' and 'Sweetheart'. *Plant Cell, Tissue and Organ Culture* 78, 173-181.

Bors RH (2005) Dwarf sour cherry breeding at the University of Saskatchewan. *Acta Horticulturae* 667, 135-140.

Brown SK, Iezzoni AF, Fogle HW 1996. Cherries. In: Janick J, Moore JN (Eds) *Fruit Breeding (Vol 1) Tree and Tropical Fruits*, John Wiley and Sons, New York, pp 213-256.

Cai YL, Cao DW, Zhao GF 2007. Studies on genetic variation in cherry germplasm using RAPD analysis. *Scientia Horticulturae* 111, 248-254.

Caprio JM, Quamme HA 2006. Influence of weather on apricot, peach and sweet cherry production in the Okanagan Valley of British Columbia. *Canadian Journal of Plant Science* 86, 259-267.

Christensen JV (1995) Evaluation of fruit characteristics of 20 sweet cherry cultivars. *Fruit Varieties Journal* 49, 113-117.

Faust M, Suranyi D 1997. Origin and dissemination of cherry. *Horticultural Reviews* 19, 263-317.

Iezzoni AF 1984. Sour cherry breeding in Eastern Europe. *Fruit Varieties Journal* 38, 121-125.

Iezzoni AF, Schmidt H, Albertini A 1990. Cherries (*Prunus* spp.). In: Moore JN, Ballington JR (Eds) *Genetic Resources of Temperate Fruit and Nut Crops*,

International Society for Horticultural Sciences, Wageningen, Netherlands, pp 110-173.

Inchlag C, Hoffman-Sommergruber K, O'Riordian G, Ahorn H, Ebner C, Scheiner O, Breiteneder H 1998. Biochemical characterization of Pru a2, a 23 kd thaumatin-like protein representing a potential major allergen in cherry (Prunus avium). *International Archives of Allergy and Immunology* 116, 22- 28.

Kaack K, Spayd SE, Drake SR 1996. Cherry processing. In: Webster WD, Looney NE (Eds) *Cherries: Crop Physiology, Production and Uses*, CAB International, Wallingford, UK, pp 471-483.

Kappel F, Fisher-Fleming B, Hogue E 1996. Fruit characteristics and sensory attributes of an ideal sweet cherry. *Hort. Science* 31, 443-446.

Kappel F, MacDonald R, McKenzie DL 2000. Selecting for firm sweet cherries. *Acta Horticulturae* 538, 355-358.

Matthews P, Dow P 1983. Cherries. John Innes Institute 72nd Report for the two years – 1981 – 1982, 151.

O'Rourke D 2006. World Sweet Cherry Review (2006 Edn), Belrose Inc., Pullman, Washington, pp 1-92.

Olden EJ, Nybom N 1968. On the origin of *Prunus cerasus* L. *Hereditas* 59, 327-345.

Schuster M, Schreiber H 2000. Genome investigation in sour cherry, P. Cerasus L. *Acta Horticulturae* 538, 375-379.

Šimunic V, Kovaþ S, Gašo-Sokaþ D, Pfannhauser W, Murkovic M 2005. Determination of anthocyanins in four Croatian cultivars of sour cherries (Prunus cerasus). *European Food Research Technology* 220, 575-578.

Song GQ, Sink KC 2006. Transformation of Montmorency sour cherry (*Prunus cerasus* L.) and Gisela 6 (P. cerasus X P. canescens) cherry rootstock mediated by *Agrobacteruium tumefaciens. Plant Cell Reports* 25, 117-123.

Srinivasan C, Padilla IMG, Scorza R 2004. *Prunus* spp. Almond, apricot, cherry, nectarine, peach and plum. In: Litz RE (Ed) *Biotechnology of Fruit and Nut Crops*, CAB International, Wallingford, UK, pp 512-542.

Vasar V 2000. The influence of antioxidants on in vitro rooting of sweet cherry (*Prunus avium*) microshoots. Transactions of the Estonian Agricultural University (Estonia), 1-5.

Wang H, Nair MG, Strauburg GM, Booren AM, Gray JI 1999. Novel antioxidant compounds from tart cherries (*Prunus cerasus*). *Journal of Natural Products* 62, 86-88.

Webster AD 1996. The taxonomic classification of sweet and sour cherries and a brief history of their cultivation. In: Webster WD, Looney NE (Eds) *Cherries: Crop Physiology, Production and Uses*, CAB International, Wal- lingford, UK, pp 3-24.

Webster AD, Schmidt H 1996. Rootstocks for sweet and sour cherries. In: Webster WD, Looney NE (Eds) Cherries: *Crop Physiology, Production and Uses*, CAB International, Wallingford, UK, pp 127-166.

Wolfram B 1996. Advantages and problems of some selected cherry root-stocks in Dresden-Pillnitz. *Acta Horticulturae* 410, 233-237.

Wünsch A, Hormaza JI 2002. Molecular characterization of sweet cherry (*Prunus avium* L.) genotypes using peach [*Prunus persica* (L.) Batsch] SSR sequences. *Heredity* 89, 56-63.

Chapter 6

Hazelnut

The hazelnut is the nut of the hazel and therefore includes any of the nuts deriving from species of the genus *Corylus*, especially the nuts of the species *Corylus avellana*. It also is known as cobnut or filbert nut according to species. A cob is roughly spherical to oval, about 15–25 mm (0.59–0.98 in) long and 10–15 mm (0.39–0.59 in) in diameter, with an outer fibrous husk surrounding a smooth shell. A filbert is more elongated, being about twice as long as its diameter. The nut falls out of the husk when ripe, about 7 to 8 months after pollination. The kernel of the seed is edible and used raw or roasted, or ground into a paste. The seed has a thin, dark brown skin, which sometimes is removed before cooking. Hazelnuts are used in confectionery to make praline, and also used in combination with chocolate for chocolate truffles and products such as Nutella and Frangelico liqueur. Hazelnut oil, pressed from hazelnuts, is strongly flavoured and used as a cooking oil. Turkey is the world's largest producer of hazelnuts. Hazelnuts are rich in protein, monounsaturated fat, vitamin E, manganese, and numerous other essential nutrients. Turkey and Italy dominate world production and produce over 70 per cent of the world's hazelnuts. It is considered that as far as Australian growers are concerned there is no real value in the production of out-of-season produce for export to the northern hemisphere, mainly because of the generally non-perishable nature of hazelnuts and the ability to store them for long periods. The major markets will be for kernels with an attractive appearance that can complement overseas imports

Botany

Corylus is one of the six genera belonging to the birch family. The hazelnut (*Corylus avellana* L.) forms the basis for the more important commercial cultivars. The terms 'filbert' and 'hazelnut' are often used interchangeably to include all plants in the genus *Corylus*. *C. maxima* and *C. colurna* (Turkish hazel) also possess desirable characteristics. The latter has possibilities as a non-suckering rootstock.

In its natural form the hazelnut is a deciduous, monoecious, multi-stemmed bush, but commercially should be grown as a single trunk tree. Tree sizes are up to 6 m tall in the commercial state. The leaves of *C. avellana* are 5–12 cm long, flat, almost round, long pointed, hairy on both surfaces, with the margin doubly serrated. Hazelnuts have a number of unique characteristics which include flowering periods of weeks or even months between pollination and fertilisation. They can also tolerate temperatures down to –10°C during the period of pollination and are usually unaffected by late spring frosts.

Nutritive Value

Hazelnuts are high in energy and kilojoules, are sealed by nature and contain no additives.

Table 6.1: Nutritional Composition of the Hazelnut (per 100 g fresh fruit)

Water (per cent)	5.8
Kilojoules	2655
(Calories)	(634)
Protein (g)	12.6
Fat (g)	62.4
Carbohydrates (g)	16.7
Vitamin A (IU)	–
Thiamine (mg)	0.46
Riboflavin (mg)	–
Niacin (mg)	0.9
Vitamin C (mg)	–
Calcium (mg)	209
Phosphorus (mg)	337
Iron (mg)	3.4
Sodium (mg)	2
Potassium (mg)	704

Varieties

The many cultivars of the hazel include 'Atababa', 'Barcelona', 'Butler', 'Casina', 'Clark', 'Cosford', 'Daviana', 'Delle Langhe', 'England', 'Ennis', 'Fillbert', 'Halls Giant', 'Jemtegaard', 'Kent Cob', 'Lewis', 'Tokolyi', 'Tonda Gentile', 'Tonda di Giffoni', 'Tonda Romana', 'Wanliss Pride', and 'Willamette'. Some of these are grown for specific qualities of the nut, including large nut size, or early or late fruiting, whereas others are grown as pollinators. The majority of commercial hazelnuts are propagated from root sprouts. Some cultivars are of hybrid origin between common hazel and filbert. One cultivar grown in Washington, the 'DuChilly', has an elongated appearance, a thinner and less bitter skin, and a distinctly sweeter flavor than other varieties.

Description of Important Commercial Varieties

Ennis

Is a high yielding variety producing jumbo size nuts.Main producer, it flowers over a 12 week period.It is generally used as the primary producing variety, with others planted as pollinisers which also produce their own style of nut.

Barcelona

It is a popular variety, table nut with sizes ranging from jumbo, large to a multitude of sizes. Late nut pollinated by Lewis and Casina.

Butler

Also produces large to jumbo nuts. Early polliniser. Ennis and Butler cross pollinate each other.

Lewis

Smaller medium a specialty nut. Early polliniser

Casina

It produces a small nut that blanches well and is used in the confectionary industry, esp in chocolate.(Speciality Nut- Small Size).Mid-season polliniser. Ennis and Casina cross pollinate each other.

Willamette

Produces a Medium size nut, blanches well. Mid polliniser of Ennis

Halls Giant

Large nut but not high yielding. Late polliniser of Ennis

Pollination

Hazelnuts for commercial purposes should be considered self-sterile because peak periods of male and female flowering may not coincide for any one variety. It is important for growers to have genetically compatible pollinators and pollinizers that shed pollen at the time female flowers are receptive. It is recommended that a range of varieties be planted to ensure the dissemination of cross-compatible pollen. The hazelnut cross-pollination guide overleaf indicates the pollen compatibility of newer hazelnut varieties, some of which are yet to be released from quarantine. One polliniser tree to eight of the main variety should ensure wind cross-pollination, but in unfavourable seasons a 1:5 ratio may be beneficial. Older orchards may require a smaller number of pollinisers. The blooming season can take up to three months from initial pollen shedding of early varieties to full bloom of female pistillate flowers.

The male flowers can be identified as catkins which, when shedding pollen, elongate to approximately double their dormant length. Female flowers emerge at a similar time and the stigmas appear as red hairs through the tips of the bud.

Male and female flower buds are borne on the laterals of one-year shoots.

The Turkish tree hazel (*Corylus colurna*) has particular value as a non-suckering rootstock. Its deep taproot results in increased drought tolerance in the drier Australian summers. However, seeds of this species are difficult to germinate and, because of the taproot, seedlings are difficult to transplant. Only vigorous varieties of hazelnut should be grafted onto the Turkish tree hazel as the rootstock tends to outgrow all but the most vigorous varieties. More research is required in Australia into the use of C. *colurna* as a rootstock to identify any compatibility problems and eventual size of trees. *C. colurna* seedlings are useful as a rootstock for ornamental hazels where nut production is not important.

Rootstock selections Newberg and Dundee have been imported into Australia as rootstocks, as have a number of interspecific hybrids 'trazels' which may have potential as rootstocks.

Harvesting

Hazelnuts are harvested annually in mid-autumn. As autumn comes to a close, the trees drop their nuts and leaves. Most commercial growers wait for the nuts to drop on their own, rather than using equipment to shake them from the tree. The harvesting of hazelnuts is performed either by hand or, by manual or mechanical raking of fallen nuts.

Hazelnut orchards may be harvested up to three times during the harvest season, depending on the quantity of nuts in the trees and the rate of nut drop as a result of weather.

Two different timing strategies are used for collecting the fallen nuts. The first is to harvest early, when about half of the nuts have fallen. With less material on the ground, the harvester can work faster with less chance of a breakdown. The second option is to wait for all the nuts to fall before harvesting. Although the first option is considered the better of the two, two or three passes do take more time to complete than one. Weather also must be a consideration. Rain inhibits harvest and should a farmer wait for all the nuts to fall after a rainy season, it becomes much more difficult to harvest. Pickup also varies with how many acres are being farmed as well as the number of sweepers, harvesters, nut carts, and forklifts available.

Drying

Hazelnuts should be collected promptly after falling as rain can cause discolouration of the shell. Nuts left on damp ground for over a week will gradually darken, become less attractive and be prone to fungal attack.

Following collection, the nuts should be cleaned and dried to approximately 8 per cent –10 per cent moisture. In the case of some confectionery companies, 6 per cent moisture is the maximum required. Temperatures of 32°C–38°C are commonly used for drying, and equipment for other purposes, such as prune driers or small dehydrators, can be adapted to dry the nuts. The amount of heat required is relatively small and the speed of drying is not critical.

Processing

Technical development is a key factor of labor efficiency and productivity. Small farmers in have generally no or very little mechanization. Collection is usually done manually. After being collected from the fields, hazelnuts are laid in 10-15cm thickness to be predried under the sun until their leaves turn brown. After drying, hazelnuts are separated from their leaves manually or by using harvesting machine (thresher or husker) and laid over canopies in thin layers to be dried under the sun. The total period of drying, including the pre-drying, can be a maximum of 15-20 days depending on the weather conditions. After drying, the hazelnuts are moved in to a vacuum based gravity separator that removes the rest of the husk material, broken shells and other unsuitable material. From here, hazelnuts are transferred to the sorter which helps to separate them according to size categories. This allows the cracker to be set to the specific size of hazelnuts for more uniform cracking. In the cracker, about 70 per cent of hazelnut shells are broken on the first pass. Then hazelnuts are moved back into a separator to remove the whole nut kernels from the shell material. The cracking process is repeated several times. Lastly, hazelnuts go for hand sorting to remove, by hand, unwanted kernels and any shell material or whole nuts.

Yields

Hazelnuts begin to bear at approximately three years old and at six years yields should approach 2 to 2.5 kg/tree. At the lower limit this translates to1000 kg/ha, assuming 500 trees/ha. Yields of 2.5 to 3.5 t/ha sustained annually would be considered an acceptable level of production. Mature trees grown without irrigation or fertilisers can produce 20–25 kg/tree. With intensive management, yields can be up to 40 kg/tree.

REFERENCES

Allen, A 1987, *Growing Nuts in Australia,* Night Owl Publishers, Shepparton.

Jaynes, RA (ed.) 1979, *Nut Tree Culture in North America,* Northern Nut Growers Association Inc., Connecticut.

Westwood, MN 1978, *Temperate Zone Pomology,* W.H. Freeman and Company, San Francisco.

Woodroof, JG 1979, *Tree Nuts: Production, Processing, Products,* 2nd ed., AVI Publishing Company Inc., Westport, Connecticut.

Chapter 7

Peach and Nectarine

Introduction

Peach (*Prunes persica* (L.) Batsch), along with its smooth-skin mutant nectarine is one of the most important temperate stone fruits grown in the world, though its culture has found a reasonable place in the subtropical too, despite the quality of the poor fruit. Peaches are an ancient fruit, believed to originate in China over 4,000 years ago. During the silk trade between China and Persia, the fruit migrated to the Middle East. From Persia's soils, the Romans and Greeks took the juicy peach to Europe sometime around 300 and 400 BC. According to the book, "A History of Sino-Indian Relations," peaches likely came to India during the Eastern Han period from 25-200 AD, though they were perhaps here as early as Zhang Qian's visit to Bactria circa 126BC. The Mughals were especially fond of peaches. As written in "Mughal Gardens," records dating from 1650 mention the fruits growing in the royal gardens of Sirhind and Lahore. Commercial cultivation of peaches, however, only came about during the latter half of the 19th century. Peaches have much more to offer than their amazingly delicious taste. Being loaded with vitamins, minerals, antioxidants and other chemical contents, peaches provide lot of health benefits. Three wild species are still grown there, namely, *P. davidiana* in the north, which is used as rootstock; and *P. mira* and *P. fergamensis,* which are indigenous to the Tibetan plateau and Sinkiang Province, respectively. The large-sized fruit varieties are grown mainly in the temperate regions of the world, where it find congenial environmental conditions for the healthy development of the plant and fruit. Important centers of commercial production of peaches lie between latitudes 30 and 45° North and South. The leading peach-producing nations are Italy, the Unites States, France, Japan, Argentina, Germany, Spain and Greece. The subtropical growing conditions required for the fruit relegate its location to the northern regions of the country. A few of the southern hill stations grow peaches, but on a small, limited scale. Some peach varieties grow well in the drier, temperate areas—such as Jammu and

Kashmir, Uttarakhand and Himachal Pradesh—while other sub-tropical variants, like nectarines, grow best in J&K, Punjab and Uttar Pradesh. Peak peach season is from early April to late June, and a few exceptional varieties grow a few weeks before and after these months.

Botany

Taxonomy

The peach belongs to the family Rosaseae, subfamily Prunoideae, genus Prunus and subgenus Amygdales. The scientific name for the peach is *Prunus persica* (L.) Batsch. Other peach species are *P. davidiana, P. mira, P. fergamensis* and *P. kansuensis* (1). Another species, *P. behmi,* considered a natural hybrid of almond and peach, occurs extensively in the dry temperature zone of Kinnaur, India and is commonly used as rootstock for almond, peach and Plum. The subgenus Amygdalus to which peach belongs is characterized by sessile or short stalked flowers, open before leafing and commonly borne on 1-year-old shoots. The flowers are simple, perfect, complete and perigynous and are always found in a lateral position. The fruit is drupe with a thin epidermis, a fleshy mesocarp, and a stony endocarp containing the seed. The fleshy mesocarp may adhere t the stone (cling-stone) or be separate from it (freestone). The fruits are variable in size, shape, color of skin and flesh. Most of the cultivars have bitter kernels, while a few have sweet kernels. A nectarine is a fuzzless peach and is usually smaller in size. The nectarine is considered botanically a variety of *P. persica* peach and is classified as *P. persica*, variety *nucipersica* (Schneid). The lack of pubescence is controlled by a single recessive gene.

Cultivars

Most of the peach cultivars in use evolved through breeding programs. The choice of peach cultivars for any region is governed by three factors; (a) the type of market to be served (b) the distance of market and (c) adaptability to local soil and climatic conditions. Cultivars for table purpose should be yellow fleshed freestone and regular annual producers. For canning purpose the fruit should be yellow-fleshed, clingstone, with a small non splitting pit, uniform size, no red color at the pit and it should mature evenly. However, in recent years, many freestone varieties have been canned. If the fruit is to be dried, it should be freestone particularly sweet in addition to the characteristics given above for canning. Freestone cultivars show less isoenzyme variation than clingstone types. Variation in the morphological characteristics of the stone in peach varieties could be used as a marker and varieties with common origin had similar stones.

Type of Peaches

Yellow Peaches

They tend to have more of an acid tang than their white-flesh counter parts. As with all peaches, you want to look for yellow peaches that feel heavy for their size, have a bit of give when held in the palm of your hand, and, most importantly, smell like peaches when you take a wiff.

White Peaches

Favored in Asia and increasingly available in the U.S., white-flesh peaches taste even sweeter than yellow peaches, in part due to their low acidity. They also tend to have a smoother, more luscious texture than yellow-fleshed peaches. They don't necessarily look terribly different until you cut into them or peel them, but since they are relatively rare and prized, they tend to be well labeled.

Freestone Peaches

Freestone peach flesh does not stick to the pit, so they are preferred for eating out-of-hand. They tend to be larger and less juicy than clingstone peaches. They bake and preserve well, too. Like yellow peaches over white peaches, the vast majority of peaches for sale to retail customers are freestone peaches. Freestone peaches come in many varieties with seasons ranging from May to October.

Clingstone Peaches

Clingstone peaches have flesh that clings to their pits. They are softer, sweeter, and juicier than freestone peaches and hence sought after for canning and preserving (commercially canned peaches are all clingstones). They are also great for baking. One advantage of shopping for peaches at farmers markets is that you can sometimes find clingstones. Most grocery stores only carry freestones.You may also find semi-freestone peaches, a hybrid of clingstone and freestones that attempts to combine the easy eating and pitting properties of freestones with the juicy sweetness of clingstones.The clingstone peach season ranges from May to August.

Donut Peaches

Donut peaches are an heirloom variety. They are flat, white-fleshed, and low in acid. They are generally available at farmers markets and specialty markets in July and August.

Nectarines

Nectarines are, botanically speaking, a variety of peach—in fact, they are so closely related, that if you plant dozens of peach or nectarine pits, a few of each will grow into the other. Some people claim that nectarines have a slightly lighter flavor, and peaches tend to be a bit more deeply flavored. As far as most people can tell, though, the only difference is the name and the lack of peach fuzz on the smooth skin of nectarines.

Peach and Nectarine Cultivars

These brief notes on peach and nectarine cultivars include only comparative characteristics needed for decision making. Special qualities or limitations are presented but, unless otherwise noted, the cultivar is commercially satisfactory in such characteristics as tree growth, hardiness, cropping ability, fruit size, appearance and quality.

Peaches and nectarines are among the most sensitive of the tender fruits to winter injury. Care must be taken when selecting the orchard site. The root systems of the trees require a well-drained soil to achieve orchard productivity and longevity.

Fresh Market (Yellow Flesh)

Bounty: It has more colour compared to Loring, is rounder in shape and has better flavour. Bounty has had light crops under extreme cold winter conditions in Ontario.

Brighton: Has shown promise as an early-ripening peach. The fruit are medium in size, attractive red in colour and of good quality, but have a clingy flesh. Brighton ripens 5 days after Garnet Beauty.

Cresthaven: Firm, large fruit. It is moderately susceptible to bacterial spot. The fruit lacks sufficient colour to compete against other cultivars.

Early Redhaven: This Redhaven sport ripens 2 days after Garnet Beauty, but with similar characteristics. The fruit may be more highly coloured and smaller, and the firmer flesh tends to be clingy. Early Redhaven has been widely planted during recent years.

Garnet Beauty: This bud sport of Redhaven ranks second in number of trees to its parent in Ontario. Garnet Beauty is a good-quality peach, ripening about a week after Harrow Diamond. It is attractive, usually not subject to split-pits, but not fully freestone.

Harbrite: It is bud-hardy and a good peach for its season. Fruit is medium to large, round, with an attractive red colour. The fruit are resistant to bacterial spot and brown rot.

Harcrest: Ripens just before Redskin and promises to be a late-season cultivar with good disease resistance. The fruit of Harcrest are medium large, quite firm, good quality, and have good winter hardiness and disease resistance but no better blush than other cultivars in this late season.

Harken: Ripens 2 days after Redhaven. The large, attractive peach has a good red colour covering most of its surface and a bright yellow ground-colour. The flesh is firm and of good quality. Because it oxidizes slowly, Harken freezes well. Resistance to bacterial spot and brown rot is good.

Harrow Beauty: Ripens with Loring and Canadian Harmony but is more winter-hardy. The very firm, highly attractive, medium-sized fruit ships well. The rich yellow flesh has a red pigment around the pit cavity. Leaves and fruit have good resistance to bacterial spot and brown rot.

Harrow Diamond: It is winter hardy, disease resistant and has few split-pits. The fruit have an attractive red blush over a bright yellow background; the deep yellow, low-oxidizing flesh is of good quality and is nearly freestone when fully matured. Because the fruit is small-to-medium sized, this cultivar must be thinned early and adequately to obtain suitable size.

Jim Wilson: Ripens 1 day before Veeglo at Vineland and may provide a good alternative for that season. The large, round fruit are attractive, and the firm flesh is freestone. The vigorous, strong trees are capable of sizing the fruit with full crops. The cultivar is more resistant to bacterial spot than Veeglo.

Loring: This late-season peach is large, firm, yellow-fleshed, freestone and known for its good quality. Loring lacks winter hardiness and should not be planted on marginal sites. Once an industry standard, it now lacks sufficient red skin colour to compete with newer cultivars.

Redhaven: This older mid-season peach is an attractive red colour with good fruit quality. Trees frequently set heavy crops and must be adequately thinned to attain size. The crop ripens unevenly, and trees must be harvested several times. When well grown and properly handled, Redhaven is a superior cultivar with good winter hardiness.

Redskin: A medium-sized, good-quality, late-ripening freestone with fairly good colour. Trees tend to be somewhat willowy but are very productive.

Springcrest: Ripens 2 days prior to Harrow Diamond and is considered an early peach for the local fruit stands and fruit markets. Fruit size is small and has several early split-pit fruit. Skin can be very deep purple as it matures. This cultivar is also winter sensitive.

Veeglo: A cultivar between Redhaven and Loring. The round fruit have a bright yellow under colour and are suitable for freezing and canning. Veeglo is only moderately resistant to bacterial spot.

Vivid: Has bright red, attractive, firm fruit that are above average in size and of good quality. The trees are vigorous and productive. Vivid has become an important cultivar to follow Redhaven. It is less winter hardy than Redhaven. Do not plant on marginal sites.

Fresh Market (White Flesh)

White Lady: Ripens 5 days after Redhaven. Round, firm and very attractive. Low acid flavour with moderate aromatic character. It is also moderately susceptible to bacterial spot disease.

Processing Type

Babygold 5: Once considered a standard in Loring season due to its large size and quality, tree hardiness, productivity and winter hardiness. Due to its susceptibility to bacterial spot, brown rot and the upright tree growth habit, it is being replaced by cultivars with greater disease resistance such as Venture.

Babygold 7: Ripens 9 days later than Babygold 5 and extends the season for processing peaches. Like Babygold 5, the large fruits are of exceptional quality for processing. The upright trees, although productive and winter hardy, are difficult to train. The fruit is susceptible to brown rot disease and tends to drop at maturity.

Veecling: Ripens 17 days before Babygold 5 (2 days after Redhaven), The strong, productive trees produce large, good-quality fruit. Veecling is tolerant to bacterial spot but is subject to split-pits and red colouration in the flesh. The processor is not supporting any new plantings as superior cultivars are introduced.

Vinegold: Ripens 8 days before Veecling and is a good choice for an early-season canning peach. The strong and spreading trees are productive and moderately disease resistant. The medium-large fruit are of a uniform round-blocky shape and process into a richly coloured product. Split-pits have been observed during some fruiting seasons.

Virgil: Ripens 5 days before Veecling. The large, round and uniform fruit has firm flesh and good quality. The fruit is also free of red colour and resistant to split-pits. The trees are moderately resistant to bacterial spot, brown rot and Leucostoma canker. The uniformity of harvest makes this cultivar a good choice to precede Veecling.

Vulcan: Ripens 12 days before Veecling, this commercial cultivar is the earliest maturing processing peach. The medium-sized, round fruit have a red over colour and a firm golden flesh. Vulcan is surprisingly free of split-pits for the early season. The trees are vigorous, winter hardy and resistant to bacterial spot and Leucostoma canker, but moderately susceptible to brown rot disease.

Nectarine Cultivars

While nectarine and peach trees are of the same species and the same in appearance, nectarine fruit, which can be yellow or white fleshed, have smooth skin, a distinctive flavour and texture, and are usually smaller. The fruits tend to be more susceptible to aphid damage and brown rot disease. Cultivars may be described as clingstone, freestone or intermediate. Nectarines must be carefully thinned to attain marketable size. Since the trees and fruit buds tend to be less winter hardy than peach, take care to select a good site for nectarine orchards.

Fantasia: Ripens late in the season with Cresthaven peach. The fruits are medium-to-large, attractive, bright red with a yellow ground-colour, freestone and firm-fleshed. The trees are moderately hardy and moderately resistant to bacterial spot.

Flavortop: Ripens just after Loring. Fruits are large, ovate and freestone with excellent quality. Skin is highly blushed over an attractive under-colour. Flesh is yellow, firm and smooth textured. Trees are vigorous but produce light crops and are tender to winter cold. Fruit are also susceptible to bacterial spot.

Harblaze: This cultivar has promise as a commercial-type nectarine that ripens during the late Redhaven season. The vigorous, productive trees bear attractive medium-to-large-sized fruit that are semi-freestone. The fruit tends to soften quickly near maturity during final swell. Harblaze is relatively winter hardy and has a good level of resistance to bacterial spot, brown rot and powdery mildew.

Redgold: A late maturing nectarine. The fruit is freestone with a rich red blush over a yellow ground colour. Flesh is yellow with red around the pit and has the ability to hold firmness, making it an excellent storage and shipping nectarine. Trees are vigorous but produce light crops and are tender to winter cold. Fruit are also susceptible to bacterial spot and mildew.

Commercial Varieties in India

Name of Variety	*Salient Features*	*Recommended Areas*
July Elberta	Trees are medium in vigour, hardy and productive, fruits are medium to large in size and round in shape, skin dull red blushed over yellow base, free stone and early to mid season maturity.	Jammu and Kashmir, Himchal Pradesh, Uttarakhand, Arunachal Pradesh, Manipur, Nagaland, Jharkhand, Meghalaya
Red June	Trees are medium in vigour, fruits are large in size, roundish with rounded beak in shape, distinct suture, yellow with red blush on the shoulder, free stone and early season maturity.	Jammu and Kashmir, Himchal Pradesh, Uttarakhand, Arunachal Pradesh, Mizoram, Nagaland, Manipur, Meghalaya
Sun Heaven	Trees are medium in vigour, fruits are medium to large size, yellow fruit skin, semi free stone and early season maturity.	Himchal Pradesh, Uttarakhand, Jammu and Kashmir, Arunachal Pradesh, Mizoram, Nagaland
Paradelux	Trees are medium in vigour, fruits are large in size and oblong flat with prominent beak in shape, yellow skin and flesh and late season maturity.	Himchal Pradesh, Jammu and Kashmir, Uttarakhand, Arunachal Pradesh
Crest Heaven	Produces top notch freestone fruit with golden yellow skin and flesh. Mid to late season variety, blooms late, fruit lasts well on the tree. Excellent for freezing and canning.	Himchal Pradesh, Jammu and Kashmir, Uttarakhand, Arunachal Pradesh
Glo Heaven	A large peach with yellow freestone flesh has mostly red skin with no fuzz, milder flavour, excellent for canning and fresh eating and freestone.	Himchal Pradesh, Jammu and Kashmir, Uttarakhand, Arunachal Pradesh
Snow Queen	Trees are spreading, vigourous, fruits are small to medium in size, bright red colour on cream white background having smooth surface, flesh white, cling stone, maturity during mid June,	Himchal Pradesh, Jammu and Kashmir, Uttarakhand, Arunachal Pradesh
Red Heaven	Trees are medium in vigour, fruits are medium in size with prominent suture, distinct apex, golden yellow skin with red blush, free stone and early in maturity.	Himchal Pradesh, Jammu and Kashmir, Uttarakhand, Arunachal Pradesh
J.H.Hale	Trees are medium in vigour, fruits are medium to large in size, roundish ovate in shape, red purple yellow skin colour, free stone, mid to late season maturity.	Himchal Pradesh, Jammu and Kashmir, Uttarakhand, Arunachal Pradesh
Sun Red	Trees are low in vigour, fruits are small to medium in size with bright red skin, semi free stone and early season maturity.	Himchal Pradesh, Jammu and Kashmir, Uttarakhand, Arunachal Pradesh, Meghalaya, Mizoram, Manipur
Fantasia	Vigorous tree, fruits are large ovate in shape, bright yellow with red blush over the major part of fruit skin and early to mid season maturity.	Himchal Pradesh, Uttarakhand, Arunachal Pradesh
Partap	It matures in the third week of April. The fruits are yellow with red blush. Flesh colour is also yellow with red colouration. It has better firmness and keeping quality than Parbhat and Flordasun, the average yield being 70 kg/plant.	Punjab
Shan-e-Punjab	It matures in the first week of May. Fruits are very large, yellow with red blush, juicy, sweet, excellent in taste and free stone. Since fruits are firm in texture, they can withstand transportation. These are suitable for canning, the average yield being 70 kg/plant.	Punjab, Haryana, Mizoram

Name of Variety	*Salient Features*	*Recommended Areas*
Florda Prince	Its fruits ripen in the last week of April. Fruits are medium to large, round with little or no tip, red blush with yellow ground colour, flesh melting, yellow with some red colour, semi cling. On an average it yields 100 kg/plant.	Punjab, Madhya Pradesh
Early Grande	It ripen in the first week of May. Fruits are large with red blush surface. Flesh yellow, firm with some red colour next to pit, semi free when fully ripe. The average yield is 95 kg/tree. The fruits posses excellent shipping qualities.	Punjab
Prabhat	It is earliest maturing peach (mid April) and fetches good income to the growers. Fruits are medium in size, roundish with an attractive red blush. It is white fleshed when fully ripe, the average yield being 50 kg/tree.	Jammu and Kashmir, Uttarakhand, Jharkhand, Haryana, Madhya Pradesh
Florda Sun	It matures in the last week of April. Fruits are medium to large in size, roundish and yellow with red blush. Flesh is yellow, juicy and sweet. With free stone, on an average it yields 75 kg/ plant.	Uttarakhand, Himachal Pradesh, Jharkhand, Madhya Pradesh, Tamil Nadu
Flora Red	An excellent, mid season table peach, it matures in the beginning of June. Fruits are large in size, almost red at maturity, juicy with white flesh and free stone. Its average yield is 100 kg/tree.	Uttarakhand, Himachal Pradesh, Jammu and Kashmir, Arunachal Pradesh, Tamil Nadu
Sharbati	Fruits are large in size, greenish yellow with rosy patches, very juicy with excellent taste and flavour. Fruits ripen during June end to first week of July, the average yield being 100-120 kg/tree.	Uttarakhand, Haryana, Jharkhand, Madhya Pradesh, Mizoram

Fruit Development

The life of a fruit begins with initiation of the fruit bud, which after fertilization is ultimately converted into the fruit. The growth of peach fruit exhibits a double sigmoid curve with three rapid growth periods and two phases of growth. The first one is very rapid and continues until the pit begins to harden. In the second stage, the growth of the flesh is low as the seed develops and the seed coat hardens. In the last stage, the growth of fruit again increases. The young fruit just formed goes through a period of rapid cell division and growth.

During the early stages of fruit growth, the developing ovules are matured by the ovary and are its ancillary parts, which are normally green and thus capable of photosynthesis. Sucrose is the major sugar transported from the leaves to the fruit, except in the phloem where it is the sorbitol. The role of the sucrose synthelate and other related enzymes in sucrose accumulation in peach has been reported. Part of the sugar is normally used in the synthesis of pectic substances and cell wall materials or is covered into the usual storage products such as starch.

Changes in peroxides, IAA oxidize, IAA decarboxy lating and protease activity in developing peach fruit have been reported. In the development of peach fruit, ethylene evolution also varies, showing a peak after 78 days of ull bloom in pericarp and is associated with a concurrent increase in ACC (animo-cyclopropane-1-carboxylic acid), which is responsible for ethylene synthesis and ethylene –forming enzyme. Fruit set, stone weight, total soluble solids and Brix-acid ratio increases, while acidity decreases with the advancement of maturity-the greatest changes are in sucrose and sorbitol contents. Changes in some other physiochemical characteristics also take place. Quinic acid, which decreases with increase in malic acid, registered an increase upon maturity. The total pectin content of the fruit also shows a sigmoidal curve.

Maturity, Harvesting and Handling

A large number of maturity indices, such as days after full bloom (DAFB), calendar date, fruit size, firmness, sense of touch, pit discoloration, freeness of pit, taste, ground, sugar, acidity, starch, sugar-to-acid ratio, and multiple linear regression analysis, have been assessed.Ground color, texture, solids-acid ratio, and titratable acidity of peach were found to be good maturity indices. The days required for flowering to maturity vary in different cultivars, rangimg from 78-100 days after full bloom to up to 127 days.

Ratio of soluble solids to acids can be used as reliable index of peach maturity for harvest, whereas eating quality of Elberta peaches could be determined on the basis of chlorophyll content (0.06-18 mg/g fresh weight). Color, flesh, firmness, the total soluble solids content of juice, and light transmission (OD 700-740 mm) could also be used as reliable maturity indices. Nondestructive light measurement and pressure of the fruits are also interrelated for deciding the maturity stage of peaches. The critical figures for flesh firmness with a pressure tester is about 16-17lb. Onset of softening coincides with an increase in red fruit color (about 100 days after fruit set), when the respiration rate and ethylene production reach a maximum. Various quality characteristics of peach fruits have also been found to be related to maturity.

Peachs should be harvested after the attainment of harvest maturity. Since peaches do not mature uniformly. Several pickings are needed to produce good-quality fruits based on ground color, size, and suture. Peach harvesting coincides with the prevailing high seasonal temperatures, which accelerate ripening process during marketing, but it declines to half with each reduction of 5.6°C from 21.1°C. After picking, the fruits should immediately be stored under cool conditions. Peaches ripen with good flavor and aroma at temperatures above 15°C but with undesirable flavor at 10°C, and there is internal breakdown instead of ripening at 4.4°C.

Hand harvesting is the standard method for the fresh market, but machines have been developed for harvesting peaches. Mechanical harvesting however, results in considerable fruit loss due to mechanical injury, bruising, and the picking of immature fruits due to uneven ripening. Low tree profile trunk and limb damage from shaker action and subsequent invasion by fungi result in increased decay and rotting. For successful mechanical harvesting, the overall mechanical system and orchard practices have to be properly planned and managed. However, use of trunk shakers, padded catch frames, and improvements shaker attachments reduces thwe damage. Deffuzing treatments, however, which are frequently used to improve the surface of peaches, may lead to a purple-black discoloration.

Nectarines and peaches require careful handling to avoid damage, as some cultivars are softer and are more difficult to handle than others. The use of foam plastic in picking baskets and bins, shallow bluk bins, manual transfer of fruit fom bins to size graders, and cold storage all enable fruit of very soft cultivars to be marketed fresh. Vibration injury gives more rapid weight loss, decay, and ripening in peaches and nectarines in warm fruits (26°C) than in cool fruits (5°C); to prevent this, tight packs are recommended.

Grading, Pakaging and Trasportation

Grading

Grading and packaging of fruits is essential, as properly graded fruits fetch better prices in the market. Grading is done both for size and quality. Peach fruits are graded in terms of end use. Different agencies have established separate standards for grading.For example, the U.S. grade standard for "U.S.Fancy" stipulates that, except for Jhon Rivers cultivar, each nectarine shall have not less than one-third of its surface showing the red color characteristics of the variety Similarly, four grades or groups of quality of peach for harvesting have been prescribed. There are other regulations for packing peaches, including the indication "Peach" on the sides of the box, the number of pieces, the variations in diameter, *etc.*

Selective grading based on the fruit color, firmness, using high absorbance and vibration techniques have been developed, but lack of universality limits their application. A computer aided vision system has been developed to inspect the color and grade of fresh- market peaches automatically, consisting of solid-state camera, computer vision hardware, an illumination chamber and a conveyor. The sorter system could easily be implemented into an on-system. Another technique called image analysis of color for grading of peaches has also been developed.

Packaging

Packaging is the final step in the preparation of fruits for the fresh market and involves placement of fruits in scales or shipping containers or in convenient units for proper handling as well as protection during marketing and transportation. The choice of package and packaging materials depends on the specific requirements of the fruit to be packaged, especially its susceptibility to mechanical injury such as cuts, compressions, impacts and vibrations. Further, it needs to be ensured that the individual specimens do not move with resopect to each other or to the walls of the package.

The earliest packages were constructed mostly of plant materials, but a wide range of packages constructed of wood, fiber, juite or other plastic is now available. To with stand the pressure of handling and transportation, the package should be strong enough, nontoxic, allow rapid cooling, remain unaffected by moisture, and be presentable, and low in cost. Further the package should not seriously impede exchange of O_2 and CO_2 nor the removal of vital or sensible heat. Pads of paper-covered excelsior, corrugated paper, and plastic film with entapped air bubbles are widely used for protection from pressure bruising.

The wooden peach box, with a two-layer place pack, has been a standard shipping container for western fruit for many years, but labour and material costs have now forced a substantial change. Now, the modern crates (a combination of wood and fiber board) and volume- fill corrugated fiber board boxes have largely replaced veneer baskets and nailed wooden boxes. Western peaches are commonly filled in fiber board boxes holding 8.2-8.6 kg, although peaches are also packed in molded trays or cups in two and three- layer packs in corrugated boxes.

Packaging of peaches and nectarines in sealed film liners and their subsequent storage resulted in fermented fruit flavors. This was prevented by putting holes in each liner, which holds about 10 kg of fruit, which also reduced weight loss compared to unlined controls. However, high RH in the ventilated film liners stimulated decay organisms, necessitating treatment of fruits with fungicides before packaging. There is no evidence of commercial use of sealed or perforated film liners in shipping containers for peaches and nectarines.

Loss of moisture can also be a problem with peaches and nectarines. It has been observed that, under similar conditions of low humidity, peaches lose more weight than nectarines. The use of polyethylene film liners in packing trays and fruit waxing reduced moisture loss of nectarines during storage. Use of suitable packaging to reduce moisture loss from fruits is important. Packing peach fruits in wooden boxes with paper linings pretreated with $KMnO_4$ followed by storage reduces physiological weight loss, decay, and increase in TSS, acidity, and ascorbic acid with increase in storage. Wooden or cardboard boxes with shredded paper linings or egg trays as cushioning material have also been tried. Cardboard boxes with egg trays was the best packaging system for peach, as revealed by the physicochemical characteristics of fruits after storage for 6 days. Of two types of wrappers tried, paper bags and dry paddy, the later proved better for wrapping mature fruits of Flordasun peach for storage.

Nectarines can be packed to tight or loose fill, in 38 and 25 lb volume-fill boxes, or in count lugs, and tray boxes. The additional costs of count packing are cost effective in comparison to many peach packs on large sized nectarines because of their premium prices. Prepackaging in consumer packaging or simply unutilization for in the retail store at some point before the product reaches the checkout counter has been practiced too. Advantages of prepackaging to the shelf-service store include, labor savings at the checkout counter due to unit prepricing, and reduction in product losses due to costumer handling during inspection and selection from bulk display. The most widely used material for prepackaging include bags, plastic film bags, mesh bags, consumer trays, shrink-film wraps, stretch wraps, boxes, baskets, and cups. Emphases has been on prepackaging, which provides better physical protection for firm ripe peaches than the usual bulk packs in crates, cartons, or baskets.

Transportation

Large quantities of fruits are transported over long distances on land and sea under refrigeration in the developed countries. The vehicles for road transportation or frail transport are refrigerated by mechanical means. All or part of a ship can be insulated and refrigerated. The containers are designed in economize on space and maintain the temperature of the precooled fruits during transportation. However, in the less developed countries transportation is mostly without refrigeration, which leads to considerable post-harvest losses. Liquid and solid carbon dioxide could also be used to provide refrigeration.

The importance of handling more than one field or shipping container at a time and with minimum of manual labor has been recognized in the industry for many years. Wheeled clamp trucks have long been used for moving single tracks of contaniers around the packaging house. More recently, modern trucks have come into general usage for handling palletized containers at several stages from orchard to wholesale store. Wooden plates and forklift trucks are in general use at all modern packing houses. According to Martin, the size, the weight and cost of wooden pallets will lead to a substantial reduction in their use during transport, although in-house use of wooden pallets will continue at both shipping points and warehouses.

Fruit Ripening

Peaches and nectarines are the climacteric fruits that exhibit a rise in respiratory activity surge of ethylene production, and accumulation of abscisic acid at ripening of fruit. Changes in organics acids during ripening of peach have also been recorded. Melting and non melting peaches behave differently with respect to ethylene production and response to ethylene treatment. Increased cellulose activity is related directly to the degree of softening. Amylase, invertase, polyphenol oxidase, and polygalactouronase enzyme activities have been found to increase during ripening. The change in polyphen ol oxidase and peroxidase activities could be seen and correlated with browning in the developing fruits using a computer video image analysis technique.

Controlled ripening after harvest, used widely for other fruit crops, could be used effectively for peaches. However; the metabolism of fruit with abnormal ripening (ripening mutants) can provide insight into the normal fruit-ripening process as revealed by behavior of several slow-ripening genotypes. Interestingly, these fruits exhibit no increase in carbon dioxide, ethylene production, flesh color or firmness during 1 month or longer storage in air at 20°C.

Sensory descriptive analysis can be employed to evaluate maturity and ripening of peach fruits. Among the sensory qualities, aroma is most important; changes in the aroma compounds also take place on ripening. Esters, alcohols, acids such as benzaldehyde, aldehyde, lactones with C_6, C_8 and C_{10}, together with deca-lactone, are among the compounds contributing aroma. The primary aromatic compounds which are present in the intact tissue of the cell produce the secondary volatiles after their disintegration. These differences are attributed to the stage of maturity and the synthesis of enzymes. The concentration of most of these compounds increase during maturation of fruit. The process of ripening is followed by senescence, during which growth of the fruit ceases and the processes of ageing replace those of ripening.

Growth regulators such as alar (SADH) and ethephon have similar effects on peach maturation and ripening. The former is sprayed at pit-hardening stage, thus affecting fruit set, and the later is applied just before harvest, thus affecting ripening. The ultimate effect of both chemicals are uniform ripening, fewer picking improvement in internal flesh and skin color, and better fruit quality including increased TSS. Spraying trees with paclobutrazol after full flowering followed by storage (20 days at 8°C) reduced ethylene production after 12 days of storage without affecting the fruit firmness and total acids. Peach fruits harvested at 50 per cent pink half-color stage gave the best keeping quality with $KMnO_4$ (1000 ppm) in paper lining as ethylene absorbent or after treating the fruit with cycocel at 4000 ppm. Use of calcium nitrate (1 per cent) as a pre harvest spray on peach leads to minimum loss of weight, lower respiration, and lower disease incidence, with maintenance of the edible quality of fruits for 6 days. However, use of more than 2 per cent calcium chloride damaged the fruit skin.

Mineral nutrition of peach can also influence post-harvest fruit quality. Amelioration of K deficiency increases the titratable acidity of the fruit and altering the quality. Application of Mg without K accentuated the K deficiency, but their combined use improved yield and fruit size with a delay in maturity of 2-4 days. Potassium tended to increase redness on skin, where as Mg tended to decrease red pigmentation, as was true with the use of N application. However, the later also increased split-pit disorder. Enhancement of N or Mg decreased the self life and firmness of fruit but increased browning resistance and soluble solids content of peach fruit with a lowering of titrtatable acidity.

Chemical Composition and Health Benefits

Peach fruit has an outer exocarp, a mesocrap, and a stony endocarp known as seed. The major portion of the fruit is water. It is either in free or bound form and maintains the turgor of the cells and indirectily the texture of the fruit, besides being a good solvent. More than 90 per cent of total fruit is flesh at harvest maturity, with a plup-to-stone ratio of 8.5-19.50 depending on cultivar.

Peaches have much more to offer than their amazingly delicious taste. Being loaded with vitamins, minerals, antioxidants and other chemical contents, peaches provide you with the following health benefits.

Cardiovascular Benefits

Being rich in iron and vitamin K, peaches help in keeping your heart healthy. Vitamin K prevents blood clotting, thus providing protection against several heart diseases. Iron is required for the formation of red blood cells. It keeps your blood healthy and prevents anemia which occurs due to iron deficiency. Lycopene and lutein in peaches also reduce the risk of heart disease. They also contain potassium, an electrolyte that regulates blood pressure and maintains fluid balance, thus controlling your heart rate and providing protection against stroke.

Anti-Cancer

Peaches are high in antioxidants, particularly chlorogenic acid which is concentrated in their skin and flesh and is known to have anti-cancer properties. They are also a good source of vitamin A which provides protection against lung and oral cavity cancers.

Eye Health

Peaches are a good source of beta carotene, a compound that is converted to vitamin A in the body. Vitamin A is vital for the health of retina and its deficiency can cause weak eye sight. Thus, consuming peaches is a great way of improving your beta carotene levels to maintain healthy eyes. It also prevents night blindness and age related macular degeneration.

Detoxifies Your System

One of the benefits of peaches is that they help in removing toxins from your system. The high content of potassium and fiber reduces the risk of stomach ulcers, inflammation and kidney related diseases. Peaches are also beneficial in curing digestive problems like gastritis and colitis. In China, peach tea is used as a natural remedy for kidney cleansing and detoxification. Peach can help remove worms from the stomach and cure stomach upsets.

Stress Reliever

In Hungary, peach is referred to as the "Fruit of Calmness" due to its ability to relieve stress and anxiety. It helps in restoring the calmness of mind and its flowers are particularly beneficial for treating restlessness.

Dental Health

Peaches also contain fluoride and small amounts of calcium, the compounds that are present in bones and teeth. Minerals like iron and fluoride prevent cavities and other dental issues. So consumption of peaches can help cure bone diseases and several dental problems.

Weight Loss

Peaches are low in calories and fat free which makes them ideal for weight loss. Moreover, they contain natural sugars which do not raise the blood sugar or insulin levels. So including this fruit in your meals can make you slim by burning extra fat deposits in the body.

Other Benefits

Peaches have diuretic and laxative effects which help in relieving pain due to gout and rheumatism. They can inhibit tumor growth through their antimicrobial and antioxidant activity. The phytochemicals in peaches known as phenols can fight obesity related diabetes. A medicinal tea prepared from the bark and leaves of peach tree is effective in curing chronic bronchitis, coughs and gastritis.

Peach Fruit Benefits for Skin

Peaches play an important role in skincare, thanks to their rich variety of skin friendly nutrients and antioxidants. They offer the following benefits for your skin.

Removes Dead Cells

Peaches contain very high levels of vitamin C which is great for your skin.

Rubbing a slice of peach on your skin as face mask sloughs away dead skin cells and the enzymes present in it nourish and refresh your skin.

Anti-Aging Benefits

Vitamin C in peaches is a powerful antioxidant that slows down the ageing process, thus reducing wrinkles, fine lines, dark circles and blemishes. Peach is often used as an ingredient in anti-ageing face masks. It tightens the skin's pores and rejuvenates a tired skin.

Great Moisturizer

Peach is a great moisturizer for dry skin due to the presence of vitamin A and C. If you have dry skin, you can apply a mixture of peach paste and yoghurt for some time and rinse off with lukewarm water. This will not only moisturize your skin but also make it soft and supple. Rubbing peach juice on your skin will impart a natural glow to it. Peach can also be used in oily skincare regimes as well as cosmetics. Vitamins A and C help regenerate skin tissue.

UV Protection

Being rich in vitamin C, A and K, beta carotene, potassium, magnesium and selenium, peaches protect your skin from the harmful ultraviolet rays. To treat sunburns, you can boil this fruit, mash it with a fork and apply it on your face after returning from outdoors. The peel of fruit can also cleanse your skin effectively.

Healing Qualities

The proteins contained in peaches facilitate tissue repair in case of cuts or severe lacerations. Regular consumption of this fruit improves your immune system, thus warding off skin infections.

Peach Fruit Benefits for Hair

Hair health also depends upon the supply of vital nutrients to the hair follicles. Peaches are, thus, beneficial for your hair due to the presence of these nutrients.

Prevents Hair Loss

Being loaded with plenty of nutrients, vitamins and antioxidants, peaches make your hair healthy and silky. Their positive effects on your scalp include providing protection against hair loss.

Scalp Cleanser

This fruit is often used in hair masks to treat scalp problems. A dirty scalp can also lead to hair fall. Topical application of peach in the form of hair mask cleanses your scalp and makes your hair soft and shiny.

Nutritional Value

These fruits have a rich nutritional value incorporating vitamins, minerals and antioxidants. Their nutrition can be explained as follows.

Nutritional Value of Peaches (*Prunus persica*), Fresh, Nutritive value per 100 g

Principle	*Nutrient Value*	*Percentage of RDA*
Energy	39 Kcal	2 per cent
Carbohydrates	9.54 g	7 per cent
Protein	0.91 g	1.5 per cent
Total Fat	0.25 g	1 per cent
Cholesterol	0 mg	0 per cent
Dietary Fiber	1.5 g	4 per cent
Vitamins		
Folates	4 µg	1 per cent
Niacin	0.806 mg	5 per cent
Pantothenic acid	0.153 mg	3 per cent
Pyridoxine	0.025 mg	2 per cent
Riboflavin	0.031 mg	2.5 per cent
Thiamin	0.024 mg	2 per cent
Vitamin A	326 IU	11 per cent
Vitamin C	6.6 mg	11 per cent
Vitamin E	0.73 mg	5 per cent
Vitamin K	2.6 µg	2 per cent
Electrolytes		
Sodium	0 mg	0 per cent
Potassium	190 mg	4 per cent
Minerals		
Calcium	6 mg	0.6 per cent
Copper	0.068 mg	7.5 per cent
Iron	0.25 mg	3 per cent
Magnesium	9 mg	2 per cent
Manganese	0.61 mg	3 per cent
Phosphorus	11 mg	2 per cent
Zinc	0.17 mg	1.5 per cent
Phyto-nutrients		
Carotene-ß	162 µg	—
Crypto-xanthin-ß	67 µg	—
Lutein-zeaxanthin	91 µg	—

Source: USDA National Nutrient database.

☆ **Calories:** They are extremely low in calories with one large fruit of 6 pounce providing just 70 calories.

- **Carbohydrate, Fat and Proteins:** Most of the calories are provided by 17 grams of carbohydrate contained in one large peach. It also contains 1.5 grams of protein and negligible fat of 0.4 grams.
- **Dietary Fiber:** One large peach contains about 2 grams of fiber which is mostly concentrated in its skin.
- **Vitamins:** One large peach provides 570 International Units of vitamin A which is equivalent to more than 10 per cent of the Recommended Daily Allowance (RDA) of this vitamin. It is high in vitamin C, providing 12 mg or around 20 per cent of the RDA. It also provides 1.4 mg or 14 per cent of the RDA of niacin or vitamin B3. It contains other B complex vitamins and vitamin E in small quantities.
- **Minerals:** Peaches do not contain sodium but are particularly rich in potassium with one large peach providing 332 mg or around 10 per cent of the RDA of this mineral. It also provides 10 mg calcium, 30 mg potassium, 16 mg magnesium, 0.3 mg zinc and 0.4 mg iron.

The carbohydrates are the most abundant group of constituents and are present as low-molecular-weight sugars or high- molecular-weight polymers. Sugars in free or bound form are the important constituents of peach and are dependent on the cultivar as well as the root stock. Sucrose is the major non reducing sugar, glucose and fructose are the main reducing sugars, and other sugars such as galactose, xylose, glucitol, and sorbitol are also found in peach. Cellulose, pectic substances, and hemicelluloses are the carbohydrate polymers that constitute the fiber. Glucose and fructose as the predominant sugars in equal amounts in the beginning of fruit development, but later on sucrose started accumulating. However, sorbitol remained at low in the developing peach fruits.

Starch is commonly in the outermost cells of the fruit occurs in the seed of the peach. The mean amino acid content (essential) of the peach has also been reported like proteins; the fat content of peach fruit is also low and is mostly associated with cuticular layers on the surface of the fruit. The fatty acid content of the pistal of different peach varieties is similar, but the reverse was the case with anthers. Malic acid is the dominant acid, with equal amounts of citric acid and quinic acid. Ripe peaches are characterized by high Brix-acid, ratios.

The ripe peach has phenolic compounds having significance from pharmacology, color and flavor development points of view. These include chlorogenic acid and its isomers, neo-chlorogenic acid (highest in skin and peels), and catechin. North Chinese crisp peach (nonfibrous) has the highest concentration of catechin and leucoanthocyanins. Chlorogenic acid, a caffeic acid derivative epicatechin, catechin and cyanidine are the prominent phenolics found in the mesocrap of Cresthaven peaches. Phenolic compounds are significant in the quality of both fresh and canned peaches, especially their implication in red color of exocrap, astringency and enzymatic browning of thermally processed peaches.

The browning potential of peaches depends on the amount of phenolic compounds present in the fruit and the levels of activity of polyphenol oxidase (PPO) enzyme. Normally, these compounds are physically separated from PPO enzyme in the intact tissue, but with damaged (*e.g.*, bruising), PPO has access to phenolic compounds and chemical reactions leading to browning taking place. Sensory properties of fresh and processed peaches are affected to a great extent by cultural and environmental factors.

The visual impact of carotenoids in peaches is the primary basis for color grading. The yellow- orange color of peaches is attributed primarily cyanidin-3-glycoside. Leucoanthocyanin content, however, depends on the variety.

Approximate Composition of Peach Kernals

Constitutent	*Content*
Protein (per cent)	23.9-31.2
Albumins (per cent)	60.0
Globunis (per cent)	9.0
Prolamins (per cent)	6.0
Glutamins (per cent)	8.0
Nonprotein nitrogen (per cent)	17.0
Fats (per cent)	39.2-54.5
Oleic acid (per cent)	63.8
Linoleic acid (per cent)	15.4
Saturated fatty acid (per cent)	20.7
Lipid composition	
Triglycerides (per cent)	98.0
Polar lipids (per cent)	1.1
Sterols (per cent)	0.5
Starch (per cent)	3.6
Sugars (per cent)	2.2
Crude fibers (per cent)	14.8
Minerals (per cent)	2.7
Hydrocyanic acid (mg/100g)	102.2

The Predominant carotenoids and carontenol fatty acid esters in peach fruit were found to be zeaxanthis and β-cryptoxanthin, identified as xanthophylls, lycocopene, β-carotene, α-ceratolene (Hydrocarbon carotenoids), while β-cryptoxanthin, lutein and zeaxanthin were identified as carotenol fatty acid esters. The carotenoids are important for color and their role in prevention of cancer. Exposure to solor radiation, even for 3 days, markedly stimulates anthocyanin development in peach fruit, reaching a maximum in 8 days of exposure.

The main contribution of fruits including peach and their processed products to nutiration is undoubtedly the supply of vitamin C, among other vitamins present in the peach fruit.The vitamin C content depends on the variety, ripeness stage and location. Peach fruit is also a good source of minerals.

Peach kernel is a source of fats, protein, fiber and minerals. The protein are rich in lysine, leucine, isoleucine, valine, threonine, basic and acidic amino acids, but poor in methionine. The kernel oil is utilized in food, cosmetics and pharmaceutical preparations. It can be used as a fertilizer or as a cattle feed after debittering. However, peach kernel oil had marked toxic effect when fed to mice. Peach flowers and leaves are purgative and anthelminitic.

Storage

Peaches and nectarines do not grow sweeter once picked from the tree—fruits bought from the market will be as good as it gets. Peaches *can* grow softer and slightly riper if placed in a paper bag with bananas and other ethylene-emitting fruits, but it's best to purchase peaches that are already somewhat ripe. In refrigeration, ripe peaches last for an extra four days. Place the fruits in a plastic bag to capture the much-needed humidity and store in the crisper. Although cool storage prolongs its longevity, let the fruit sit at room temperature before consuming—the peach will taste better. Peaches freeze well, though they require preparation that can be laborious. First, remove the skin by slitting a small "x" near the groove at the top. Blanch by dropping the fruits into boiling water for no more than 30 seconds, and then transferring them into a bowl of ice water—this will loosen the skin. Peel the skin, starting at the cut slit. Once peeled, use a paring knife and slice the fruit in half to remove the pit. Cut into chunks or wedges, and coat them in lemon juice to prevent browning. Many choose to sprinkle sugar over the peaches as well. Freeze the chopped pieces on a baking tray with parchment first, and then move the pieces to a plastic freezer bag. They will keep for 8 months or, until the next peach season.

Post-harvest Losses

Peaches are highly perishable fruits and exhibit a high rate of deterioration due to their high moisture content. Unless the process of ripening is controlled, the fruits become spoiled within 2-3 days after picking. Because they lack hard texture, peaches are highly susceptible to bruising. Losses also occur due to microbial spoilage during harvesting, transportation and storage of fruits. Post-harvest losses of peaches were estimated to be of the order of 15-24 per cent due to microbial decay. Fungi are the major cause of deterioration of fresh fruits, while crops which have been damaged during cultivation, harvesting, storage, transportation, or marketing are infected by weak pathogens. Physical injury induced by mechanical means, natural openings on the surface of fruit, or physiological injury or injury after harvesting due to high or low temperatures or chemicals are preconditions for development of weak pathogens.

Diseases of peach fruit at post-harvest stage occur due to either contamination or infection that occurs during the growing season or during harvesting, handling, packaging and transportation. Bacteria are responsible for most of the soft rots of fruits during transport or in storage, such as soft rot by coliform bacteria, *Erwinia coratovora*, and *Pesudomonas*. Bacterial spot, a serious disease in several countries is caused by *xanthomonas compestris* pruni; it can be controlled by hygienic conditions and antibiotic spray. Post-harvest spoilage can be prevented or retarded by use of chemical inhibitors, fungicides, hot water washes and refrigeration. Blotches or spots developing on the peak skin after harvest have been associated with metal cations in water used to wash fruit, rubbing or transit injury, preharvest application of fungicides such as captan and ammonia leaks from refrigeration equipment. Nectarine pox is another disorder of commercially grown nectarines (*Prunus persica* L.). It is characterized by superficial warty outgrowth, which may be due to deficiency of boron or infection with virus.

Disinfestations of fruit insects is of particular concern when the fruit is being shipped across agriculture quarantine. Use of methyl bromide fumigation of stone fruits also centered around quarantine measures to assure freedom from codling moths, which were not generally attracted to peaches or nectarines. Fumigants can be phytotoxic. Methyl bromide-treated nectarines were found to develop deep pitting, which tended to be greater at lower temperatures (4.5°C) than at higher (26° C); damage becomes more severe at higher concentration of fumigants (48 g/m^3). However fumigation with methylene bromide (48g/m^3) for 2 h at 21° C controlled the incidence of *Cydia pomonella* significant phytotoxic effect but with altered texture. The use of Bacillus subtilis and related organisms as biofungicides has been reported. A pilot test on simulate commercial peach packing lines showed that *Bacillus subtilis* as a biological agent was equal to treatment of fruits with benomyl (1-2 mg/kg) in controlling brown rot caused by *Monilinia fructicola*.

Precooling

Fruits are cooled immediately after harvesting to reduce the field heat and thus lower the temperature and retard the process of ripening. Clingstone peaches for canning are often hydrocooled in orchard pallet bins and then top-iced before transportation. During hot weather when peach temperate is high, cooling requirement increases.

Nectarines can be hydrocooled with 100-120 ppm chlorine in the hydrocooler and hydrodumper. If hydrocooling is to be followed by extensive storage, fungicides are recommended for protection. Nectarines respond very well to waxing and fungicide protection.

Post-harvest Diseases and Control

Fruit losses caused by post-harvest diseases are among the main concerns of the peach fruit growers and marketers. Depending on weather conditions and post-harvest handling, other high-incidence post-harvest diseases of stone fruit are gray mold (caused by *Botrytis cinerea* Pers.:Fr.), sour rot (caused by *Geotrichum candidum* Link), rhizopus rot (caused by *Rhizopus stolonifer* (Ehrenb.:Fr.) Vuill.), mucor rot (caused by *Mucor piriformis* E. Fischer), alternaria rot (caused by *Alternaria alternata* (Fr.: Fr.) Keissler), and blue mold (caused by *Penicillium expansum* Link). Effective post-harvest decay control depends on an integrated management approach based on appropriate preharvest fungicide treatments, adequate harvest and handling practices, effective sanitation of fruit and facilities in the packinghouses, appropriate post-harvest antifungal treatments, and maintenance of the crop. Comprehensive detail of important Post-harvest diseases of Peaches and their control Measures are given in table.

Important Post-harvest Diseases of Peaches and their Control Measures

Name of Disease	*Symptom*	*Causal Organism*	*Control Measures*
Browm rot	✰ Small, water-soaked spots on fruits appears initialy ✰ Affected flesh becomes brown/ black in 24 hours ✰ Ashy tiffs of fungus, powdery mildew mass in rings; entire fruit decays, shrivels and mummufies	*Monolinia* sp.	✰ Refrigenation or hydro-cooling after harvest ✰ Dipping the fruit in the water at 50°C for 3 minutes ✰ Botran (DCNA (250500 pm), benomyl (150-250 ppm), or bavistin (750 ppm)
Whisker's rot .	✰ Circular tan area around healthy skin ✰ Whole skin becomes tan to brown. Later fungus grows near the center 2-3min	Fungi	✰ Precooling in storage and transit. ✰ Hot water dip at 50-54°C for
Blue mold	✰ Occurs on injuried or overripe fruits with small tan coloured spots. ✰ Skin over the spot breaks or slips ✰ Whole surface of the fruit becomes blue	*Penicillum expansum*	✰ Rapid precooling and storage at °C ✰ CA storage of fruits at 0c with 1 per cent O_2 + 5 per cent CO_2
Black mold	Small, tan, slightly unsunk Mold produces abundant black spores with sooty appearance	*Aspergillus niger*	✰ Careful handling to avoid fruit injury ✰ Avoid long storage
Green mold	Found on inkjured or softened fruits Infected spot green, affecting skin, but later covering entire fruit, including flesh	*Alternaria Alternate*	Careful handling during harvesting, packing and transportation
Grey mold	Light brown spots, usually after lonf storage	*Botrytis cinerea*	

Low-Temperate Storage

The storage life of any fruit is influenced by temperature which is known to affect the rate opf respiration. Lowering of the temperature, therefore, considerably reduces the respiration rate and extends the storage life. Refrigerated storage has made possible the marketing of fruits after their harvest season. But peaches have a shorter storage life than the majority of temperate fruits. Cold storage is the common practice to prolong the canning season. Peaches and nectarines are seldom stored except for a short period to overcome the glut in the market or to extend the processing season. In general, sound, well-matured fruits can be stored for 2-4 weeks at -0.5 to 0°C and 90 per cent relative humidity, depending on the cultivar and the growing season. However, peaches stored at 0°C for more than 4 weeks tend to have a dry and mealy flesh and may show marked browning around the pit. Some late-season peaches, such as Rio Osa gem, can be stored for 4-6 weeks at -0.6 to 0°C and 90-95 per cent relative humidity before marketing. Transferring fruit to 5°C after storage for 1 or 2 weeks at 0°C causes severe internal breakdown.

Peaches harvested at firm ripe stage and destined for processing often need 4-6 days for ripening and storage for 10-14 days at -0.5to 0°C reduces rhizopus decay during subsequent ripening without affecting brown rot. Generally, peaches are of better quality and have less decay if ripened before storage. Peaches and nectarines ripened after storage of 4 weeks at 0°C retained a significantly better internal appearance, with less internal breakdown, than those ripened after storage. Those softened at room temperature almost reharden when returned to low temperature storage.

Nectarines have higher susceptibility to shriveling than peaches and should be stored in low air velocity at proper temperature with useful supplement to protect the nectarines, including packing. Chilling increases the membrane permeability in tissue from various chill sensitive species. Intermittent exposure to room temperature has been advocated as a means of controlling wooly breakdown of peach in cold storage. Mealiness of peaches is associated with reduced electrolyte leakage and internal conductivity, and can be used as a nondestructive method to find out the presence of mealiness intact fruit. A relationship between mealiness in nectarines tissue and Ca uptake, ion leakage and internal air space has also been observed. The incidence of mealiness has been attributed to the presence of insoluble low-methoxypectic substance of high molecular weight which are formed by the action of pectinase increase during chilling and which reduce activity of polygalacturonase during subsequent ripening.

Controlled-Atmosphere Storage

The preservative effect of low temperature becomes more pronounced when it is combined with control of the composition of the storage atmosphere than when either method is used alone. A reduction in oxygen concentration and/or an increase in carbon dioxide concentration of the storage atmosphere reduces the rate of the storage atmosphere reduces the rate of respiration of fresh fruits and vegetables and also inhibits microbial and insect growth. Either controlled-atmosphere (CA) storage or modified-atmosphere (MA) storage can be used. The storage life of peaches has been extended by storing them in controlled atmosphere in recent years as CA facilities become more common. Improved equipment is now available to establish and maintain CA conditions within storerooms and during rail and road transport.

Subatmospheric-Pressure Storage

Storage of fresh produce under reduced pressure has also received attention. Subatmospheric pressure (102 mm Hg) significantly delayed the flesh softening of firm/mature peaches and extends their storage life up to 93 days, compared to 66 days for fruit stored at the subatmospheric pressure. Changes in sugar and acid content were significantly altered by the subatmospheric pressure, through at the end of the storage time both were comparable.

There is a problem of decay in abnormal ripening of peaches after 5 weeks of low-pressure storage (10 kPa) at 20°C, although low pressure (10-25 kPa) retarded softening, discoloration, and sugar depletion in peaches during storage of 2 weeks, indicating that wilting of fruits was a more serious problem than accumulation of ethanol.

When peaches are stored under unfavorable conditions for too long, they lose brightness of skin color, develop considerable off-flavor (dry and mealy), and exhibit red to brown discoloration of flesh. In addition, fluctuation in storage temperature often causes condensation of moisture on fruit, favouring the growth of microorganisms, particularly mold, resulting in decay of the stored fruits.

Irradiation

Irradiation is another method of reducing the post-harvest decay of fruits. After the end of World War II, the uses of ionizing radiation for food preservation were extensively investigated, involving both perishable and non perishable foods such as fresh fruits and vegetables. Among the radiations, gamma rays are well known for their lethal properties. At very high doses, radiation kills all forms of life, but at lower concentrations, the radiation can check processes such as sprouting and mycelial growth, reduce the number of microorganisms and hence reduce the extent of spoilage. Irradiation can be successful only if, after inactivating the decay-causing organisms, the radiation does not injure the fruits by altering the flavor, texture or appearance. Peaches inoculated with active cultures of *Monillinia fructicola* remained free of infection for 10 days at 27°C after irradiation at 182,000 rad, while peaches inoculated with *Rhizopus nigricans* remained free for only 6 days with similar treatment. Firm ripe peach fruit, when irradiation to different radiation doses and temperatures, had higher taste preference after 10 days of storage than after 1 day of storage. This was probably due to better ripening of treated fruits. However, after 10 days of storage, there was substantial increase in adjusted preference scores at the 3×10^5 radiation level. This was attributed to progressive reduction in mold growth. Gamma radiations at 40-50 Krad from cobalt -60 of California peaches and nectarines infested by Mediterranean fruit flies (*Ceratitis capitata*) prevented the hatching of eggs and better control of harvesting, shipping and handling than controls.

Processing

Peaches are principally canned; only a small proportion if frozen, dried, or used to prepare other products such as jams, preserves or beverages, the details of which is given as under:

Puree

Peaches puree is the starting material for preparation of other products. Various steps involved in the preparation of puree are washing, trimming, cooking the fruits and passing the pulp through a continuous rotary unit with perforation, followed by finishing with addition of ascorbic acid (0.14 per cent). The puree is pasteurized and filled aseptically into bulk-capacity drums or large-size cans, which are closed using vacuum and nitrogen. A sensor mechanism for pitting of peach fruit has been developed. Initial color and viscosity of puree are important quality characteristics. Peach pulp flow behavior index ranged from 0.20 to 0.39, indicating the pseudoplastic nature of the pulp.

Jam

The fruits of peach can be used in the preparation of jam. Most of the time, it is combined with other fruits to produce mixed-fruit jam. The fruit is converted into pulp and mixed with the required amount of sugar. If needed, pectin is added for proper setting of jam. Jams for domestic use are commonly packed in glass containers and usually have sufficient sugar content to preserve them against spoilage due to microorganisms. Long storage of jam affects its quality. The rate of deterioration on quality of plum was found to be about the same whether the jam previously had been stored at -20, 2, 32 or 47°F.

Nectar

Nectar is another product prepared from peach. Procedure for making nectar includes washing, halving and pitting the fruits, followed by passing over the inspection belt to remove any damaged fruits. Peeling is done by lye peeling (1 per cent sodium hydroxide at 100°C for 15 s), followed by immersion in cold water. After removal of the skins by rubbing, halves are cooked and passed through a disintegrator and pulped in pulper. Syrup and citric acid are added and the product is passed through a deareator to prevent deterioration in color and flavor. It is pasteurized in jars at 100°C for 20-30 min or after filing hot in cans at 88°C, inverting the can and holding temperature for min followed immediately by cooling.

Juice and Beverages

Preparation of beverages including ready-to-serve beverage, ginger appetizer *etc.* is another outlet for utilizing peach. Because of the difficulty of extracting juice from the fruit, which is pulpy in nature, a process for juice extraction using pectiolytic enzyme has been reported. The addition of enzyme increased the juice yield, causing a slight change of TSS, pH and acidity and a drastic decrease in apparent viscosity. Enzymatic extraction of the juice improves the color and clarity without affecting the flavor. Differences in types of sugar and acids in peach and nectarine juices have also been reported.

Baby Food

Clingstone peaches are prefered to freestone for preparation of baby food because of the white-yellow color of the puree, better storability and non melting flesh and thicker consistency. The peach pulp is blended with sugar syrup, sterilized at 110°C and deareated. It can be either canned or preserved in jars. The final peach baby food has soluble solids of 21-22 per cent, a pH of 3.0-4.1, an acidity of 0.35-0.45 per cent (citric acid) and a Bostwick consistometer reading of 7.0-8.5 cm.

Canned Peaches

Commercial canning of peaches is an important industry. Yellow-fleshed cultivars are the most common and are preferred both for processing and the fresh market. Peaches can be canned whole, as halves, quarters or sliced, in water, syrup or juices of various color, tender and of good cooking quality. They should be picked at or near the optimum maturity. Fruit for processing should have proper maturity, a diameter of 6.03 cm or more and freedom from blemishes

For canning, peaches are lye peeled (1.5-2.0 per cent hot lye). The halves are washed once then dipped in 1.0 per cent citric acid. The peeled halves are filled into empty cans which are automatically filled with syrup by a vacuum syruping machine. The peaches may receive a syrup of 40,25, 10°Brix or water, depending on the brand, but most choice-grade canned peach halves have 21-22° Brix. The filled cans are exhausted in a steam-filled exhaust box for 5-6min at 93-96°C. Coding of cans is then done by specifying the cultivar grade and content. The processing time of canned peach depends on the maturity and cultivar of fruits' vacuum indicates satisfactory processing. The cans are double seamed and stylized, usually for 20-25 min at 100°C, followed by cooling in water to 36°C. Canned peaches are stacked, labeled and stored until disposal in a warehouse at 20°C with good ventilation. The quality factors for canned peaches are proper vacuum, headspace, drained weight, color, uniformity of size, absence of defects, texture and flavor. Nitrogen fertilization increased peach flavor and texture and decreased astringency and tartness of canned. However, canned peaches with low nitrogen and high potassium regime were poor in quality. Consumers prefer canned clingstone peaches having a cutout TSS of 22-23 per cent, a TSS-acid ratio of 34-40 and titratable acidity of 0.35 per cent. Less mature peaches gave canned product with lower drained weight than more mature peaches. Drained weight and cutout syrup Brix are inversely related and increase rapidly during the first week and then slowly reach an erquilibrium 90 days after canning. Canned peach cultivars of July Elberta and Shimizu with citric acid in covering syrup and after 0-100 days of storage gave the best product.

Enzymatic firming of fruits using vacuum infusion of exogenous citrus pectin methylesterase (PME) may obviate low endogenous levels of PME for firming of peaches, which could be especially useful for freestone peaches, which have higher levels of endopolygalacturonase activity and are prone to rapid softening. The important factors affecting HCN content in canned stone fruits include initial amount of glycoside and conditions of heat processing and enzymatic hydrolysis of glycoside.

High-vacuum flame sterilization (HVFS) of fruits produces a canned product which is closer to fresh than conventionally canned product. The process is advocated to use 30 per cent lesser energy in processing than the conventional sterilization procedure. In HVFC processing, the fruits are packed with little or no water, deareated to a high vacuum level and given a minimal thermal treatment which enhances the attributes of canned fruits. The process required 7.5 min total heating time and 5.5 min for diced fruit to achieve biological stability. Comparison of cutout analysis of conventional canned and high-vacuum flame-stylized fruits retained original flavor attributes and texture of fresh fruit for 18months. Omission of covering syrup combined with necessary blanching treatment allowed packing of one-third more fruit into a standard can.

Conventionally, peaches are canned in sugar syrup of 40-42°Bricks with 0.1 per cent citric acid. To make the product nutritionally better, instead of covering sugar syrup, juice may replace a part of the covering medium. The product is claimed to have similar physicochemical characteristics but improved sensory qualities compared to conventional product.

Diced Peaches

The fruits used for preparation of diced peaches are of canning ripeness. The fruits are halved, peeled in 2 per cent lye solution at 103°C for 20-38 s, and diced to give 1.27-cm cubes. Diced peaches and pears are sometimes mixed. Filling of the cans and scraping is done, followed by exhausting and processing for 20 min at 100°C in an atmospheric cooker.

Steep Preservation

Steep preservation is simple technique which can be used for preparation of peaches. It does not require costly equipment or packaging materials. Steep preservation method can be used for peaches but nor for nectarines.

Marmalade

Marmalades have the characteristics of jellies and preserve combined, contain the fruit pulp and may contain the skins suspended in jellied juice. They are made from slightly under ripe fruits that are rich in pectin and acid, chiefly from citrus fruits, alone or in combination with other fruits. Popular fruit combination includes orange and peach.

Peach Butter

Peach fruit butter is made by cooking he pulp to a smooth consistency, thick enough to hold its shape but soft enough to spread easily. Fruit butters are different from jams in that the product is clear but more concentrated. Since, butters are very thick, special care is needed to avoid scorching. Usually, sugar is added at the rate of $^{2/3}$ units of sugar to 1 part fruit. For more flavor, cider, cinnamon and mace are added as desired. The mixture is cooked to a heavy consistency, packed in hot sterile jars and corked quickly.

Peach Conserve

Conserve is similar to jams with chopped nut (pecans/walnuts) added for texture and flavor. Conserves may be a mixture of two or more fruits, with the chief ingredient in specific conserve being peaches. The product contains higher proportions of fruit than marmalades.

Fruit Bar

Fruit bars of leather is a product obtained by dehydration of fruit purees into leathery sheets. Peaches with high sugar and flavor but low fiber are suited t make leather, especially from culls and overripe fruits. Since, fruit leather is poorly packed and unrefrigerated, it loses color and flavor gradually and quality standards are difficult to maintain.

Pickle

Peach produces one of the most popular and acceptable pickles. Small freestone peaches from the grading tables of fresh or canned fruits are utilized. The fruits are peeled by dipping in boiling 2 per cent lye, rinsed with water and dipped in 1 per cent citric acid solution, often with the addition of apple cider vinegar or brown sugar. Sweet peach pickles are made with 50° Bricks syrup having a cutout of 30° Bricks, through really sweet pickles have 65-67° Bricks of syrup with cutout of 40° Bricks. The shrinkage is more in heavier syrup than in lighter syrup. Pieces of peach pickles tend to float in the syrup until they attain equilibrium.

Refrigerated Peach Slices

The use of low-temperature preservation is a very promising technique, but the primary reason for the limited shelf life of refrigerated peach slices is the loss of fresh natural flavor and development of a seedy flavor and gradual softening of texture. Various steps involved in the preparation of refrigerated peach slices include harvesting freestone peaches at firm ripe stage, mellowing for 1-2 days, grading fruit for maturity and uniformity, followed by peeling in boiling 2-5 per cent lye solution. The peel is then removed, followed by neutralization. The slices are packed in convenient size cans. The process is completed by coating 8 parts peaches with 1 part sugar containing 0.15 per cent ascorbic acid and 0.06 per cent sodium benzoate, mixing gently, filling into sterile glass or plastic jars, closing with vacuum or screw caps, cooling to 4°C and storage at 0°C for 20 weeks.

Dried Peaches

Dehydration is the oldest and most common form of food preservation. Potential spoilage of dried fruit depends on how much water is available to the spoiling microorganisms, which is expressed in terms of water activity (aw); Yeast needs more water (0.85) than fungi (0.80) but less than bacteria (0.90) and is, therefore, a very stabilizing factor against microbial deterioration. Freestone peaches are used almost exclusively for drying.

The steps involved in common pre drying treatment applied to the fruits are (a) selection and sorting for size, maturity and soundness (b) washing (c) peeling by hand, lye solution or abrasion (d) cutting into halves, slices, cubes or segments and (e) sulfuring. The quality of dehydrated products depends on many factors, which as raw materials, drying temperature, process time, moisture content and sulfur dioxide concentration.

For many years, sulfur dioxide (SO_2) an antioxide has been used for preservation besides protection of carotene and ascorbic acid contents of dried fruits. Packing materials and packaging atmosphere can be used to control SO_2 loss from dried peaches during extended storage Nitrogen packing also reduces the loss of SO_2 from fruit. However, marked reduction of SO_2 in fruit before consumption can be introduced by impressing the fruits in hot water. Dehydration of sun-dried peaches, after a 3-min dip in 1.0 per cent ascorbic acid, had more attractive color and flavor than non treated fruits. Peaches have drying ratios varying from, 3.5:1 to 7:1, depending on cultivar and maturity. Peaches can be dried having 4 per cent moisture under vacuum only. The dried products are utilized as pie, tart and turnover filling, while the powder provides excellent purees, spreads or glazes after proper dehydration and preparation. The drying of the fruit could be achieved by solar or mechanical drier. In the latter case, forced air at high temperature is also used. It has been found that 75 per cent air recirculation in the drying process required the least energy.

Osmotic dehydration has received attention as an intermediate step in drying, dehydro freezing and freeze-drying of fruits. The method is based in removing only part of the natural water content of the fruit to produce products of high quality, especially flavor. During the process, exchange of sugars or other components take Place. Under this process sugar exchange dynamics between the syrup and the fruits, including peach is related not only to the flux of sugars from the syrup into the fruit, but to the individual sugars present in the fruit. The exchange is dependent on the species of fruit, the relative defeasibility of the sugars, and enzymatic activity within the fruit, peach was found to have complex behavior. According to osmosis laws, the fruit should be penetrated by all the sugars present in syrup, as their concentrations are higher in the syrup than in the fruit.

REFERENCES

Amoros, A M. Serranto, F. Riguelme and F. Romoyaro. 1989, Levels of ACC and physical and chemical parameters in peach development, *J. Hort. Sci.* 64(6): 673.

Anderson, R.E., 1982, long term storage of peaches and nectarines intermittently warmed during controlled atmosphere storage, *J.Am. Soc. Hort. Sci.* 107(2): 214.

Baugher, T.A and S.S. Miller.1990. Nectarine pox: A disorder of nectarine fruit. *Hortscience* 26: 210.

Bhargava, J.N, D.R. Thakur, M.P.Dwivedi and S.N.Tripathi.1986. Determination of maturity indices in July Elberta peach fruit, *Advances in Research on Temperate Fruits* (T.R.Chadha, ed.), Parmart University of Horticulture and forestry, Horticulture publication, Solan, India.

Brecht J.K., and A.A. Kader., 1982, Post-harvest physiology of non ripening nectarines selections. *Hort-Science* 17: 490.

Brecht, J.K, and A.A. Kader. 1984. Ethylene production by fruit of some slow ripening nectarine genotypes, *J.Am. Soc. Hort.Sci.* 109: 763.

Byrne, D.H, A.N. Nikolic and E.E.Burns. 1991. Variability in sugar and acids, firmness and colour characteristics of 12 peaches genotypes, *J. Am. Soc.Hort. Sci.* 116(6): 1004.

Chaparro, J.X., R.E Durham, G. A Moore and W. B. Sherman. 1987. Use of isoenzymes techniques to identify peach x *Nonpareil" almonds hybrids. *Hort Sci.*22: 300.

Chapman, G. W, R. J. Horvat and W.R. Forbus. 1991, Physical and chemical changes during the maturation of peaches *J. Agr. Food Chem.* 39 (5): 867.

Chapman, G.W and R.J Horvat. 1990. Changes in nonvolatile acid, sugar, pectin and sugar composition of pectin during peach (Cv. Monero) maturation. *J. Agr. Food Chem.* 38: 383.

Crisosto, C.H, A.N. Miller, P.B. Lombard and S.Robinson. 1990. Effect of fall ethephon application on bloom delay, flowering and fruiting of peach and prune, *Hortscience* 25(4): 426.

Delwiche, M. L., 1987. Grader performance using peach ground colour maturity chart, *Hortscience* 22 (1): 87.

Dhillon.W.S and S.S. Cheema. 1991. Physico -chemicals changes during development of peach cv. Flordasum, *Indian Food Packer.* 45(4); 56.

Dineer, I, M. Yildiz, M. Loker and H. Gun. 1992. Process parameters for hydrocooling apricots, plums and peaches. *Int. J. Food Sci. Technol.* 27(3): 347.

Erez, A and J.A. Flore. (1986). The quantitative effect of solar radiation on Redhaven peach fruit skin color, *Hortscience 21*(6): 1424.

Fourie. P, E.C.F, Hansman and G.L. Wium. 1992. Effect of fresh fruit characteristic and cold storage on the quality of dried apricots and peaches, *J. Hort. Sci.* 67(1): 59.

Galleschi, L, E. Scieno, R. Izzo, M.F Quartacci and A. Masia. 1991. Changes in protease activities during developing of peach mesocrap. *Plant Physio. Biochem* 29(6): 531.

Giangiacomo, R, D. Torreggani and E, Abbo, 1987. Osmotic dehydration of fruit. part.1: Sugar exchange between fruit and extracting syrup. *J. Food Preserve,* 11(1): 183.

Hansche, P.E. 1986. Heritability of juvenility in peach. *Hort Science* 21: 1197.

Hansche, P.E. 1989. Three brachytic dwarf peach cultivars: Valley gem, Valley red and Valley sun. *Hort. Sci.* 24: 707.

Hasse, C O., 1975. Peaches. *Advances in Fruit Breeding* (J. Janick and J. N. Moore, ends) Purdue University Press, West Lafayatte, in. p.285.

Horvat, R.J, G.W. Chapman, J.A. Robertson, F.J. Meredith, R.Scorza, A.M.Claham and D.Morgens. 1990. Comparison of volatile compounds from several commercial peach cultivars, *J. Agr. Food Chem.* 38: 234.

Ibarz, A and J.E. Lazano.1992. Rhehological characteristics of concentrated plum and peach pulps. *Rivista Exponala de Ciencia Y Techonology Alimants* 32(1); 85.

Ibraz. A, Gonzels, S. Eslugas. M. Smith and A.A. Kadar. 1992. Composition and characterization of prunes and juices, *J. Agri. Food. Chem.*, 40: 784.

Javeri. H, R. Toledo and I. Wicker.1991. Vacum infusion of citrus pectinmethylesterase and calcium effects on firmness of peaches. *J. Food Sci.* 56(3): 739.

Joshi.V.K, S.K. Chuhan and B.B .Lal. 1991. Extraction of juice from plum, peach and apricot by pectolytic enzyme treatments, *J. Food. Technol.* 28(1): 64.

Kader, A.A., and A. Chordas. 1984. Evaluating the browning potential of peaches. *Calif. Agr.* (3 and 4): 14.

Kamal, B.S and Y. Kakuda. 1992. Characterization of the seed oil and meal from apricot cherry, peach and plum, *J. Am. Oil Chem. Soc.* 69(5): 493.

Kawano, S, H. Watanabe, and M. Iwamoto., 1992, Determination of sugar content in intact peaches by near infrared spectroscopy with fibre optics in interactive mode, *J.Jpm. Soc. Hort. Sci.* 61 (2): 445.

Khachik, F, G.R. Beecher, and W.R. Lubby.1989. Separation, identification and quantification of the major carotenoids in extracts of apricots, peaches, cantaloupe and pink grape fruit by liquid chromatography, *J. Agr. Food Chem. 37*: 1465.

Kuezynski, A, P.Varoquax and F. Varoguex. 1992. Reflectometric method to measure the initial color and browning rates of white peach pulps. *Science des Aliments* 12(2): 213.

Lill, R.E, E.M. O'Donoghue and G.A. King. 1991. Post-harvest Physiology of peaches and nectarines, *Hort. Rev.*11 (1): 413.

Luh, B, C.E. Kean and J.G. Woodroof. 1986. Canning of fruit, *Commercial fruit processing* (J.G. Woodroof and B.S. Luh, ends), AVI West port C.T. 6, P.163.

Lyon, B.G, J.A. Rohartson and E.I. Meredith. 1993. Sensory descriptive analysis of cv. Cresthaven peaches, *J. Food Sci.* 58(1): 177.

Machado, C.A.E, B.H, Nakasu, E.A. Olivveira and E. Karsten.1986. Superoxide dismutase isoenzymes patterns in some peach genotypes. *Pesquisa agropecuaria brasileria* 21(11); 1193.

Maness, N.O, G.H. Brusewitz and T.G. McCollan. 1992. Internal variations in peach fruit firmness. *HortScience* 27 (8): 903.

Meredith, F.L, R.G. Leffler and C.E. Lyon. 1990. Detection of firmness in peaches by impact force response. *Trans. ASAE* 33 (1): 186.

Mikel, W.B and S.L. Olds. 1993. Peach and nectarines. *Encyclopedia of Food Science Technology of Nutrition* (R.K. MaCre, R.K. Robinson and M.J. Sadler, eds) Academic Press London., p. 3464.

Miller A, N and C.S Walsh. 1990. Indole-3-acid and concentration and ethylene evolution during early fruit developing in peach. *J. Plant Growth Regulation* 9(1): 37.

Miller, B.K and M.J. Delwiche. 1989. A color vision system for peach grading. *Am. Soc. Agr. Eng.* 32(4): 1484.

Moore, G.A. 1988. Non-convectional technique for peach breeding and genetic studies. The peach- world cultivars to marketing (N.F. Childers and W.B. Sherman, eds) Horticultural Publications, Gainesvillla, FL, P. 122.

Moriguchi, T.Y. Ishizawa, T.sanada., Teramoto and S. Yamaki. 1992. Role of sucrose synthease and other a related enzymes in sucrose accumulation in peach fruit. *J. Jpn Soc. Hort.Sci.* 60(3): 531.

Morris, J. R, L.D. Ray and D.L. Cawthon., 1978, Quality and post-harvest behavior of once-over harvested clingstone peaches treated with daminozide, J. *Am. Hort. Sci.* 103: 716.

Pathak, R.K and R.A. Pathak. 1991. Peaches. *Temperate Fruits* (S.K. Mitra D.S Rathore and T.K. Bose, eds). Horticulture and Allied publishers, Calcutta, p.179.

Phillips, D.J. (1988). Reduction of transit injury associated with black discolouration of fresh peaches with EDTA treatments, *Plant. Dis. Rep.* 72: 118.

Randhawa, S.S. 1987. *Wild Germplasm of Pome and Stone Fruit*. Indian Agriculture Research Instuite Reg. Sta. Shimla 1987, p.51.

Robertson, J.A, F.I. Meredith, W.R. Forbus and B.G .Lyon. 1992, Relationship of quality characteristics of peaches (Cv. Loring) to maturity, *J. Food Sci.* 57(6): 1401.

Ryall, A.L., and W.T. Pentzer., 1982, *Handling, transportation and storage of fruit and vegetables. 2. Fruits and Tree* Nuts, AVI, Westport, CT.

Salunkhe, D.K and B.B. Desai., 1984, *Post-harvest Biotechnology of Fruits,* Vol. 1, CRC Press, Boca Raton, FI, p. 157.

Sandhu S.S, B.S. Dhillon and W.S. Brar, 1986. Changes in phenolics and anthocyanins in developing the fruit of early and late maturing peach cultivars, *Indian J. Hort.* 43(1 and 2): 69.

Sandhu, S.S, J.S. Randhawa and B. S. Dhillon. 1986. Storage life of Flordasun peach fruits. *Indian Food Packer* 36: 48.

Sandooza, J.K, R.K. Sharma, R. S. Singhrot, and J.P. Singh., 1986, Studies on shelf life of different maturity stages of peach cv. Flordasun as affected by $KMnO_4$ and $AgNO_3$. *Haryana J. Hort. Sci.* 15(3/4): 210.

Shah, G.H and G. S. Brains, 1991. Flow behaviour of peach and apricot pulp and concentrate of some Indian varieties. *J. Food Sci Technol.* 28(5): 308.

Sharma, R.K, J.K. Sandooza, R.S. Singhrot and J.P. Singh., 1986, Studies on the shelf life of peach cv. Flordasun as affected by various packing materials, *Haryana J. Hort. Sci.* 15 (3/4): 188.

Singh, B.P, O.P. Gupta and K.S. Chauhan. 1982. Effect of preharvest calcium nitrate spray on peach on the storage life, *Indian J. Agr. Sci.*52: 235.

Singh, N, M.J. Dewiche and R.S. Jhonson., 1993, Image analysis methods for real time colour grading of stone fruit, *Computer Electronics Agr.* 9(1): 71.

Stosic, D, M. Gorunovic and B. Popovic. 1987. Preliminary toxicological study of the kernel and oil of several species of genus *Prunus, Plants Medicinales et Phytotherapie 21*(1): 8.

Stutte, G.W. 1989. Quantification of net enzymatic activity in developing peach fruit using computer video image analysis, *Hortscience 24*(1): 133.

Tonutti, P, P. Casson and A. Ramina, 1991. Ethylene boisysthensis during peach fruit development. *J. Am. Soc. Hort. Sci.* 116(2): 274.

Tsay, L.M, S.Mizuno and N.Kozukhe. 1984. Changes in respiration, ethylene evolution and abscisic acid content during ripening and senescence of fruits picked at young and mature stage. *J. Jpn. Soc. Hort. Sci.* 52: 458.

Ueda, Y., M. Nakamoto, and K. Ogate., 1980, keeping quality and control of ripening in various fruits and vegetables in low pressure storage, *J.Am. Soc. Hort. Sci. Technol. 27: 149.*

Upchurch, B.L., M.J. Delwiche, and D.L. Peterson. 1990. Evaluation of an optical measurement for peach maturity sorting. *Appl. Eng. Agr.*6 (2): 189.

Valpuesta, V, M. Quesada, C. Sanchez –Roldan, H.A Tigier, A. Heredia and M.J. Bukovac. 1989. Changes in indole -3-acetic acid and peroxidease iso enzymes in the seeds of developing peach fruit, *J, Plant Growth Regulation* 8(4); 255.

Visage, T.R and G.J. Eksteen. 1981. Picking maturity of stone fruits, *Deciduous Fruit Growers* 31(9): 358.

Vyas, K.K and V.K Joshi. 1982. Canning of fruits in naturally fruit juices I. Canning of peach in apple juice. *J. Food. Technol.* 19(1): 39.

Walker, T.H and L.R. Wilhelm, Drying fruit with circulated air for energy saving. *Am. Soc.Agr.Eng*. 92: 6535.p.20.

Werner, D.J, D.F. Ritchie, D.W. Cain and E.I. Zehi. 1986. Susceptibility of peaches and nectarines, plant introduction and other Prunes species to bacterial spot, *Hortscience* 21: 127.

Wills, R.B.H and M.S. Mahendra. 1989. Effect of post-harvest application of calcium on ripening of peach. *Austral. J. Exp. Agr*. 29(5): 751.

Wolf Gardt, P.J. 1983, Marketing of Kakamas type of peaches for fresh consumption, *Deciduous Fruit Grower* 33: 31.

Woodroof J.G. 1986. Other products and processes, Commercial fruit processing (J. G Woodroof and B, S. Luh, eds.), AVI Westpot CT.

Woodroof, J.G. and Luh, B.S. 1986. *Commercial Fruit Processing*. 2nd ed. AVI West Port CT.

Chapter 8

Pear

Introduction

Pear (*Pyrus communis* L.), a typical fruit of temperate climates, with delicate pleasant taste and smooth, has a wide acceptance throughout the world. Pyrus species are native to the Northern Hemisphere of the old world. European and West Asian species are native to Eastern Europe and South Western Asia. East and North Asian species (oriental group) are native to Eastern Asia including China, Japan and ManAuria. Patharnakh (*Pyrus pyrifolia*) (Burm. F. Nakai) originated in China from where Chinese merchants and settlers brought it to Amritsar's village Harsa Chhina during the time of Lord Kanishka (120-170 AD). From here patharnakh spread to other areas. In Himachal Pradesh and Uttar Pradesh, Patharnakh is cultivated under the name of Gola pear. These are deciduous, cold-requiring fruit crops which can grow in tropical; lowlands but never set fruit there, since they need a certain number of hours with temperatures below 7°C to replace the dormant period of the temperature climates. Hence, the production of pear is concentrated mostly in temperate regions of the world. Extracts of different parts of the plant have shown variable antibacterial action. Fresh pear juice exhibited good activity against Micrococcus Pyrogenes and Escherichia coli. An aqueous extract of the leave was active against some strains of Escherichi. Chlorogenic acid is present in the vegetative parts of the tree. The leaves contain arbutin, isoquercitin, sorbitol, astragalin and tannin. Nutritionally rich pears are considered to be fairly a good source of fibre and can provide substantial quantity of potassium in the diet.

Botany

The requirement of cold climate is necessary to break down the growth inhibitors which accumulate during the season of active growth in leaves and bud scales. With sufficient chilling, development is hindered by physiological disorders known as prolonged rest and delayed foliation. Close to the equator, at altitudes where temperatures below 7°C occur regularly, the day temperature is too low for normal growth and fruit ripening. The pear fruit classified as a pome, the pericap, the fleshy edible part of it arises form the fusion of the receptacle stamens, ovary wall and calyx. It has five carpels, each containing two seeds. The fruit is typically in a pyriform shape with a wide variation. Pears are often called the European buffer pear because of the smooth, buffery texture of their flesh. Pears are classified by maturity. Summer, autumn and winter, summer types mature from July to August on the Northern hemisphere and from January to February in the Southern hemisphere; they usually have a short storage life 103 months. Autumn types mature from September to October in the Northern hemisphere and from March to April in the Southern hemisphere; they can be stored 2-4 months successfully under proper conditions. Winter types mature in October to November in the Northern hemisphere and April-May in the Southern hemisphere; under proper storage, some varieties can have a storage life of 6-7 months., Bartlett is the most widely grown cultivars of the summer and fall groups of pears. A few other varieties are hardy, Clapp, Favorite and winter groups including Anjous, Bosc, Comice, Winter Nelis and Seckel and Florida Home.

The pear cultivars of India fall into two groups, the first having the soft-fleshed fruits with inconspicuous grit cells of the European pears (*Pyrus communis*) and the second giving the firm, hared flesh with prominent grit cells of the oriental pears (*P. pyrifolia*). A few cultivars, such as Kieffer, Le Conte and Smith, of hybrid origin with intermediate characters, also have been introduced. The introductions of soft pears possibly made along with apples constitute about 171 cultivars, of which the most popular is Bartlett. Very early-ripening cultivars, particularly china pear and red strains of Bartlett fetch premium prices. The recently introduced Flemish Beauty, Devoe, Max Red Bartlett and Manning Elizabeth have been found promising. Junska Zlato from Yugoslavia and some red-fleshed pears such as Sanquinole and varieties of Sand pears from Japan and China are being introduced.

Important Species

1. *Pyrus pyrifolia* (Burm) F. Nakai Sand Pear

Trees are spreading in nature and can grow upto 9 m high. The bark of the trunk is light brown, rough and shreading. Branches grow fast and become drooping. Leaves 10-12 cm long with Serrated margins, round base and acuminate apex. The upper side of the lamina is dark green and lower side of light green in colour. Fruits are round to pyriform in shape and have deciduous calyx. Fruit tastes sweet. Flesh white in colour, fruits usually hard at maturity, can with stand transportation.

2. *Pyrus communis* L. (European Pear)

Trees are of pyramidal in shape, upright growing. Bark of the trunk light brown to grey in colour. Twigs show lenticles young shoots light brown in colour. Leaves medium in size, ovate 5-10 cm long with fine serration. The base of leaves is obtuse with acuminate apex. Leaves are dark green from upper surface and light green from lower surface. Fruits variable pyriform. The length of neck differ from cultivar to cultivar. Fruits with fleshy pedicels and persistent calyx. Fruits sweet, soft fleshed and creamy white.

3. *Pyrus serotina* Rehd. (Shiara)

Tree is spreading. The bark of the trunks is dark brown. Leaves just like Pyrifolia but smaller in size ovate with pointed type. The spurs are stout and spring fruits round, hard and brown in colour. Fruits have bold seeds. The taste is sour. Seeds are used as rootstock for pyrifolia and communis species.

4. *Pyrus pashia* Buch-Ham (Kainth)

Trees are spreading and upright. Tree size variable may grow 5 m to 9 m high. The bark is dark-brown. The young shoots end in spines. The small spurs also spiny sharp. Leaves like 'patang' elliptical with fine serration on the margins and acute apex. Fruits are brown in colour hard and round in shape. It grows wild. Fruits size variable. 3-5 seeds per fruit brownish black. It is an important rootstock for pear cultivars.

Suitable Varieties/Cultivars

A. Hard Pear (*Pyrus pyrifolia*)

1. Patharnakh

Very important cultivar of hard pear group is commonly known as gola pear is some areas. It is commonly cultivated cultivar in Punjab. Patharnakh requires only 250-300 chilling hours for flowering and fruiting. It is heavy bearer with well sized, firm fruits. Fruits can withstand long- distance transportation. Trees are spreading and vigorous. The scaffolds grow upright and then spread. It starts bearing at the age of 5- 7 years depending upon the rootstock used. The spurs are stout and can bear fruit for 15 years, if not broken. It takes 2-3 years for taking re-growth of the spurs to bear fruit. Fruits are of medium size, round and yellowish green with prominent dots on the surface (lenticles). Flesh white in colour, gritty due to the presence of grit cells. Flesh is crispy and juicy. Keeping quality very good, can be kept at room temperature for three to four weeks. Fruits mature in the last week of July. A fully grown tree can bear upto 3 quintals of fruit. Average yield 1.5 quintals per tree. TSS of juice 13 per cent with 0.5 per cent acidity.

B. Semi-soft Pears

(These are natural hybrids of the *Pyrus pyrifolia x Pyrus communis* species).

1. Punjab Beauty

The trees are upright in growth forms pyramid shape, medium vigour, regular bearer. It resembles 'Le conte' cultivar in tree shape and foliage. Fruits are medium in size, yellow in colour with red blush at full maturity. Flesh is white and juicy. TSS of juice 14 per cent and acidity 0.3 per cent. The fruits become soft at ripening. Fruits mature in end July. Average yield 80 kg per plant.

2. 'Le Conte'

Tree is of medium in vigour with upright and spreading habit. Fruits are of medium in size mostly pyriform in shape with small necks. The colour of the fruit is greenish yellow. Flesh white, juicy and sweet. Fruits mature in first week of August. TSS of juice 14 percent and acidity 0.3 per cent. Average yield per tree 80 kg in on year and 30 kg in off year.

3. Punjab Nectar

This cultivar is similar to the above cultivars of soft pear group. It also requires about 250-300 chilling hours. Trees grows upright and spreading at maturity. Fruits are of medium in size with 14 per cent TSS of juice and 0.31 per cent of acidity. Fruits ripen in end July. Average yield 80 kg per tree.

4. Punjab Gold

This is also a selection from semi-soft pear group. The young plants grow upright and are less spreading. Fruits become yellow at ripening. The TSS of juice 14.5 per cent with 0.3 per cent acidity. The fruits ripen in end July. Average yield 80 kg per plant.

5. Baggugosha

The trees grow similar to 'Le Conte'. The trees are taller and more spreading. It is an alternate bearer. The fruits are little smaller than 'Le conte' but are more pyriform in shape with long neck. Fruits ripen in August. Very popular in foot hills of North India. Average yield 80 kg/tree. The TSS of juice 13 per cent with 0.3 per cent acidity.

6. Kieffer

It is a hybrid between two species *Pyrus communis* and *Pyrus pyrifolia*. The trees are very hardy upright growing and spreading at maturity. The foliage is dense, leaves are smaller than patharnakh. Tree is very productive just like patharnakh. The spurs are stout. Fruit size is equivalent to patharnakh with prominent points at the calyx and fruits develop red bluish near maturity. The flesh is gritty. TSS of juice 12.5 per cent with 0.3 per cent acidity. It is late maturing cultivar. Fruits ripen in 2nd week of August. Average yield 100 kg/tree, regular bearer.

7. China

Trees are of medium vigour. It bears small-sized fruits, which ripen in end June. Fruits develop red colour partially. Due to early maturity it is grown in the foot hills for early marketing. The tree yields up to 55 kg of fruit. The TSS of juice 12 per cent with 0.32 acidity.

8. Max-red Bartlette

It is a bud sport of Bartlette. The chilling requirements are little less, that is why performs well in Punjab conditions. Tree is medium size with very high density of spurs. It is a regular bearer. Fruits ripen in August and develop red colour which makes the fruit very attractive. Average yield per tree 100 kg.

Rootstock

Many rootstocks are being used to propagate pear trees. Promising characters of rootstock are given below:

Pear Root Suckers (*Pyrus calleryana*)

This is a very old rootstock which is in use since the introduction of pear cultivar patharnakh by the Chinese. Initially in rootstock trials at P.A.U. Ludhiana it has been considered to be *Pyrus pyrifolia*. A sucker was planted in the old orchard of the P.A.U. in 1976. It grew as a tree quite different from *P. pyrifolia* and did not flower for twenty years, but remained vegetative. Its roots continued to give out rootsuckers, whereas no rootsuckers developed on *Pyrus pyrifolia* roots. Hence it has been identified as root suckers of oriental pear *Pyrus calleryana*. Old pear orchards produce root-suckers. The root-suckers are uprooted during September or October. These are planted in the nursery beds for grafting. Only the established suckers are grafted with desired cultivar during December and January. Plants get ready to transplanting in the field within a year. The root-suckers develop a double tier root system. One, feeder roots develop in the upper 15-25 cms of soil depth and second, lower at 50-60 cms depth. Very stable rootstock for plain. The plants produced on this rootstock are semi-vigorous.

Pyrus pashia (Kainth)

The fruits are harvested in October. Seeds are extracted immediately and are sown directly in beds without any treatment. Seeds germinate within a week and become budable in April or may be grafted in December-January. Kainth cuttings can also be used as a rootstock for achieving the uniformity in a rootstock. For this cuttings are prepared during December and treated with 100 ppm of IBA solution for 24 hours and then planted for rooting. Over 90 per cent cuttings root. The shoots can be budded in April-May or grafted in December.Kainth forms a vigorous rootstock for all the pear cultivars. Kainth roots also do not give out suckers. Its plants only produce collar suckers, which should be removed.

Pyrus serotina (Shiara)

It is very important rootstock for pear in hilly areas. Fruits are larger in size than kainth. The seeds are extracted from fruits in the same way as in kainth. The seeds are bolder than kainth. Plants propagated on this rootstock are as vigorous as in kainth. Seedlings are healthier than kainth.

Pyrus calleryana (D6)

The rootstock is commonly used in Australia. It is also a vigorous rootstock for pears. The tree growth is similar to kainth. The foliage also resembles like kainth.

Cydonia oblonga (Beedana)

This is a dwarfing rootstock for pears. In a trial at fruit research station, Gangian of the P.A.U. the Patharnakh and soft pears propagated on quince died within 7-8 years of age. The bud union configuration bulged out and partition developed between scion and rootstock in xylem wood. It is compatible with high chilling requiring pear cultivars *viz.* Anjou, Old Home, Hardy and Flemish Beauty. Rootstock can be raised both through seeds and cuttings. Quince roots are resistant to pear root aplids and nematodes. Some Quince strains which are compatible to low chilling requiring soft pears should be tried.

Patharnakh (*Pyrus pyrifolia*)

Patharnakh cuttings root well. The own rooted cuttings produce precocious plants. Patharnakh form a good rootstock for soft pears. There are no root-suckers originate from the roots. Rooting of cuttings can easily be promoted by dipping cuttings in 100 ppm IBA solution for 24 hours before planting in the beds in December. Sprouted patharnakh cuttings can be budded in April-May or grafted in December. Own rooted plants of Patharnakh should not be planted since the vigorous trees are prone to tree felling during rainy season. Cuttings are good rootstock for semi soft pears due to semi vigorous plants.

Suitable Varieties

Anjou Pears

Anjou pears have a mild flavor and a firm texture, while still being sweet and juicy. Anjous can be green or red (which are often labeled "Red Anjou"), but the color doesn't indicate any major flavor or texture difference, and they can be used interchangeably in recipes. Look for Anjous that feel heavy for their size, with bright, taut skins and relatively firm texture when lightly squeezed with the palm of your hand. Never pinch or poke pears with your fingers to test for ripeness, since it will in all likelihood bruise the fruit. Anjous are great for eating out of hand or for cooking. Anjou pears are usually available October through May.

Asian Pears

Asian pears are super crunchy - more like crisp apples than other pears. They look more like apples than pears, too. While there are many varieties of Asian pears, the ones most commonly available in the U.S. are a very matte, tan color with a bit more texture and roughness to the skin than other apples or pears.Asian pears are great for eating raw, especially when sliced or diced into salads. They have more of a crisp-apple texture than soft, grainy pear texture. They are so good raw, in fact, that there is no real need to cook Asian pears, but if you have a glut of them, know that they work wonderfully in tarts and crisps, like this Asian Pear Crisp.

Bartlett

Are the juiciest pears when eaten raw. They are so delicious when ripe and juicy, it's worth the drip marks on your shirt, but you may feel like you should have a bib on when biting into one of these.

Since Barletts are so juicy, they lose their shape when cooked. They turn to mush at the slightest mention of heat. As with all fruit, look for Bartletts that feel heavy for their size. They will have some give with squeezed when ripe, but avoid checking them too much since they bruise easily. Barlett pears are harvested July through October. Bartlett pears are also known as Williams pears.

Bosc Pears

Bosc pears are crisp when raw and hold their shape beautifully when cooked. They have the best example of the soft-yet-grainy texture classically associated with pears. Their brownish russet skin is easy to recognize, as is their classic pear shape, not to mention their heady pear aroma, especially when sniffed at the stem. Ripe Bosc pears will have a bit of give as you hold them firmly in your hand. Avoid pressing down with your fingers into the pear to check for ripeness, since that can bruise the fruit. They will also smell beautifully of pears. So stick your nose down into the stem end and give it a solid sniff. If you don't smell a whole lot of pear, move along. Use Bosc pears in any recipe calling for cooking pears in which you want the whole, halved, sliced, or chopped pear to hold its shape while cooking. Use them to make desserts like this Pear Tart. Bosc pears are also delicious raw, especially sliced into salads. Luckily, the Bosc pear season is a long one. Look for Bosc pears starting into September and running through winter, with pears from some orchards going into April.

Comice Pears

Comice pears are perhaps the best pears for eating raw. They have a great fruity aroma and flavor and a slightly finer, less grainy texture than other pear varieties. They are almost a cross between a bosc pear (itself a fine variety) and an Asian pear (super crisp and apple-like).Comice pears are slightly more rounded and apple-shaped than other pears. Look for Comice pears that feel heavy for their size and test the fruit around the stem to see if the pear is ripe: it should be quite tender and smell like, well, a ripe pear. They won't fall apart like Bartlett pears. Comice pears are usually available in September through February.

Concorde Pears

Concorde pears have beautiful long, tapered necks. They are also ideal combination of Comice pears and Bosc pears. They are juicy, smooth, and don't brown too much when cut, so they're great eating pears like Comice pears. Yet their dense flesh holds its shape when cooked like Bosc pears, so they work well as poached pears or in pear tarts.Concorde pear season starts in the fall and runs through February. They can be eaten when just harvested, when they are crisp, and allowed to ripen further as they develop a softer texture and more mellow, vanilla-scented flavor.Concorde pears can be completely green or develop a red or russet blush, which is a sign of sun exposure, not ripeness. To determine if they're ripe, look for Concorde pears that are just a tad soft right at the neck.

Forelle Pears

Forelle pears are bit bigger than seckel pears, but not much. Like seckels, they are an excellent snacking fruit because of their size (they are just a few bites big). Forelle pears are available in season from October into February.

French Butter Pears

French butter pears are great for cooking with and, as long as you let them get fully ripe, have a wonderfully soft and rich texture for eating raw, too. They can be a pretty green color but some turn a more golden color when ripe (like all pears, true ripeness can be told by the tenderness of the flesh around the stem more than from the color).

Red Anjou Pears

Red Anjou Pears are pretty much exactly like Green Anjou Pears, but they are the striking and glorious rusty red color.

Seckel Pears

Seckel pears are tiny—sometimes just a bite or two—with extremely firm flesh. They are ideal for baking, canning, and poaching. Their firm texture and acidic taste make them a bit trickier for eating out of hand, especially if you're expecting the soft, sweet experience of a ripe anjou or bartlett pear. Seckel pears are available from August u[to December.

Maturity and Harvesting

Pears are the only temperate tree fruits that cannot be on the tree to ripen. Theh are clamactric fruits that require harvest before the climactric rise in respiration occurs if storage life is to be maximized. If they are harvested too late, they develop off-flavors and core break down, a post-harvest physiological disorder. If harvested too early, they do not develop good eating quality and are more susceptible to scald, another post-harvest physiologica disorder.Harvesting pear at optimum maturity is important to ensure a high quality product. Relaible harvest indicators enable growers to determine when optimum pear maturity occurs. Flesh firmness, ground color, corking of lenticels, fruit finish, ease of separation, days from full bloom, heat units, and starch-iodine test are all methods used to indicate pear optimum maturity.

The percentage of soluble solids is not used as a maturity indicator, although pears must contain levels of 10 per cent or more if they are to be harvested. Barttlett pears with 13 per cent soluble solids do not require any indicator to determine maturity. Fruit flesh firmness is the most common method used to determine maturity in pears. The number of days from full bloom to maturity varies with variety and growing region.Bartlett, a summer type, has a date indicator of 110-135 days from full bloom to harvest, while Anjou, a winter type, requires 140-165 days from full bloom to harvest date. This indicator does not take into account yearly weather difference, such as abnormally cool summers or unusually warm harvest seasons. Days from full bloom gives the grower an idea of when to start using more precise indicators, such as flesh firmness.

Heat unit accumulation takes into account annual weather differences in a way that days from full bloom does not.This method sums or accumulates the no. of degrees above a given temperature observed during all or part of the growing season.Heat units offer a more precise measure of harvest maturity than days from full bloom. Maturity standards for ground color, corking of lenticles, and fruit finish depend on the variety. In Bartlett, ground color is defined as a lightening of the green ground color to yellow. The starch-iodine test is similar to that used for apples. Ease of fruit separation is not a precise as flesh firmness and heat units, but if it is not used, the impact on tree productivity can be serious. If pears separate easily from the tree's fruting spur is often removed with the fruit, greatly reducing the the tree's future fruting potential.

All commercial cultivators of pears are harvested while hard and green in color.However the time of harvest is critical, because if the fruits are picked too early, their dessert quality is poor, and if they are picked too late, their storage life is much shortened. The range of the maturity for optimum quality for either fresh market or prossecing is rather narrow. Color, flesh firmness, and elapsed period from full bloom to 'readiness for harvest' provide useful indices of harvest maturity of Bartlett pears. A pattern of green to yellow color changes in japanease pears can be used to evalivate variations in the degree of maturity of fruit on the trees, the rate of the maturation of individual fruits, the dropping tendency of mature fruit, and the harvest dates.

Grading and Packing

Harvested fruits should be kept under shade in an airy 'Varandah.' Fruits should not be wet at the time of packing. For sending the fruit to a distant market it should be packed properly in wooden boxes. A box may contain 17-18 kg of fruit. Place the fruits in layers in the boxes by placing some grass/rice trash at the bottom and at the top and cover it with paper before closing the box with its head. Packed boxes are transported by trucks. Grade the fruits before packing.

Post-harvest Handling

If the pears are going into long-term storage, they need to be cooled down or precooled to storage temperature within 48 hrs of harvest. Forced air cooling is the most common method, since chloride used in hydro-cooling may have a detrimental effect on the pear skin. Once precooled, the pears can be either stored in bulk bins or proceesed through a packing line. A mechanized packing line washes, dries, sorts, grades, and packs the pears into smaller containers. During this process, small and defective fruits are culled, while sound fruits are graded by size and packed into 20- 22 kg boxes with 70-165 pears per box depending on size. Boxes may be volume filled and tray-packed boxes usually have polyethylene linears. The fruit may be individually wrapped in tissue-type paper to cushion them in transit. Wrapping has become less common in recent years, because the pool of workers skilled in this practice is getting difficult.

Ripening

Although the pears can ripen while still attached to the tree, a generally accepeted commercial practice is to pick the fruit before the onset of respiratory clamateric rise. This practice aids in increasing shelf life of fruit during storage and transport. The biochemical and physical changes associated with the visible process

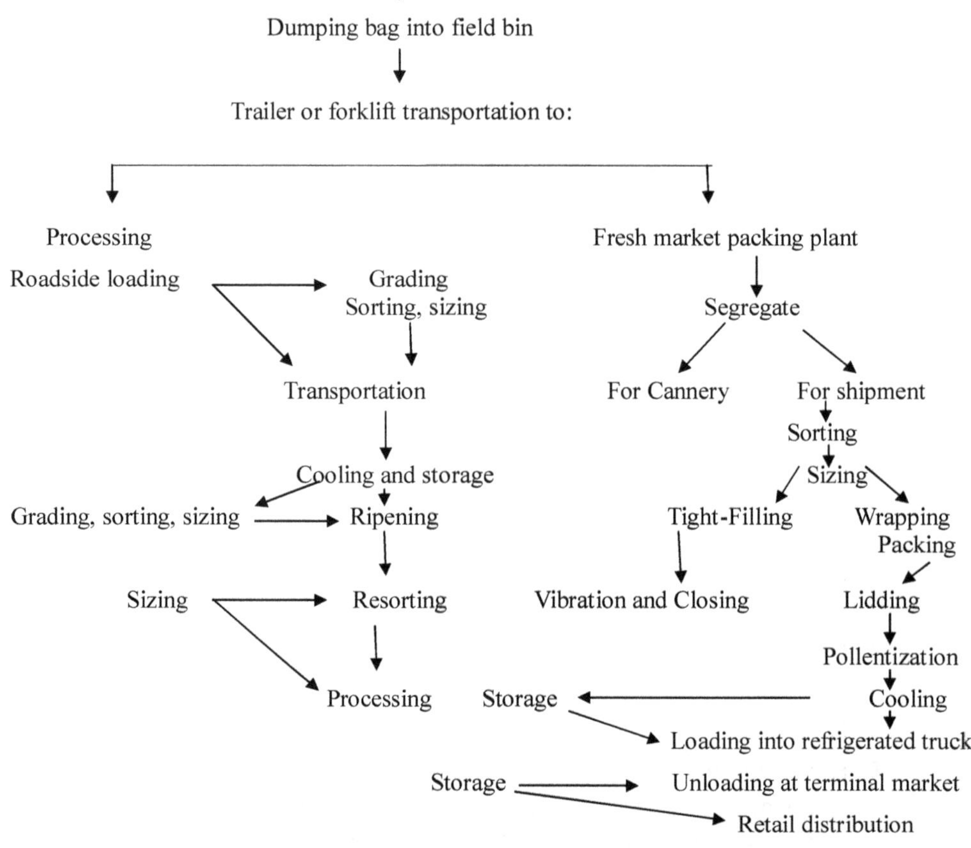

Post-harvest Handling System for Bartlett Pears (Kader, 1992.).

of ripening include changes in color, texture, sweetness and astringency, and the development of the characteristic flavor of a given cultivar. Fruit respiration begins to rise before visible ripening sets in. During this period, profound biochemical and metabolic changes take place which initiatethe ripening process. Pears may require special ripening conditions for best quality. They ripen best and with more aromatic flavours at 18.3-21.1°C than at either lower or higher teoperatures. In fact, some cultivars will not ripen if the temperature is above 24° C. Once a fruit is eating ripe, its shelf life is very short but may be prolonged by refrigerated storage.

Flavor, texture and color of pears are dependent on the stages of ripeness; Ripening is the transformation of the physiologically mature fruit from an unfovarable stage of consumption to a favorable state. In pears ripening takes place after harvesting. Ripening proceeds under the control of natural growth regulators, *i.e.*, auxins, cytokinins, gibberllins, abscisic acid, and ethylene, and in pears precededby a shap increase in respiration rate.

The ripening process involves a wide array of physiological and biochemical changes in the fruit. Flavor changes remarkably during this phase. Starch is converted to sugar and sorbitol to frustose. Fruit acidity and tannin content drop. Aromatic esters that give pear fruit to their distinct flavor are produced. Pectins increase, cell walls soften and cell membrane becomes leaky. Softening of the flesh and the rise in the ethylene evolution and respiration assiociated with ripening were delayed when mannose was vacuum infiltrated into intact fruit of cv. Bartlett. Holding the fruits at 30°C between harvest and storage suppressed the subsequent normal ripening processes of skin yellowing and flesh softening. At higher concentration of ethrel, fruit firmness decreased more rapidly than in the controls. Total solube solids (TSS) tend to increase during ripening, the increase being greater with high ethrel concentration, while acidity decreases. Starch content decreases sharply during the initial 5 or 10 days. Pigments increase while phenolics decrease during ripening.

Most early-ripening pears do not have a long storage and break down quickly in cold storage. Late maturation pears usually have a better storage life. In fact, many of these late maturation pear cultivars, such as Anjou, Bosc, Comice, Packham's Triumph, and Passé Crasane, require a period of cold storage to develop maximum flavor.

Chemical Composition

Nutritionally, pears are considered to be a farily good source of fiber and can provide substantial amounts of potassium in the diet. The major organic acid in pears is malic acid. Various amounts of citric, tartaric, and oxalic acids may also be found. Pears are also known to contain sugar, alchohol, and sorbitol, which have wide applications in the manufacture of sugar-free foods, but also cause gastrointestinal discomfort in sensitive individuals at high concentration. Reducing sugars of which fructose is the major component, constitute over 80 per cent of the total sugars present in pears. These fruits also contain maltose, galactose, xylose and probably arabinose. Other Carbohydrates in pears include starch and cellulose. During ripening; the starch of the fruit is converted into sugars. High quality pear flavor is associated with high sugar content.Values for total sugar in common pear cultivars range from 8 to 11 per cent per unit weight of edible pulp. The impact of sugar is affected by the juiciness of the fruit. Fruit that is dry trends to taste less sweet than fruit that is juicy.The presence of juice within the fruit appears to enhance the impact of sugar on the taste.

The pectic conustitutents in pears depend on the type, the stage of ripening, and the storage conditions. During ripening, protopectin is largely hydrolyzed into soluble pectin, and this conversion is the principal factor concerned in the softening of pears. The soluble pectin content of Bartlett pears was found to be 0.77 per cent at the time of picking (unripe stage) and 0.7-0.8 per cent in the ripe fruit.

Nutritional Composition of Raw and Dried Pears (100g edible basis)

Constitutent	*Content*	
	Raw	*Dried*
Water (per cent)	87.7	26.0
Food energy (kJ)	264.60	1369.80
Protein (g)	0.7	5.4
Fat (g)	0.4	3.2
Caarbohydrate total (g)	15.8	64.1
Fiber (g)	1.4	—
Calcium (mg)	13.0	61.0
Phosphorus (mg)	16.0	84.0
Iron (mg)	0.3	2.3
Vitamin A (j u)	20.0[a]	120.0
Thiamin (mg)	0.02	0.2
Riboflavin (mg)	0.04	0.03
Nicotinic acid (mg)	0.1	1.1
Ascrobic acid (mg)	4.0	12.0

Sugars of Pear Fruit and their Concentration

Sugar	*Dry Weight (per cent)*
Fructose	7
Glucose	4
Sorbitol	3
Sucrose	2
Xylose	*
Galactose	*
Arabinose	*

* Minor amount.

Pears have a very low nitrogen content. Protein nitrogen constitutes only about 8 per cent of the total nitrogen. Lysine, phenylalanine and leucine are the most abundant amino acids in the protein of pears, especially during maturation. Pears also contain biotin, pantothenic acid, folic acid, and vitamin B_{12}. The concentration of ascorbic acid and biotin is fond to be much greater in the peel than in the pulp of pears.

Malic and citric acids are the principal organic acids present in pears. Table pears contain more of malic acid, while as in some juicy pears, citric acid accounts for up to 45 per cent of total acids. The acidity level of pears varies from pH 2.6 to 5.4. The organic acids present in pear fruits are shown in Table. The enzymes present in pears are amylase, catalase, peroxidase, xanthineoxidase, pectin galacturonase, protopectinase and pectase.A high pectin polygalacturonase activity has been demonstrated in ripe Bartlett pears. Polyphenolase is considered to be responsible for the browning of pears. When cut, chlorogenic acids being the chief substrate. Ripe pears contain appreciable amounts of acetaldehyde, which is believed to be the causative agent in the production of scald and break down in the fruits.The extent of scalds is stated to be related directily to the concentration of acetaldehyde in the tissue.

Composition of Edible Portion of Important Types of Pear Grown in India

Constituent	*Kashmir*			
	Bagugosha	*Nakh*	*Country*	*Kieffer*
Moiture (g/100 g)	86.5	83.6	86.0	86.3
Protein (g/100 g)	0.4	0.2	0.2	0.2
Fat (g/100 g)	0.1	0.3	0.1	0.2
Minerals (g/100 g)	0.3	0.4	0.3	0.2
Fiber (g/100 g)	2.1	0.6	1.0	1.4
Other carbohydrates (g/100 g)	10.6	14.9	12.4	11.7
Calcium (mg/100 g)	20.0	20.0	6.0	10.0
Phosphorus (mg/100 g)	20.0	20.0	10.0	10.0
Iron (mg/100 g)	1.5	1.0	1.0	0.4
Vitamin A (I.U./100 g)	14.0	9.0		
Thiamin (mg/100 g)	—	—	0.02	0.03
Riboflavin (mg/100 g)	—	—	0.03	0.02
Nicotimic C (mg/100 g)	0.02	0.2		
Vitamin C (mg/100 g)	1.0	3.0	7.0	

In general, extreme bitterness and astringency is undesirable in dessert pears. Astringency and bitterness is associated with the skin in some pear cultivars. Peeling the fruit often helps to remove the astringency and bitterness factors. If masked by sugar and other flavor. Bitterness and astringency is attributed to the presence of polyphenolic and phenolic substances (tannis). Polyphenolics of high molecular weight tend to be more astringent, where as polyphenolics of low molecular weight tend to be bitter.

Although sweetness, acidity, astringency and bitterness are important in diterming flavor, the aromatic volatiles define the various distinctive fruit flavor. In Bartlett pear as many as 77 volatile components have been found. The principal volatile compounds in fully ripoe pears were ethyle, propyl, butyl and hexyl acetates, which accounted for 70.6 per cent of the total volatiles. The typical aroma of the Bartlett pears has been attributed mainly to the presence of esters. An unsaturated acid, Trans -2, 4-decadienonic acid, has been identified.

Storage

Low-Temperature Storage

Copmpared to apples, pears have a shoter storage life. This fruit has a higher respiration rate at all temperatures, even at optimal storage temperatures. Summer and Autumn pears have a shoter storage life than witer types. Pears grow in cooler areas are also know to have a shorter storage life and develop core breakdown sooner than those grown in warmer areas. Pears can be held longer than the recommended storage term without visible damage, but they loose their Ability to soften properly and this ius an important part of the ripening processes. The metabolic processes required for softening no longer function.The pear looses abilityto synthesize the enzymes required to soften the fruit. Bartletts are best stored at -1.7° to 1.1° C, while other varities are best stored at -1.1to - 0.6°. The relative humidity should be at least 90-95 per cent. If the air flow in storage is high, then the relative humidity should be at least 95 per cent to compensate for the drying effect of the air. Post-harvest dipping of pears in 5 per cent calcium choloride greatly improved their keeping quality. This treatment reduces fruit softening and gave firmer fruits during storage.

Controlled- and Modified-Atmosphere Storage

The storage life of pears can be successfully extended with controlled-atmosphere (CA) storage. Pears are sensitive to carbon dioxide. At levels higher than 3 per cent, core and flesh browning may develop, while at 5 per cent serious brown core damage occurs. Any efforts to control the atmosphere during pear storage must consider the phytotoxic effects of the corbon dioxide. Optimum corbon dioxide levels for pears are 0.8-1.0 per cent and optimum oxygen levels are 2.0-2.5 per cent.

Recommended Storage Conditions,
Storage Pife and Physical Characteristics of Pear Fruit

Storage Temperature (°C)	
Low	–2
High	–0.6
Relative humidity	90-95
Approximate storage period (days)	14-28
Highest freezing point (°C)	–0.94
Water content (per cent)	89.1

Firmness of pear slices can be maintained by storage in 12 per cent Co_2 and in 0.5 per cent O_2, respectively, or by dipping in 1 per cent calcium chloride and storing in air or under CA conditions. When stored in air at 0° C, barttlett and Anjou pears coated with either Pro-long or Multi-save can be maintained firmer, higher in titratable acidity, and greater in color. Polythylene nox liners provide an inexpensively generated modified atmoephere around the fruit in shopping containers; this can extend the storage life of the pears.

Subatmospheric-Pressure Storage

Subatmospheric pressure significantly (at 1 per cent level) Softened and extended the storage life of pears. Pears can be stored for 3.5 months under normal refrigeration. At 461mm Hg, they stored successfully up to 5 months, at 278 mm hg for up to 7 months, and at 102 mm Hg for up to 8 months. Decreases in firmness were especially delayed at 102 mm Hg. The color of the pears were retained fairy well up to 5 months at 120 mm Hg. Degradation of cholorophyll was delayed by subatmospheric pressure, the longer chlorophyll was retained. Subatmospheric pressure delayed losses of sugar in the pears. At the end of the storage the storage content of pears from subatmospheric pressure were significantly (at the 1 per cent level) less than that of control. However, the treated pears were still marketable at that time.

Irradiation

Pears were irradiated with y-rays at 500, 1000 and 1500Gy or x-rays at 40, 60 and 100 Gy before cold storage in CA or normal atmosphere. Ripening was accelerated after irradiation, as shown by respiration and ethylene activity tests. Pear fruits could be stored satisfactory for 7 months at 0-0.5° C, 92 per cent RH, in 2 per cent CO_2 and 2 per cent O_2.

Post-harvest Diseases and Disorders

The skin of pear turns brown when bruised by rubbing against a hard surface such as a conveyor belt or the side of paller bin. Such friction bruises can also occur in transit in loose packs of fruits and may appear as continuous bands around the fruit. The common physiological disorders of pears are core breakdown, scald (superficial and senescent), and loss of ripening capacity. Like apples, pears are affected by grey mold rot and blue mold rot. In addition, a type of decay called bulls eye rot also occurs late in the pear storage season, causing considerable losses of pear. Senescent scald can develop in pears that have been stored beyond their potential post-harvest life. Scalded fruits often change their background color during storage and lose their capacity to ripen.Fruit from very early or late harvest, fruit suffering delayed cooling, and fruit held at too high a storage temperature are also more susceptive to senescent scald. Symptoms begin on the fruit surface but can progress into the flesh during ripening, proper harvest maturity, good temperature management, and avoiding too long storage are all important in minimizing senescent scald of pears.

Processing

Pears are consumed primarily as fresh fruit. However, a good portion of the crop is crushed to produce juices for beverages and wines. Large quantities are canned and some are dried. Pears are a good source of pectin and also contain appreciable amounts of sugar and thiamin. They are reported to help in maintaining a desirable acid/base balance in the human body. Pears have been recommended for patients suffering from diabetes because of their low sugar content.

Canned Pears

Bartlett pears are preserved by canning. Pears are washed in weak acid or alkaline solution to remove pesticides residues, then washed in water, drained and held in cold storage prior to canning. Pears are peeled manually, mechanically or by lye peeling method (5 per cent NaOH at 60°C for about 1 min). They are washed and then halved in two pieces lengthwise and cored. The halves are preserved 20-40 per cent sugar syrup, depending on the type of pack. The filled cans are exhausted in water at 79.4°C for 10-12 min and sealed. The sealed cans are heat processed in boiling water for about half an hour and cooled in water. Pears are also chopped into smaller pieces and canned in fruit cocktail. The standard of identity for canned fruit cocktail allows not less than 25 per cent and not more than 45 per cent pears in the product.

Canned Pear Products

Pears are pureed for baby food and for the making of pear butters and jams. Pear concentrates especially charcoal-filtered, no-essence-returned products have a bland taste and are finding wide applications as natural sweetening agents in no-added-sugar fruit spreads and beverage products. Perry is a fermented beverage similar to cider but produced from pears.

Dried Pears

A small percentage (less than 1 per cent) of the pears produced are dried. The process usually involves sulfuring and requires 24-30 hr for a final moisture content ranging from 15 to 25 per cent.

Candy

A food-quality candy can be prepared from sand pear and Bagugosha cultivars by steeping the slices in 40 per cent Brix sugar syrup which was increased by 10 per cent Brix and dried in open sun. Candy packed in polythene bags retained the characteristics pear flavor of about 16 weeks under ambient temperature. The flesh Sand pear contained 85 per cent moisture as compared to 20.1 per cent in the candy-made fruits. Ascorbic acid retention was 50 per cent after 40 week of storage at room Temperature. The varieties Leconte, Smith, Stone Pear or Naspati are found to be suitable for jam and chutney.

Juice and Concentrate

A clear juice of excellent flavor can be prepared from Bartlett pears, the process requiring treatment with pectic enzymes. It is recommended for use in jellies and sherbets and after acidification as a beverage. In California, the Bartlett is the principal type used for making pear nectar. The nectar is the pulpy liquid food which contains not less than 40 per cent fruit by weight and additional optimal ingredients including sweeteners, acidifiers and ascorbic acid.

Composition of Sand Pear and Bagugosha Pear Fruit and Fresh Candy

Composition	Fruit		Fresh Candy	
	Sand Pear	Bagugosha	Sand Pear	Bagugosha
Moisture (per cent)	85.0	83.0	20.1	19.2
TSS (per cent)	10.8	14.5	74.0	76.0
Reducing sugars (per cent)	4.2	8.6	34.7	39.4
Nonreducing sugars (per cent)	6.3	10.1	38.4	35.5
Acidity (per cent)	0.10	0.20	0.17	0.21
Ascorbic acid (mg/100 g)	1.9	2.1	19.9	-
pH	3.7	3.5	3.7	4.0

Utilization of Fruit Waste

Pear waste obtained during the peeling, coring and stemming operations in the canneries amounts to 30-35 per cent. It can be used in the preparation of vinegar, brandy or denatures alcohol. Some pear waste is employed in making syrup suitable for canning of pears or for table in use. It is also fried and processed to a useful stock feed that is comparable in nutritive value to beef pulp mixed with beet molasses. Pear waste can be used as a soil conditioner and in composting.

Chemical Compositions of Pear Fresh Waste

Constituent	Constituent (per cent)
Total solids	14.8-16.3
Proteins	0.5-0.6
Crude fiber	2.2
Total Sugar	7.1-8.3
Ash	0.3-0.4

REFERENCES

Al-Bachir, M and P. Sass. 1989. Effects of ionizing radiations on the respiration intensity of pears during storage, *Acta, Agron. Hung.* 38(1-2): 49.

Cappellini, R.A, M. J. Ceponis and G.N. Lightner. 1987. Disorders in apple and pear shipments to the New York market, *plant Dis. Rep.* 71: 852-856.

Downs, C.G, A.E, Pickering and M. Reihana. 1989. Influence of temperature between harvest and storage on the ripening of "Packhams Triump" pears. *Sci. Hort.*39 (3); 235.

Geeson, J.D, P. M. Genge, R. O. Sharples and S.M.Smith. 1991. Limitations of modified atmosphere packaging for extending the shelf- life of party ripened Doyenne de Comice pears. *Int.J. Food. Sci Technol.* 26(2): 225.

Haggag, M.N. 1987. Effects of preharvest and post-harvest calcium treatment on storage behavior of "Le Conte" pears. *Alexandria J. Agr. Res.* 32(3): 175.

Jennings W.G and R. Tressl. 1974. Production of volatile compounds in the ripening Bartlett pear, *Chem. Microbiol. Technol. Lebensm.* 3.52.

Kader, A.A. 1992. Post-harvest Technology of horticulture Crops.p: 296.

Kvale, A. 1990. Maturity indexes for pears, *Act Hort.* (1990).p.285: 103.

Mann, S. S and B. Singh. 1990. Effects of ethrel on ripening of fruits of Patharnakh pear harvested on different dates. *Act Hort.* 297; 529.

Meherink, M and O. L. Lau. 1988. Effects of two polymeric coatings on fruit quality of "bartlett" and Anjou pears. *J. Am. Soc. Hort. Sci.*113 (2): 222.

Publications and Information Directorate, *Wealth of India*, Vol. 8, *Raw materials,* Council of Scientific and Industrial Research (CSIR), New Delhi, 1969, pp.330-333.

Rani, U and B.S. Bhatia. 1985. Studies on pear candy processing. *Indian Food Packer 39*(5): 40.

Rosen J.C and A. A. Kader. 1989. Post-harvest physiology and quality maintenance of sliced pear and strawberry fruits. *J. Food Sci.*54 (3): 656.

Salunkhe, D.K and B.B. Desai. 1984. Post-harvest Bio technology of fruits. Vol. I, CRC Press, Boca Raton, FL., p.123

Salunkhe, D.K., and M.T. Wu. 1973. Effects of sub-atmospheric pressure storage on firmness, cholorophyll and total sugars of pears. *J. Am. Soc. Hort. Sci.* 113.

Schwarz, A. 1991. A trial of cold storage of pears in controlled atmosphere. V.Beuree Bosc. *Revue Suiss de Viticulture, d'Arboriculture et d' Horticulture* 23 (3): 157.

Shita, H. 1990. Changes in the volatile composition of "La France" pear during maturation, *J. Sci. Food agriculture.*52 (3): 421.

Watkins C.B and C. Frenkel. 1987. Inhabitation of pear fruit repining by mannose. *Plant physiol.*85 (1)56.

Westwood M.N. 1987. *Temperate Zone Pomology,* W.H. Freeman, San Francisco, 1978.

Williams, A.A, A. G. H. Lea and C.F. Timberlake. 1977. Measurement of flavor quality in apple, apple juices and fermented ciders. Flavor quality: Objective measurement, *ACS symp.ser.51: 71.*

Yadav, I.S. 1988. Germplasm collection of pome, stone and nut fruits in India, Central Institute of Horticulture Northern Plains, Lucknow, Indian Council of Agricultural Research (ICAR), 1988, p.74.

Chapter 9

Pecan Nut

Introduction

The pecan nut (*Carya illinoensis*), like the walnut, belongs to the family Juglandaceae. Names formerly used for the nut are *Hicoria pecan* and *C. illinoenensis.* The tree is indigenous to North America where it grows wild in the states along the Gulf of Mexico and around the Great Lakes. Some 400-year-old trees still bear nuts. Pecan-nut trees grow very fast and become very tall unless growth is controlled. It is regarded as one of the biggest fruit/nut crops. The nut has a very high protein content and its nutritional value is one of the highest of all fruits grown in South Africa. The nuts are rich in vitamins, carbohydrates and nut oil.

Pecan nut is the one of the most important temperate nuts grown in India. In India, it is mainly grown in Jammu and Kashmir, and Himachal Pradesh. The total area under pecan nut production is increasing due its high economic returns and adaptation to intermediate zone of Jammu.

Before European settlement, pecans were widely consumed and traded by Native Americans. As a food source, pecans are a natural choice for pre-agricultural society. As a wild forage, the fruit of the previous growing season is commonly still edible when found on the ground.

Pecans first became known to Europeans in the 16th century. The first Europeans to come into contact with pecans were Spanish explorers in what is now Louisiana, Texas, and Mexico. These Spanish explorers called the pecan, *nuez de la arruga*, which roughly translates to "wrinkle nut". They were called this for their resemblance to wrinkles. The genus *Carya* does not exist in the Old World. Because of their familiarity with the genus *Juglans,* these early explorers referred to the nuts as *nogales* and *nueces,* the Spanish terms for "walnut trees" and "fruit of the walnut". They noted the particularly thin shell and acorn-like shape of the fruit, indicating they were indeed referring to pecans. The Spaniards took the pecan into Europe,

Asia, and Africa beginning in the 16th century. In 1792, William Bartram reported in his botanical book, *Travels*, a nut tree, *Juglans exalata* that some botanists today argue was the American pecan tree, but others argue was hickory, *Carya ovata*. Pecan trees are native to the United States, and writing about the pecan tree goes back to the nation's founders. Thomas Jefferson planted pecan trees, *C. illinoinensis* (Illinois nuts), in his nut orchard at his home, Monticello, in Virginia. George Washington reported in his journal that Thomas Jefferson gave him "Illinois nuts", pecans, which Washington then grew at Mount Vernon, his Virginia home.

Pecan is a 32 chromosome species that readily hybridizes with other 32 chromosome members of the *Carya* genus, such as *Carya ovata*, *Carya laciniosa*, *Carya cordiformis* and has been reported to hybridize with 64 chromosome species such as *Carya tomentosa*. Most such hybrids are unproductive. Hybrids are referred to as "hicans" to indicate their hybrid origin.

Botany

The pecan tree is a large deciduous tree, growing to 20–40 m (66–131 ft) in height, rarely to 44 m (144 ft). It typically has a spread of 12–23 m (39–75 ft) with a trunk up to 2 m (6.6 ft) diameter. A 10-year-old sapling grown in optimal conditions will stand about 5 m (16 ft) tall. The leaves are alternate, 30–45 cm (12–18 in) long, and pinnate with 9–17 leaflets, each leaflet 5–12 cm (2.0–4.7 in) long and 2–6 cm (0.79–2.36 in) broad.

A pecan, like the fruit of all other members of the hickory genus, is not truly a nut, but is technically a drupe, a fruit with a single stone or pit, surrounded by a husk. The husks are produced from the exocarp tissue of the flower, while the part known as the nut develops from the endocarp and contains the seed. The husk itself is aeneous, that is, brassy greenish-gold in color, oval to oblong in shape, 2.6–6 cm (1.0–2.4 in) long and 1.5–3 cm (0.59–1.18 in) broad. The outer husk is 3–4 mm (0.12–0.16 in) thick, starts out green and turns brown at maturity, at which time it splits off in four sections to release the thin-shelled seed.

The seeds of the pecan are edible, with a rich, buttery flavor. They can be eaten fresh or used in cooking, particularly in sweet desserts. One of the most common desserts with the pecan as a central ingredient is the pecan pie, a traditional Southern U.S. dish. Pecans are also a major ingredient in praline candy.

Pecan wood is used in making furniture and wood flooring as well as flavoring fuel for smoking meats.

Descriptions of Cultivars

A large number of pecan varieties have been developed through the selection of desirable seedling trees. Selection of disease resistant varieties is a major consideration in areas experiencing high humidity and poor air circulation. Description of some of the promising pecan cultivars is given below.

1. Stuart

It is a popular old cultivar which has originated as seedling in Mississippi.

Nuts are large, of high quality and with well defined markings and a fairly thick shell. The tree is a moderate bearer.

2. Desirable

It is a hybrid between Success x Jewett or Russell. The nut is large with a semi thin shell which cracks easily. The tree comes to bearing early, very prolific and regular bearer and is also resistant to scab. The planting of this variety with other varieties facilitates pollination.

3. Nellis

Nuts are long and shells are thin. Kernels are of good quality.

4. Mahan

The tree is vigorous, precocious, prolific bearer with good foliage. The nuts are extra large shell, thin and tend to poor filling on older trees. It is a good variety for the areas which receive abnormal rainfall.

5. Western

It is the most commonly planted cultivar in the U.S.A. and has originated from the cross between a seedling and the variety Texas. Nuts are medium sized, thin shelled and having good kernel quality. It is a prolific bearer and suitable for high density planting. It is susceptible to scab disease.

6. Wichita

It is a hybrid between Halbert X Mohan. It is very popular cultivar being one of the most precocious and prolific of all cultivars. The nuts are of medium size, attractive appearance, high quality and have high kernel percentage. It is susceptible to scab disease. For effective pollination, the cultivar should be grown with Western or Cheyenne.

7. Cheyenne

It is a cross between Clark and Adom. The tree is very precious, productive, scab resistant and recommended for high density planting. For effective pollination, it should be grown with Wichita.

8. Mohawk

It is a cross between Success X Mahan. The tree is precocious, prolific, vigorous and fairly resistant to disease. Nuts are elongated, thin shelled, very attractive and have kernel percentage of 55-60. The kernels are quite smooth, easily shelled and of high quality.

9. Burkett

Nuts are large round with a semi thin shell. They are not easily separated from the kernel.

10. Elliott

It is an early maturing with small round nut. The nuts have medium thick shell and good flavor. The trees are vigorous and regular bearer. It is resistant to scab disease.

11. Curtis

The variety is suitable for lower elevations experiencing a long growing season. The nut is small thin shelled suitable for shelling. The kernel is of high quality. It is moderately resistant to scab disease.

12. Pawnee

This is a selection from Mohawk X Starking Hardy Giant crosses. It is an early, prolific and fast growing variety in young age. Nut quality is excellent with good size, appearance and kernel percentage. The plant is tolerant to scab disease.

13. Moore

It produces good crops of small nuts but it is susceptible to scab.

14. Cape Fear

The variety produces fairly large nuts. Shell cracks well and possesses high quality. Trees are highly productive and start bearing fruit after seven to eight years of age.

15. Ukulinga

This variety has originated as a chance seedling in South Africa. It is suitable for humid tropical regions. The nuts are large, shell medium thick and kernel percentage is 52 -54. It is less prone to irregular bearing.

Composition

The nut contains 3.4 per cent water, 9.2 per cent protein, 71.2 per cent fat, 14.6 per cent carbohydrates, 2.3 per cent fiber, 1.6 per cent ash or 73mg per cent, calcium, 289 mg per cent phosphorus, 2.4 mg per cent iron, 603 mg per cent magnesium, 0.86 mg per cent thiamine, 0.13 mg per cent riboflavin, 0.9 mg per cent niacin and 2 mg per cent ascorbic acid.

Pecans (*Carya illinoinensis*), Nutritional Value per 100 g

Principle	*Nutrient Value*	*Percentage of RDA*
Energy	691 Kcal	34.5 per cent
Carbohydrates	13.86 g	11 per cent
Protein	9.17 g	17 per cent
Total Fat	71.9 g	360 per cent
Cholesterol	0 mg	0 per cent
Dietary Fiber	9.6 g	25 per cent
Vitamins		
Folates	22 µg	5.5 per cent
Niacin	1.167 mg	7 per cent

Principle	*Nutrient Value*	*Percentage of RDA*
Pantothenic acid	0.863 mg	17 per cent
Pyridoxine	0.210 mg	16 per cent
Riboflavin	0.130 mg	10 per cent
Thiamin	0.660 mg	55 per cent
Vitamin A	56 IU	2 per cent
Vitamin C	1.1	2 per cent
Vitamin E	24.44 mg	163 per cent
Electrolytes		
Sodium	0 mg	0 per cent
Potassium	410 mg	9 per cent
Minerals		
Calcium	70 mg	7 per cent
Copper	1.2 mg	133 per cent
Iron	2.53 mg	32 per cent
Magnesium	121 mg	30 per cent
Manganese	4.5 mg	196 per cent
Phosphorus	277 mg	40 per cent
Selenium	3.8 µg	7 per cent
Zinc	4.53 mg	41 per cent
Phyto-nutrients		
Carotene-ß	29 µg	--
Crypto-xanthin-ß	9 µg	--
Lutein-zeaxanthin	17 µg	--

Source: USDA National Nutrient Database.

Health Benefits of Pecans

1. A Handful Goes a Long Way

Just a single ounce of pecans (about 19 halves) is 200 calories, 3 grams of fiber, 3 grams of protein, and 20 grams of mostly unsaturated fats, making it a hearty, satisfying snack with loads of antioxidants, vitamins, and minerals.

2. They're Important for Immunity

Pecans provide phytonutrients, plant-based compounds that have powerful antioxidant benefits. They're also a good source of the mineral zinc, crucial for immune-cell development and function. Diets high in zinc are linked with a lower risk of many diseases, particularly those related to age and lifestyle. What's more, eating ellagic acid-containing foods — an antioxidant found in pecans — is associated with a reduced risk of some cancers.

3. They're Full of Flavonoids

According to the USDA, pecans have more flavonoids — a type of antioxidant found mostly in veggies and fruit — than any other tree nut. People who eat diets high in flavonoids are less likely to develop chronic diseases, such as heart disease, diabetes, some cancers, and cognitive decline.

4. They're Crucial for Heart Health

Pecans are chock-full of mono unsaturated fatty acids, a type of fat linked to improving total cholesterol levels. Another benefit to these tasty tree nuts: Pecans are filled with beta carotene and vitamin E, which protect cells from damage by mitigating the effects of chronic inflammation.

5. They're Surprisingly Low in Sugar

Not only do they contain very little sugar, but pecans may also help improve blood sugar levels overall by slowing down the rate of absorption from the bloodstream into peripheral tissues. This helps to maintain energy levels and prevent blood sugar spikes, which ultimately lead to major crashes later on.

6. They can Boost your Brain

It may not be news to you that nuts are some of the most nutritious foods, but chew on this: Diets high in vitamin E have been linked with lowering risk of Alzheimer's disease and dementia by up to 25 per cent. That's because vitamin E and other antioxidants in plant-based foods help reduce oxidative stress caused by inflammation, protecting cells and therefore tissues of vital organs like your brain.

7. They're Helpful for Weight Loss

While it's true that nuts contain lots of fat, the type found in tree nuts and fruit (*e.g.*, olives) can actually promote weight loss and maintenance. That's because nuts are a filling source of oleic acid, a better-for-you fat that makes you feel full while also boosting heart health. Eating every three to four hours, limiting added sugars and saturated fats found in processed foods, and making more of your meals plant-based are also key to successful, long-term weight loss. As a deliciously satisfying snack, pecans are a good one to keep on hand for a nutritious nosh that won't leave you feeling deprived.

8. They've got Key Minerals

Pecans are excellent sources of manganese and copper, two minerals that boost overall metabolic health, may contain anti-inflammatory properties, and potentially help reduce risk of heart disease. The added benefit: These nutrients have shown promise specifically in preventing in high blood pressure.

9. They can Satisfy a Salty-Snack Craving

Unsalted nuts are a naturally sodium-free snack, making them ideal for anyone following a lower-sodium diet or cutting back on salty foods. Diets high in sodium have been linked to increased risk of chronic disease, especially high blood pressure.

Swapping in pecans as a crunchy ingredient in any recipe can help to retain flavor with powerful health benefits.

Maturity and Harvesting

Pecan nuts usually ripen from April to July, depending on the area. As soon as the nut is physiologically ripe, the green husk goes dry, it cracks open and the nut drops out. Usually nuts are mainly collected from under the trees manually. For various reasons, a certain percentage do not drop from the trees. These nuts are called stickers and are shaken from the trees. If a large number of the nuts are stickers, it may be due to poorly filled nuts, scab or other factors such as irrigation or fertilization. A period of 4 weeks or more is often required from the time of maturation of the earliest to the latest nuts

Storage

Pecan kernels develop optimum flavor shortly after harvest and gradually deteriorate in all quality parameters during subsequent storage. Sequence of this deterioration is as follows:

1. Loss of volatile substances so that blandness tends to increase.
2. Onset of oxidation, causing color darkening and the development of a state aroma and flavor.
3. Hydrolysis of fats resulting in increases in free fatty acids and development of an acrid flavor.

Pecan kernels are especially susceptible to staleness and rancidity because of their high oil content which ranges from 55-75 per cent by weight among varieties and the high percentage of unsaturated fatty acids present in the oils (45). Deterioration in quality can be retarded if kernel moisture is reduced to 3.5-4.0 per cent immediately after harvest and if the pecans are stored at reduced temperatures until used kernel quality can be maintained for up to 3 years at a temperature of -2°C and below.

In commercial practice, pecans gradually deteriorate in quality because they are not refrigerated during the entire time between harvest and use. Pecan kernels were subjected to steam conditioning and dielectric heating treatments evaluated initially and during 16 weeks of accelerated storage to determine the effect of temperature on color.

Steam conditioning treatment, which raised kernel temperature to 93°C, caused significantly greater darkening of the kernels initially and during storage than did dielectric heating to 88, 136 and 156°C.

Processing

Pecans are purchased in the shell, shelled, salted and packed in tins, bags or cartons. They are salted, oil roasted or dry roasted and used in conjunction with almost any food, including pies, cakes, cookies, deserts, confections, fruit cups, salads, fruit, cakes and snacks.

According to Heaton and Woodroof, kernel moisture should be removed promptly after harvest in order to prevent development and discoloration. They found that most satisfactory color, texture, flavor and oil stability resulted when the pecans were dried with moderately warm (below 40°C) dry air circulating constantly around the nuts. Early harvesting and controlled drying of pecans provides a light colored, bright, uniform kernel with increased storage stability.

Waste material from the pecan shelling industry has shown commercial possibilities for the recovery and production of oil, tannin, shell flour and activated charcoal. Tannic acid produced from pecan shells is used by the training industry and by the oil industry to control the viscosity of drilling muds. The spent shells after extraction of tannin are used for making activated charcoal.

REFERENCES

Cavaletto, C.G. and H.Y. Yamamoto, 1971. Factor effecting macamdamia nut stability. 3. Effects of roasting, oil quality and antioxidants. *J. Food Sci.,* 36: 82.

Cavaletto, C.G.E. Ross and H.Y. Yamamoto, 1968. In shell storage effect on quality of proceses macdemia nuts. *Food Tecnol.,* 22(4): 172.

Crain, W O., Jr. and C.S Tan, 1975. Volatile components of roasted macdemia nuts. *J. Food Sci.,* 40: 207.

Delacruz, A., C.G. Cavaletto, E. Ross and H.Y. Yamamoto, 1966. Factors affecting macadamia nut stability Roasted kernels. *Food Technol.,* 20(9): 123.

Dhaliwal, G.S., J.S. Jawanda and D.K. Uppal, 1978. Studies on physico-chemical characteristics of almond. *Punjab Hot. J.,* 18: 72.

Guadagani, D.G., E.L. Soderstorm and C.L. Storey, 1978. Effect of controlled atmosphere on flavor stability of almonds. *J. Food Sci.,* 43(4): 1007.

Heaton, E.K., R.E. Worthington and A.L. Shewfelt, 1975. Peacon nut quality: Effect of time of harvest on composition, sensory and quality characteristics. *J. Food Sci.,* 40(6): 1260.

Higaki, T. and R.R. 1963. The changes of macademia nuts (*Macadamai integgifolia* Maidens and Betche). *Am. Soc. Hort. Sci. Proc.,* 83: 359.

Jones, W.W. and P.J. Ito, Mauka and Makai, 1977. Two new macdemia cultivars suitable for high and low elevation. *Proc. Hawaii Macdermia Producer Assoc.,* p. 34.

Kind, A.D., Jr., M.J. Miller and L.C.E. Eldride, 1970. Almond harvesting, processing and microbial flora. *Applied Microbiol.,* 20(2): 208.

Lipovec, J., 1990. Storing chestnuts species and cativars. *Agro-Survey,* 18: 30.

Lutz, J.M and R.E. Hardenburg, 1986. The commercial storage of fruits, vegetable and florist nursery stock. *U.S Dept of Agriculture Handbook,* 66(Rev.).

Resurreccion, A.V.A. and K.E. Heaton, 1987. Sensory and objective measures of quality of early harvested and traditionally harvested pecans. *J. Food Sci.,* 52(4): 1038.

Ryall, A.L. and Pentzar, W.T., 1974. *Handling, Transporting and Storage of Fruits and Vegetables, Vol. 2: Fruits and Tree Nuts*. AVI, Westport, C.T.

Senter, S.D. and R.J. Horvot, 1976. Lipids of peacan nut meats. *J. Food Sci.*, 41(5): 1201.

Senter, S.D., W.R. Forbus, S.O. Nelson and R.J Hovweert, 1970. Effects of de-electric and steam heating treatments on then pre-storage and storage color characteristics of pecans kernels. *J. Food Sci.*, 49(6): 1532.

Storey, C.L., 1976. Mortality of *Scitophilus oryzae* (L.) and *S. granaries* (L.) in atmosphere produced by exothermic inert atmosphere generator. *J. Stored Prod. Res.*, 11: 217.

Storey, W.B., 1965. Macadamia nuts bits and pieces. *California, Macdamia Soc. Yearbook*, 11: 71.

Teotia, M.S., 1991. Post-harvest technology of almonds (*Prunus amyodalus*). *Beverage Food World (Indian)*, 18: 27.

Teotia, M.S., S. Kaur and S.K. Berry, 1987. Chemistry and technology of almonds. *Indian Food Racker.*, 41(5): 49.

Vickery, J.R., 1971. The fatty acid composition of seed of oil of the proteaceae: A Chemo-taxonomic study. *Photochemistry*, 10: 1239.

Wilkinson, A.E., 1987. Pecan. *The Encyclopedia of Fruits, Berries and Nuts and How to Grow Them*. Shree Publishing House, New Delhi, pp.123-126.

Williams, F.M.G. La plante and E.K. Heaton, 1975. Evaluation of quality of peacons in retail markets. *J. Am. Soc. Hort. Sci.*, 98: 460.

Winterton, D., 1968. The macademia nut industry: Problem and prospects. *Food Technol.* (Austrial), 20: 119.

Woodroof, J.G. and E.K. Heaton, 1961. Peacans for processing, *Georgia Agr. Exp. Sta. Uni. of Ga., College of Agriculture Bull N.S.* 80.

Woodroof, J.G., 1967. *Treenuts: Production, Processing and Products, Vol. 1*, AVI, Westport, CT, pp. 59, 60, 72.

Wright, R.C., 1941. Investigation on the storage of nuts. *U.S. Dept. of Agriculture Technical Bulletin* 770.

Wright, W. R., 1960. Storage decays of domestically grown chestnuts. *Plant Disease Rep.*, 44: 820.

Chapter 10

Pistachio

Pistacia is a genus of *Anacardiaceous* plants comprised of trees and shrubs which exude turpentine or mastic. *Pistacia vera* L. bears the pistachio nut of commerce. Several of the Pistacia species are used as root stocks upon which to bud or graft the pistachio or as ornamentals. The pistachio is the only commercially acceptable edible nut of some 12 to 15 Pistacia species, two American and the rest Old World, all but *P. vera* having drupes with an indehiscent bony endocarp. *P. vera* has much larger elongated, keeled fruits with an endocarp that splits into two valves along the keel. The color of its kernel ranges from light yellow to deep green throughout; the quality from a dry, woody, to a rich, oily, nut-like flavor. Pistachios have been used for centuries in the preparation of delicacies and sweet dishes. The health benefits of pistachios are associated with improved cardiovascular health, weight management, improved digestion and protection against diabetes. It is grown on a commercial scale for export principally in Turkey, Iran, Italy, Syria and Afghanistan. Lesser amounts are produced in Lebanon and Pakistan. Scattered small lots of nuts are sometimes exported from nearby countries, but are usually consumed locally. For the most part these nuts are received unshelled. They reach the consumer as "Red" pistachios after they are roasted and salted, and the shell is colored with a vegetable dye; as "White" pistachios after they are roasted and the shell is coated with a mixture of salt and cornstarch and as "Naturals" when salt only is added after roasting.

Botanical Classification

The pistachio of commerce is the only edible species among the 11 species in the genus Pistacia; all are characterized by their ability to exude turpentine or mastic. Several are referred to as pistachios, but the name is generally reserved for the edible nut of commerce. Its Latin name is *Pistacia vera* L. A member of the family Anacardiaceae, it is related to the cashew, mango, poison ivy and oak, peppertree and sumac. The tree has a pinnately compound leaf. Each leaf subtends a single axillary bud. Most of these lateral axillary buds differentiate to inflorescence primordia and produce a nut- bearing rachis the following year; thus, pistachios bear laterally on one-year-old wood. Botanically, pistachio nuts are drupes, the same classification for almonds, peaches, apricots, cherries and plums. All drupes consist of three parts; an exocarp, a fleshy mesocarp and endocarp that encloses a seed. The difference lies in the edible portion. In pistachios and almonds the seed is consumed, rather than the mesocarp as in stone fruit. The pistachio tree is dioecious (*i.e.* two houses"), meaning the male flowers are borne on one tree and the female flowers on another. Therefore, both male and female trees are required to produce nuts. The female flower isapetalous (no petals) and has no nectarines, thus does not attract bees. The pollen is spread by wind. The pistachio tree is deciduous, so it loses its leaves in the fall and remains dormant through the winter. The rooting habit of the tree is classified as a phreatophyte. Phreatophytes have extensive root systems allowing them to mine the soil deeply. They achieve full bearing between 10 to 12 years of age. The tree has an upright growth habit characterized by a strong apical dominance and a lack of lateral vegetative buds in older trees. These characteristics have strong implications for young tree training, mature tree pruning and rejuvenation of fruiting wood in older trees.

Nutritional Profile of Pistachio Nuts

Pistachio nuts are one of the healthiest nuts that are not only dense in nutrients, but are also extremely beneficial for heart health. These nuts are high in fiber and healthy fats that help in lowering cholesterol and aid in weight management.

- ☆ Pistachios are richer in protein content compared to other nuts. 1 ounce of pistachios contains 6 grams of protein – the highest in comparison to other nuts.
- ☆ The total fat content of pistachios is the lowest compared to other nuts. 1 ounce of pistachios contains 13 grams of fat.
- ☆ A major part of the fats present in pistachios are healthy mono-unsaturated fatty acids such as antioxidants and oleic acid.
- ☆ Various important trace minerals are found in pistachios such as calcium, potassium, iron, magnesium, phosphorus, zinc, selenium and copper.

Pistachio Nutritional Composition per 100 g

Folates	51mg	13 per cent
Niacin	1.3mg	8 per cent
Pantothenic acid	0.520mg	131 per cent
Pyridoxine	1.7mg	131 per cent
Riboflavin	0.160	12 per cent
Thiamine	0.870mg	72.5 per cent
Vitamin A	553IU	18 per cent
Vitamin C	5mg	8 per cent

Vitamin E	22.60mg	150 per cent
Sodium	1mg	0 per cent
Potassium	1025mg	22 per cent
Calcium	107mg	11 per cent
Copper	1.3mg	144 per cent
iron	4.15mg	52 per cent
Magnesium	121mg	30 per cent
Manganese	1.2mg	52 per cent
Phosphorus	376mg	54 per cent
Selenium	7µg	13 per cent
Zinc	2.20mg	20 per cent

Top 10 Benefits of Pistachio Nuts for Health

Extremely delicious and packed with nutrients, pistachios have been the symbol of robust health since ancient times. These nuts were added to milk in order to enhance strength and improve overall health.

Rich Source of Nutrients

Pistachio nuts form part of a balanced diet because these nuts are a rich source of healthy fats, proteins and minerals. There are just 557 calories for each 100g of pistachios. Pistachios are loaded with vitamin E (especially gamma-tocopherol), B-complex vitamins, folate and vitamin B6. Vitamin E acts as a potent antioxidant that protects the cell membranes against harmful free radicals.

Pistachios are a powerhouse of essential minerals such as copper, potassium, iron, magnesium, zinc and selenium. Each 100 g of pistachios contain 144 per cent of daily recommended levels of copper, which is an essential trace mineral required for red blood cell synthesis, neurotransmission and enhanced metabolism.

Rich Source of Antioxidants

One of the most effective health benefits of pistachios is that it is loaded with powerful antioxidants. Even the shells of the nuts are storehouse of antioxidants. The mono-unsaturated fatty acids known as oleic acid found in pistachios are one of the most powerful antioxidants that protect the heart by lowering the level of LDL (bad) cholesterol and increasing the level of HDL (good) cholesterol in the blood. It helps in reducing the risk of stroke and coronary artery disease.

In addition, the phytochemicals such as carotenes, polyphenolic compounds and vitamin E contribute to the overall antioxidant activities and remove the harmful oxygen-free radicals.

Improve Heart Health

One of the most important pistachio benefits is that it helps in maintaining proper functioning of the heart and therefore, they are also known as heart-friendly nuts. Munching on pistachio nuts lowers total cholesterol and triglyceride levels in the blood. It reduces the level of LDL (bad) cholesterol and increases the level of HDL (good) cholesterol, and the results can be noticed within a short period of regular consumption. The antioxidants such as vitamins E and A neutralize inflammation of arteries and blood vessels and reduce the risk of developing heart diseases.

Regular consumption of pistachios raises the level of the antioxidant lutein that protects against LDL oxidization. The compound l-arginine present in pistachios softens the lining of the arteries so that they become flexible and it also prevents the formation of blood clots, there by minimizing the chances of the arteries becoming clogged with plaque.

Aids in Weight Management

Nuts and weight management usually don't go hand in hand but you can actually promote weight loss by snacking on nuts such as pistachios, provided you limit your portions. Pistachios must be included in a healthy diet for weight loss because of its high dietary fiber content and low saturated fat content. A 30 grams serving of pistachios contain almost 3 grams of dietary fiber that helps in filling you up so that you eat fewer calories through the rest of the day.

In addition, the high protein content of the nuts promotes the energy burning process of the body and speeds up weight loss. Therefore, nuts and dried fruits should be included in the snack options if you are trying to lose weight.

Protect Against Age-Related Macular Disease

Age-related macular degeneration (AMD) is a gradual breakdown of the macula of the eye that contains the light-sensitive tissues. The macula is responsible for central vision that allows the person to see fine details with clarity. The symptoms of macular degeneration include blurred vision and dark areas in central vision.

Pistachio nuts contain two carotenoid antioxidants – lutein and zeaxanthin that are not found in other nuts. These protective antioxidants protect the tissues of the eyes from inflammation caused by free radicals and oxidative stress. These antioxidants also protect against age-related macular degeneration – one of the leading causes of visual impairments and blindness in elders around the world.

Reduce the Risk of Diabetes

One of the vital health benefits of pistachios is that it helps in controlling and preventing Type 2 Diabetes, one of the most common forms of diabetes mellitus that is caused due to insulin resistance. Eating at least two servings of pistachios per day helps in supplying the recommended daily value of phosphorus that not only aids in breaking down the proteins into amino acids, but also improving glucose tolerance.

Pistachio nuts are especially effective for diabetic patients because in diabetes, the sugars form abnormal bonds with proteins and leave them unusable. This process is called glycation. But the antioxidants present in pistachios reduce the process of glycation and aid in keeping blood sugar levels under control.

Healthy Blood

The list of benefits of pistachios is incomplete without discussing its helpfulness for developing healthy blood. Pistachios are loaded with B complex vitamins, especially vitamin B6 that is an essential vitamin for the production of haemoglobin. Haemoglobin is the main component of red blood cells that helps in carrying oxygen to all the organs, muscles and tissues of the body.

The initial production of haemoglobin requires the presence of vitamin B6. Therefore, daily consumption of pistachios can help in treating blood-related diseases such as anemia. In addition, it also enhances the functioning of various organs due to the increased supply of oxygen throughout the body.

Boost the Immune System

The best way to sustain a healthy immune system is through adequate intake of vitamin B6 that is found abundantly in pistachio nuts. Deficiency of vitamin B6 leads to slowed down immune function that leads to decreased white blood cell response and shrinkage of the thymus – the critical immune system organ.

Vitamin B6 found in pistachios enhances the production of red blood cells and improves the functions of the lymph glands such as the thymus, lymph nodes and spleen and also ensures optimum production of red blood cells that protects the body against infection causing bacteria and viruses. Regular vitamin B6 intake has been shown to increase immune response even in critically ill patients.

Improved Functions of the Nervous System

Here, once again the high vitamin B6 content of pistachio nuts comes into use. Vitamin B6 found in pistachios has a wide range of benefits for the nervous system. The messaging molecules amines need amino acids for developing, which in turn require vitamin B6 for their creation. Proper transfer of impulses through the nervous system is maintained by the insulating sheath, known as myelin present around nerve ûbers.

Vitamin B6 plays a fundamental role in the formation of myelin. In addition, vitamin B6 also contributes to the fusion of melatonin, epinephrine, serotonin and gamma-aminobutyric acid – all of which regulate the transmission of messages through the nervous system. So, it is important to include pistachios in a healthy diet in order to maintain healthy nerve functions.

Reduce Skin Dryness

Last but not the least, comes the beauty benefit of pistachios. The natural oils present in the nuts have amazing emollient properties which moisturize the skin and protect it from excessive dryness. The saturated fats found in pistachio nuts also work as excellent carrier oils and is widely used in traditional massage therapy and aromatherapy. The potent antioxidants such as vitamins E and A help in maintaining the structure of the skin cells so that the skin remains supple and youthful. It also slows down the process of skin aging and prevents the appearance of wrinkles, fine lines and age spots. In addition, vitamin E also works as an excellent natural sunscreen that protects the skin from the severe UV rays of the sun and reduces the risk of skin cancer.

Varieties/Cultivars

They are classified according to their origin; their colour and size differentiates them. The main types of pistachio are the pistachio of Sicily, which is green, the smaller Tunis and the Pistachio of Levante, yellow and therefore less appreciated commercially.

The main types found in the market are the Noble or Sicily pistachio, with a green and very appreciated almond; Tunis, smaller but equally appreciated, and the Levante, with a yellow edible part and less accepted because of its flavour.

Usually, the pistachio varieties are classified according to their place of origin or culture. Each country has its own selections, whose main differences are the colour, flavour, size, period of harvesting and qualities.

Some of the most frequent cultivars in Iran, the main producing country, are Mirhavy, Momtaz, Owhadi, Safeed, Wahedi, Sefideh-Montaz, Imperiale de Dameghan, Ravzine, Bademi and many others as female cultivars. As masculine cultivars they use the native ones coming from the seed.

The most important cultivars in Syria are Achouri, Batouri, Alemi, and Lazouardi. In Turkey "Uzun", "kirmizi", " Halebí, "Abiad miwahi", " the Jalalé, " Aintaby" and " Ayimi. In Turkestan "Kouchka", " Akart-Tachecmé " and " Chor-Tchéchimé ".

Some Varieties of Pistachios "Kerman"

Pistachio nut of great size and good quality. Selected in Iran, it was introduced in the U.S.A. and it is also cultivated in Spain (in Castilla-La Mancha) where the fruit ripens during the first fortnight of September.

"Peter" It is used as a male cultivar with Kerman it has a good polen production and they partly coincide during the flowering period. Selected in California.

"Uzun" Pistachio nut of average size, long and clear green. It is cultivated in Turkey.

"Kirmizi' Pistachio nut of average size and reddish colour. Along with the cultivar Uzum, it is the most cultivated variety in Turkey.

"Abiad miwahi' Pistachio nut of average size, white colour and excellent quality. Cultivated in Turkey.

" Achouri ' Pistachio nut of average size, red colour, excellent quality and very productive. Cultivated in Syria.

"Batouri' Thick fruit of whitish colour and good quality. Important cultivar in Syria.

"Sefideh-Montaz" and " Imperiale de Dameghan"

The fruit of these varieties is round, thick and yellowish. Very appreciated in Iran.

" Kouchka '

Quite thick pistachio, cream white colour and good quality.

"Mateur" Long fruit, average size, yellow greenish colour and good taste quality. It was selected in Tunis and it gives good results in Spain. In Castilla-La Mancha it ripens at the end of August.

"Larnaka' Average size pistachio, less long than 'Mateur ". Original from Cyprus. It is cultivated in Greece and in Spain, giving good results.

"Aegina ' Medium size fruit, long and similar to " Mateur ". It comes from Greece and it also gives good results in Spain.

Effects of post-harvest factors on shell splitting No post-harvest practice has yet been demonstrated to significantly impact the percentage of split pistachio nut shells. However, pistachio shells have a high percentage of moisture.

Nytime the harvested pistachio nut is subjected to heat, during post-harvest transport, preprocessing waits anddrying, this heat will decrease the moisture content of the shells, literally shrinking the shell about the nut and increasing the width ofthe split. Thus, pistachios can leave the field with a lower percentage of wide splits thanthey have when they arrive at the processor. During processing, this increase in split width occurs very early in drying and increases as dryer temperatures increase from 125°F to 190°F. This increase can result in nut kernels dropping out of the shell

Storage

It has been observed that freshly harvested, dried and shelled pistachio are very stable in storage. Nuts with 4.5 per cent moisture and 49.2 per cent oil stored are 0°C and in N_2 at 3.3 and 23.9°C can be stored up to 20 months.

Tips for Selection and Storage of Pistachios

A few useful tips and guidelines can help you select the fresh and crunchy pistachios and extend their shelf life significantly.

- Pistachios are one such variety of nuts that are available in the market all year round.
- You can choose between shelled, in shell, salted, roasted and sweetened varieties.
- It is best to buy unshelled ones with the outer coat intact because the whole nuts are the least processed ones.
- You can buy the nuts from bulk bins or in airtight packages.
- If you are buying from the bulk bins then check the nuts in your hand and see if they have a compact shape and uniform texture.
- Reject the nuts that have spots or mold or if they emit a rancid smell.
- Raw pistachios can be stored in a cool, dry place for many months in an airtight container.

- ✰ Shelled nuts should be stored in a container and kept in a refrigerator in order to prevent them from turning rancid.

Processing

The moisture content of pistachio kernels at harvest time may be high as 45 per cent. This must be reduced to 4-5 per cent within a few days of harvest. Natural air drying can be employed in low, humidity and high-temperature conditions. The nuts can, however, be dehydrated by forcing heated air at 35-37.8°C, very effectively and rapidly. In-shell pistachio can be dried from 22-30 per cent to 5-6 per cent moisture required about 18 h in the dryer to reach 5-6 per cent moisture.

REFERENCES

Hormaza, J.I., L. Dollo and V.S. Polito. 1994. Determination of relatedness and geographical movements of Pistacia vera (Pistachio; Anacardiaceae) germplasm by RAPD analysis. *Economic Botany* 48: 349- 358.

Kader, A.A., C.M. Heintz, J.M. Labavitch and H.L. Rae. 1982. Studies related to the description and evaluation of pistachio nut quality. *J. Amer. Soc. Hort. Sci.* 107: 812- 816.

Joley, L.E. 1979. Pistachios, p. 163-174. In: R.A. Jaynes (ed.), *Nut Tree Culture in North America*. Northern Nut Growers Assn., Hamden, CT.

Maggs, D.H. 1990. The Australian pistachio Sirora. Fruit Var. J. 44: 178-179. 5. Parfitt, D.E. 1997. Pistachio, p. 581-582. In: *The Brooks and Olmo Register of Fruit and Nut Varieties*, ASHS Press, Alexandria Virginia. 744 p.

Polito, V.S. and J.G. Luza. 1988. Longevity of pistachio pollen determined by in vitro germination. *J. Amer. Soc. Hort. Sci.* 113: 214-217.

Pontikis, C.A. 1986. 'Pontikis' pistachio. *Hort. Sci.* 21: 1074. 8.

Chapter 11

Plum

Introduction

Plum is one of the important stone fruit crops of the world. A plum is the fruit of the sub genus *Prunus* of the genus prunus *Prunus*.The sub genus is distinguished from other sub genera (peaches, cherries bird cherries *etc.*) in the shoots having terminal bud and solitary side buds (not clustered), the flowers in the groups of one to five together on short stems, and the fruit having the groove running down one side and a smooth stone (or pit). Mature plum fruit may have a dusty white waxy coating and is known as "wax bloom". Dried plum fruits are called as dried plums or prunes, although in American English, prunes are a distinct type of plum, and may have pre dated the fruits now commonly known as plums. Plums may have been one of the first fruits domesticated by humans. Three of the most abundant cultivars are not found in the wild, only around human settlement: *Prunus domestica* has been traced to East European and Caucasian mountains, while prunus salicina and Prunus simonii originated in Asia. Plum remains have been found in Neolithic age archeological sites along with olives, grapes and figs. The name plum derived from Old English plume or "plum, plum tree, which extended from Germanic language or Middle Dutch, prune, and latin, prunum. Plums are generally consumed fresh, with the exception of small quantity used for canning and beverage preparation.

Botany

Types

Plum belongs to the genus *Prunus* of the subfamily prunoidate of the family Rosaceae. Nearly 200 species in the genus Prunus. Plums have a chromosome number of 8, but the somaic chromosomes number ranges from 16 to 48. Plums are related species have been placed in the subgenus Prunophora, section Euprunus and Prunocerasus. The important plums are the three types: the European (*Prunus domestica*), the Japanese (*P. salicina*) and hybrids of the latter. Other species are of interest to plum breeders for imparting desirable characters to existing. The commercially important plum trees are medium sized, usually pruned to 5-6 meter height The tree is of medium hardiness. Without pruning the trees can reach 12 metres in height and spread across 10 meters. They blossom in different months in different parts of the world.

Cultivars or for Use as Rootstock

European Plums

Prunus domestica is the most important source of commercial cultivars, European plum is said to be native Eastern Europe or Western Asia and has been cultivated in Europe for at least 2000 years. P. Domestica plums are heaxaploids and originated as a result of hybridization of diploid (*P. carasifera*) and tetraploid (*P. spinosa*) species. *P. domestica* fruits exhibit both yellow and green ground colors and also both red and blue skin colors, with an unlimited range of variation. The size, shape and flavor of the fruit also support the above origin. The cultivated varieties of European plums have been classified into four groups.

Prunus

This is commercially the most important group. The fruit is oval with bulging of the central side, blue or purple, firm, thick and freestone. Prune fruits are high in sugar content and can be successfully dried without removing the pit. Important cultivars are Agen (French), Stanley, Sugar, Imperial Epineuse, Italian, German, Giant and Tragedy.

Reine Claude (Green Gage)

The fruit is characterized by more or less round fruit with green yellow or slight red color. The flesh is sweet and juicy. Cultivars include Reine Claude, Bavay, Jefferson, Washington, Imperial Gage and Hand, used for canning and the fresh market.

Yellow Egg

Yellow egg is a small group of plums used for canning. The fruit is large, long, oval and usually with yellowish flesh. Cultivars include Yellow Egg, Red, Magnum, Bonum and Golden Drop.

Lombard, Lombard is group of large oval red or pink plums and of some what fair quality important cultivars are Pond, Bradshaw and Lombard.

Another European species, *P. insititia*, includes the small-sized fruits. Plums of this species are known commercially as Damsons, Bullaces and Mirabells. The plants of this species are compact and dwarf, having smaller and flowers and can be easily distinguished from *P. domestica* by their growth habits. The fruits are round and purple (in the Damsons) or yellow (in the Mirabelles). Damsons are excellent for preserves and jams, but are not eaten fresh. Mirabelles have excellent flavor.

Japanese Plum (*P. salicina*)

The Japanese plum is next in important to the large-fruited European plums. It is believed to have originated in china rather than in Japan. It was introduced in Japan 200-400 years ago, from which it disseminated around the world. It is readily distinguished from *P. domestica* by its rough bark, numerous persistent fruiting spurs and by mostly oblate to hear-shaped fruits with a more pointed apex than other species. Fruits are never blue. Trees of the Japanese plum come into bloom quite early and thus make it susceptible to frost. There is considerable variation in tree growth habit among cultivars. Some are spreading in habit while others exhibit upright growth. Leaves are medium sized, sharp pointed and almost free of pubescence. Flowers are produced three in a bud on either one-year-old shoots or on spur.

American Plums

Most native American plums are self-unfruitful, so pollinizers should be placed in the orchards. Some of the species and clones are described briefly:

P. American Marsh: Fruit are yellow or orange with golden yellow flesh. Leading cultivars are Desoto, Hawkeye, Wyant and Terry.

P. hartulana Baily**:** it is more resistant to rot and is used for jams and marmalades. Important cultivars are Wayland and Golden Beauty.

P. munsontana Wight and hedre: it is resistant to spring frost and brown rot of the fruit. The important variety is Wild Goose, widely planted in the lower Mississippi valley.

P. besseyi Bailey**:** it is used for hybridizing and as a dwarfing rootstock for other plums.

P. subcordata Benth: it is native to south central origin and northest California. The fruits are used commercially for preserve and jellies.

Commercial Cultivars

Japanese Plums

Early Magic –The early magic ripens in mid-July. Fruit size is medium. This medium sized fruit is purplish-red and covered with waxy bloom giving it a bluish cast. The flesh is amber-yellow, firm, juicy, sweet and very good tasting.

Early Golden - The early golden is one of the first plums of the season, ripening in the second part of July. Similar to shiro plum, it is small to medium in size, firmer than the shiro, mild tasting, and sweet and does not stick to the pit. The early golden is an excellent choice to satisfy your early season sweet tooth.

Methley – One of the first out of the orchard in mid-July, this is well known variety that has been present on the market stands for a long time. This fruit is harvested with a green shadow, but ripens to a vibrant purple with a deep red flesh at market. This small round fruit is the perfect pop-able sweet treat on a hot July day.

Shiro – The shiro needs no introduction as it is the most well-known of the Japanese varieties. Be sure to handle this petite yellow plum with care as they bruise easily. Smooth and sweet, you will surely encounter the shiro at famers markets in late July.

Santa Rosa – The Santa Rosa is a beautiful, large, red fruits with gold flesh. It is a sweet plum that is delicious when eaten fresh, cooked or canned. It is ready the first week of August.

Starking Delicious – This new variety is gaining acclaim for its great taste and ease of growing. It is disease resistant making it a very environmentally friendly option. Ripening in the second week of August, this deep red Japanese plum is a delicious summer treat.

Ozark Premier - The Ozark premier is a large, plump, roundish plum. The skin is red with a waxy bloom. Its firm flesh is yellow, fine-grained, and juicy. The Ozark premier will appease your appetite with its sweet, great taste.

Burbank – The Burbank is a well-known old variety. The fruit is medium-sized and has attractive orange-red color that covers most of the surface with a base color that is amber-yellow. The flesh is yellow, fine-grained, firm and juicy, sweet and very good tasting. The peak harvest is in the second part of August-beginning of September

Redheart – you would be fortunate to find the redheart at the farmers market in mid-august. A finicky producer, the redheart is one of the tastiest plums grown in the state. It is aesthetically appealing being large, smooth, and heart-shaped with dark purplish red skin covered with golden specks. The flesh is blood red, firm and juicy. There are few other varieties that rival the redheart's sweet aromatic goodness.

Rubyqueen – The Rubyqueen is a medium sized fruit with a firm flesh and excellent flavor. It has a beautiful dark red/black skin with a deep red flesh. This is really a jem of a plum.

Fortune – The fortune is a large bright red plum on a yellow background. The flesh is yellow, firm and juice. This attractive, good-tasting plum ripens in mid to second part of September.

Lydecker – The Lydecker is a new variety of plum harvested the first week of September. It is dark blue-black and nearly round, similar to many California dessert plums. It is said to have a superior ripe flavor to the more common varieties in its ripening season.

Simka – The fruit of the simka is large, uniform, with a very shallow suture. It is a very attractive, good-tasting plum. It is sweet with an excellent firm texture The skin is very dark red almost ebony and the flesh is very light green to slightly yellow. It's a very good choice for the first week of September.

Alderman –The Alderman is a very attractive plum that has nice brilliant orange-red skin color and orange-yellow flesh. Skin is very firm, shiny, and waxy-like. The fruit is medium to large often 2"-2 1/2" in diameter. It has good texture and taste. It is a very good late plum. It ripens in the third/fourth week of September.

European Plums

Vibrant - The first European plum to be harvested, the Vibrant is a beautiful plum with a violet-blue skin and amber flesh. It is a medium to large-sized fruit with good firmness and sweetness with a medium acid content for a nice balanced flavor.

Vanette - The Vanette is a large, purple-blue, freestone plum. It has good sugar/acid ratio that accounts for excellent taste. Most years ripens in the third week of August. It is very good dual - purpose plum and is one of the best fresh market plums. It is an excellent all-around plum. It's great for cooking, canning and fresh eating.

Castleton - The fruit of the Castleton is medium-sized, dark blue, oblong, and freestone. It has a sweet to mildly acidic taste. It is very good dual purpose plum; suitable for fresh and cooking/preserving. A known favorite when it comes to home canning, it makes an excellent burgundy jam. Find them at the end of August to the first part of September at a farmer's market or fruit stand near you.

N.Y. 6 – The N.Y. 6 is a relatively large blue-skinned, yellow-fleshed European plum. It is a favorite variety for baby food as it has a very mild taste which makes it excellent choice for mixing with other fruit. It is a great choice for baking.

Early Italian – The Early Italian is an old, well-known variety of blue plum that is also called "Early Fellenberg". With its pleasant firmness and its great taste, the Early Italian is one of the highest caliber plums Michigan has to offer.

Stanley – The Stanley is the go-to for European plums. It is medium to large fruit with dark blue skin and yellow-green flesh. It can be identified by its distinct neck.

Valor – The Valor plum is large and very good tasting. Its skin is an attractive dark purple, speckled and the flesh is greenish- yellow. It is semi-freestone and can be found starting the first week of September.

N.Y. 9 – This plum ripens in the first week of September. The fruit size varies due to the crop load and goes from small to large. The flesh is green and the skin is purple, covered with waxy bloom so it appears blue. It has mild taste and is rather sweet. Though, it is processing variety, when picked when the flesh color starts changing from green to amber, it has just enough acid to make it well eating plum.

Blufre -The Blufre is a medium-large sized European plums. It is blue-skinned and yellow-fleshed with a very distinctive flavor. The Blufre is a good choice for all your baking needs.

Long John – The fruit of the Long John is large and has an interesting shape: it is quite long and bit "flattened". The skin is dark maroon, almost black, and covered with the waxy bloom, which gives it nice blue color. The flesh is orange, firm and pleasantly tart. It is freestone and it ripens with the Stanley at the first of September, but is larger and better quality.

Autumn Sweet – The Autumn Sweet is a new European blue-skinned, yellow-fleshed plum. It is said to have superior quality to that of the traditional Italian plum. It ripens the first week of September, making it a great choice for lunchboxes that kids are sure to love.

Blue Damson – The Blue Damson is an old variety, renowned for its superb preserves and baking characteristics. It is a small blue plum with a yellow flesh. The Damson is in high demand throughout farmer's markets from individuals longing for the most mouthwatering jams and most intriguing plum bounces.

Italian – The Italian is a medium to large fruit with purple skin and yellow flesh. It is a sweet plum that stores well and is a great dual purpose plum.

Tulare Giant - The Tulare Giant is a new Japanese plum variety. It is a large reddish-grey skin plum

with an amber flesh. Eat this one fresh because its incredible sweetness is rivaled by few other varieties

Empress – The Empress is a well-known European plum variety. It has large, elliptical, symmetrical fruit of very good quality. The skin is purple and covered with heavy waxy bloom. The flesh is greenish-yellow and it is semi-cling. The Empress is a very nice late-season choice.

Fruit Growth and Development

The pattern of fruit growth in plum, as in other prune species, is the double sigmoid curve. Prior to fertilization, one of the two ovules of the plum fruit aborts. During the subsequent development of the fruit, the outer wall of the ovary gives rise to three distinct layers of pit. Two hypodermal layers ion the ovary produce the skin of the fruit, which consists of the cuticle, the epidermis and a few layers of collenchymas cells. The flesh contains most of the vascular bundles of the fruit, which are embedded in the lignified tissues of the pit. The fresh weight of the fruit and it size increases throughout the growing season. The sugar and total soluble solids content increase throughput the period of growth. Organic acids accumulate in the fruit during early stages of growth and then gradually decrease. The phenolic content of the fruit is high during the early stages, decreases subsequently and then remains constant until harvest. Volatiles which determine the flavor or aroma are also produced. Wax develops on the skin of the fruit. The fruit attains its full size and optimum maturity, although it is still unripe.

During ripening, pacific substances in the cell walls change from an insoluble to a soluble form, resulting in softening of the fruit. The chlorophyll content of the skin decreases and carotenenoid content increases. The plum is considered to be a climacteric fruit; the respiration rate is high during growth. As maturity approaches, it decreases to a preclimacteric minimum and increases irreversibly to a maximum during ripening. At this stage, the fruit is soft and sweet, with a characteristics flavor and is ideal for eating. Subsequently, senescence sets in, whereupon the respiration rate decreases and the fruit become overripe.

Maturity

Optimum maturity standards are dependent on the cultivar and the intended use of the fruit. Plums destined for long-distance transportation have to be picked at a firm ripe stage. The maturity indices for plums and prunes include skin and flesh color, flesh firmness, juiciness, soluble solids, and ratio of soluble solids to acids.

fisher found that total soluble solids (TSS) content of the flesh was more consistent index of eating quality of plums than flesh firmness. However flesh color, TSS-to-acid ratio, and firmness were found to be reliable indices of maturity of plum. The optimum picking maturity varies from cultivar to cultivar. For Santa Rosa plums it was calculated to be 94 + 3 days from full bloom. In calafornia, French prunes require an average of 158 days to mature, with a range of 145-165 days.

Spray of different chemicals affected various physicochemical characteristics of Japanease plum; for instance, with KNO_3 the size of the fruit increased, while $C_a(NO_3)_2$ enhanced their firmness. Similarlily, the application of growth regulators/retardants as such or in combination with other chemicals influenced various chemical characteristics of the treated fruits. Another growth retardant, Semifresh, prevented undesirable changes in the properties of the fruit and has the potential for enhancing the post-harvest life of plum fruits.

Harvesting

Plums usually ripen unevenly over the tree. Therefore, fruit should be harvested into two or three pickings. Plums which are picked for canning are usually harvested at one time. They are picked by hand into buckets or baskets with padded liners, or into picking bags. So far as possible, the peduncle should be allowed to remain attached to the fruit. Precautions must be taken not to puncture the fruit with fingernails *etc.* Care should also be taken to prevent direct exposure of the fruit to the sun. Attempts to harvest plums mechanically have not succeeded due to the susceptibilityof the fruit to physical damage. However prunes are now harvested mechanically. Fruther, mechanical harvesting of prunes resulted in gathering of more than 75 per cent of the crop. Skin color mature oriental plums vary from light green to yellow, red-blushed, or solid red. Each cultivar has its charecteristics color relating to "readiness" for harvest; for instance beauty having 85 per cent surface yellowish green or trace of red, and Santa Rosa, with 40 per cent of surface red color, or full light greenish yellow. Coloring and softening are the major changes. Soluble solids do not increase appreciably after harvest. Containers for prunes and plums include baskets, crates, wood lugs, and cartons. The use of fiber board is increasing for tight-fill and tray packs. Plums are packed in shallow crates about 4-5 in. deep, packing not more than tree layers of fruit. After packing the fruit should be cooled immediately to 0°C, which would stop the ripening process in plums for a period of about 12 days.

Packaging and Transportation

In packaging, the fruits are assembled in convenient units for handling and transportation to protect them during marketing and storage operations. Improvements in packaging have contributed greatly to more efficient transportation and marketing of fresh fruits. Among the packaging materials most commonly used in earlier times were woven leaves, grass stems which have been replaced by wood, cardboard and recently plastics. Plastic flim is most effective in minimizing the loss of moisture and nutrients from fresh fruits. Packages such as waxed fiber board cartons, parchment wraps, and other specially treated packing material retard moisture loss significantly. While refrigerated trains or trucks are common in some development countries for the transportation of fruits, most countries have to be satisfied with ordinary truck or trains, resulting in far greater losses.

Chemical Composition

The approximate compolsition of the fruit is given in table. The fruit is rich source of sugars. It has very little starch and no sucrose at the immature stage, but the later increases during ripening and its consentration exceeds the reducing sugar content. D-glucitol, sorbitol, galactose, mannose, arabinose, rhamnose, raffinose and xylose are the other sugars detected in plum. Although D-glucouronic acid is rare in the Free State, it is found as a constitutent of many polysaccharides such as gum of fruits. Amygdalin, a glycoside, is present in the seed of the plum fruit. Various pectic substances contribute to the texture of the fruit. Phenolic substances present in the plum fruit include *neo*-chlorogenic catechin, caeffeic acid, chlorogenic acid and rutine with phenolic acid as the predominant phenol. Some of these compounds to the astringent taste of the fruit.

Approximate Composition of Plum Fruit

Composition	*Range*
Fruit	
Fresh weight (g)	36-126
Water (g/100 g of edible poprtion)	86-88
Protein (per cent)	0.4-0.8
Fat (per cent)	0.1
Total sugar (per cent)	6.7-9.9
Fructose (per cent)	0.9-3.4
Glucose (per cent)	1.7-5.2
Sucrose (per cent)	1.0-4.2
Dietary fibers (per cent)	1.3-2.4
Organic acid (total) (meq H^{+}/100g)	1.4-2.7
Malic acid	0.14-2.54
Citric acid	0.03-0.04
Quinic acid	0.12-0.41
Ash (per cent) 0.3	
Energy (kJ)	125-187
Titratable acidity (meq H^{+}/100g)	16.3-30.5
Minerals (mg/100g)	
K	120-190
Na	0.0-3.0
Ca	6.0-8.0
Mg	4.0-7.0
Fe	0.1-0.4
Z	0.1

Composition	*Range*
Vitamins (mg/100 g)	
Vitamin C	4.0-11.0
Thiamin	0.02-0.05
Riboflavin	0.04-0.05
Niacin	0.2-0.9
Carotence	0.26-0.78
Flavanon (mg/100 g)	46.3-57.0
Anthocynin (mg/100 g)	926
Pectin (per cent)	0.81-0.98
Kernal (per cent by weight)	
Water	9.6
Protein	20.0-23.9
Fat	39.5
Lecithin	0.15
Carbohydrates	15.0
Ash	3.4
Fiber	12.2
Amygdalin	0.3
Phosphorus	0.3
HCN	0.07

Type and Concentration (mg/g dry weight) of Phenolic Compounds in Plum Fruit

Phenolic Compound	*Plup*	*Exocrap*
Neochlorogenic acid	2.6	6.25
Chologenic acid	0.11	1.37
Caffeic acid	0.055	0.036
Catechin	—	2.20
Rutin	—	6.30
Anthocyanins	0.79	0.59

The major organic acids found in are malic and quinic acid, which impart an acidic taste to the fruit. Flavanoes are also present, but due to their lack of color these are not easily detected. Though the fruit has other vitamins also, it is vitamin C which is present in a quantity to be of nutritional significance. The protein content of the fruit is low, like other fruits, and the chief amino acids present include asparagines, aspartic acid, glutamic acid, glutamine, serine, theronine, α alnine. T-amuynobutric acid, valine, leucine and proline besides traces of hydroxyl proline. The color of the fruit is mostly contributed by anthocynins, which are located in the epidermal layers. The color effects of anthocynin have been explained based on their complexes with metals. Their color shades are dependent on the PH of the medium.

Cyanidin-3-rutinoside and peonidin-3-rutinoside are the major anthocyanin of plum. Anthocyanin content increases in fruit as they approach maturity. Carotenoides also contribute to the color of some plum cultivars and are affected by drying. The fruit is good source of minerals such as K, Na, Ca, Mg, Fe and Zn.

A number of compounds have been associated with ripening of plum, including a variety of terpenes, alcohols, aldehydes and esters, which may distinguish a ripe fruit from an under ripe one. The aroma of plum blossoms is due to the pressence of benzaldehyde. The surface of the fruit contains wax layers containing D-glucose, L-arabinose, D-xylose, L-rhamnose, gluconic acid, and hexanoic acid.

Prunes contain low sodium, a higher level of dietry fiber, iron, β-carotene, and potassium, besides being a natural laxative. Prunes juice could be used as humactant. Its consumption reduces plasma low-density lipoprotein cholesterol significantly compared to grape juice.

The composition of plum pit shows that it is a source of protein and fats. A fatty oil can be extracrted from the kernel which is bitter in taste and has in it HCN—a toxic compound. It can be used as a lubricant, an illuminant, and for hair dressing.

Nutrients and Mineral Content of Fresh Plums and Dried Prunes

Nutrient	*Content per 100 g Edible protein*	
	Plum	*Prune*
Moisture (per cent)	85.2	32.4
Enerjy value (kJ)	230.0	1000.0
Carbohydrates (g)	11.0	55.5
Dietary fiber (g)	2.0	7.2
Protein (g)	0.8	2.6
Total fat (g)	0.6	0.5
Nicotinic acid (mg)	0.5	2.0
Pantothenic acid (mg)	0.18	0.46
Riboflavin (mg)	0.10	0.16
Thamin (mg)	0.04	0.08
Folic acid (μg)	2.0	4.0
Vatimin A (μg)	192.0	1194.0
Vatimin B (mg)	0.08	0.26
Vatimin C (mg)	10.0	3.0
Vatimin E (mg)	0.65	—
Sodium (mg)	0	4.0
Potassium (mg)	172.0	745.0
Calcium (mg)	4.0	51.0
Magnesium (mg)	7.0	45.0
Phosphores (mg)	10.0	79.0
Iron (mg)	0.1	2.5
Coper (mg)	0.04	0.43
Zinc (mg)	0.10	0.53

Storage

The plum is a climatic fruit and has a short shelf life. To extend the storage life of fruit, its temperature must be lowered as soon after harvest as possible. Precooling which removes field heat from the fruit should be done immediately after harvesting, packaging, and palletizing. The ideal storage conditions for plums are a temperature of -0.6 to 0°C and a relative humidity of 85-90 per cent. The storage period depends on the cultivar. Most cultivars can not be stored for longer than 3-4 weeks, but others can be stored for up to 3 months. Some varieties exhibit chilling injury if stored at 0° C for a long time. To prevent chilling injury, the temperature raised from -0.6°C to about 7-10 days, and this temperature is maintained for the rest of the storage period.

Post-harvest Diseases

Plums are to very suscitible mechanical damage resulting from fruit compression during packing. Bruising is also caused by vibration during transit. The bruised tissue first appears to be water soaked or glossy and later on truns brown. Higher temperatures also accelerate the process. Prunes may also split or crack along the sides or ends, causing faster decay. Brown rot, rhizopus rot, blue mold rot, claddosporium rot, and gray mold rot are the important diseases causing severe post-harvest losses. The amount of decay during storage and reduction of molds by chemical treatment is related directily to the method of harvest and temperature of incubation.Molds that cause decay are species of *Cladosporium, Monilinia Rhizopus,* Mucor, Penicillium Botrytis. Alternative. Prunes treated with chlorine, estconazole and potassium sorbate showed no mold growth up to 10 days and significantly reduced decay. Prunes harvested machanicially showed significantly less decay than hand picked prunes.

There is a lack of awareness among marketers of the need for post-harvest protection of fruits from fungal attack. The common methods of control include preharvest fungicidal spary, fumigation of fruits by gases and dips in antibiotics, fruit wrappers and skin coating gammaradiation. A combination of heat and chemical treatment and use of wax emulsions. Post-harvest application of food-grade fruit coating containing fungicide or O- phenylphenol (OPP) as an active ingreadent controlled fungal rots in plum caused by *Aspergillus niger, Geotrichum candidum. Monalinia fructicola, Rhizopus stolonifer, Trichoderma* and *Pennicillum digitatum.*

Low Temperature Storage

A storage temperature of about 0° C with 90-95 per cent relative humidity has been recommended to hold both European and Japanese plums for about 2-4 weeks, depending on the cultivar and the stage of fuit maturity at low storage temperature (0-4°C), fruit development is distinctly slowert than at higher temperature (8-12°C). Slighty unripe plums had better resistance to fungal infection tthan tree-ripe plums. Plums with higher soluble solids content and picked late could be stored for 4-5 months. The non-sealed plyethylene liners used in the containers further reduces loss of moisture.

During storage, loss in weight ranging from 1.4 to 2.3 per cent due to shriveling in plum fruits is commonly observed, depending on the cultivar, while the loss due to disease ranged from 0-8.9 per cent after 30 days of storage at 0-1°C and 90-95 per cent RH immediately after harvest, depending on the variety. Storage losses were found to depend on the time of picking, infection by *Botrytis cinearea* and *Monalinia* spp., and the temperature of the storage, but the effects of these factors varied with the varieties. If fruits are held longer in cold storage, there is a danger that shriveling, mealiness, internal browning, and abnormal flavor will develop. However, ripening of early Italian prunes was enhanced by a few days' exposure to low temperature. These prunes develop more color, have less acid, and are softer than prunes ripened immediately at ripening temperature without cold treatment. Treatment of plum fruit with wax emulsion was effective in reducing percent loss in weight and spoilage during storage.

Internal breakdown has been associated with maturity stage and storage temperatures of the fruits. The use of the dual-temperature storage reduces considerabily the incidence of internal breakdown compared to a single – temperature regime in Songold plum

Controlled Atmosphere Storage

Controlled atmosphere (1 per cent O_2 and less than 0.2 per cent CO_2) extended the storage life of English grown Victoria plums up to 4 weeks at -0.6°C. The interruption of storage by 2 days at 18.3°C in air after the 16th day further extended the storage life and reduced internal browning and jellying. The Exposure of Italian prunes to CO_2 concentrations of 20-60 per cent for 2 days at 7.2° C prior to storage at 0°C reduced softening and decay. Sealing of plums in polyethylene bags extended the storage life of fruits, most probably by accumulation of a high concentration of CO_2 in the sealed polyethylene bags. The CO_2 also retarded color and ripening of plums in transit and increased their marketable life. Exposure to a low –O_2 atmosphere inhibited ripening, including a reduction in ethylene production rate, retardation of skin color, and flesh softening with msintenance of titratable acidity. However, the most important detrimental effect of low –O_2 treatment has been the development of an alcoholic off-flavor that had a logarithmic relationship with ethanol content of fruit skin. Plums cv. Santa Rosa and Songold could be stored under controlled atmosphere at 0.5°C in 4 per cent O_2 + 5 per cent or 7 per cent CO_2 for 14 days in excellent condition with out internal browning.

Hypobaric Storage

Ripening of plum was greatly delayed at 100 mm Hg and 20°C, but at 6°C there was no difference in ripening between normal and low-pressure storage. Hypobaric storage is also a form of controlled- atmosphere storage in which the produce is stored in a partial vaccum. The vacuum chamber is vented continuously with water- saturated air to maintain O_2 levels and minimize water loss. Reductions in partial pressure in Oxygen and reduction in ethylene in hypobaric storage results in retarded ripening of fruits.

Irradiation

The ionizing radiations have been found particularly useful to disinfect the produce from fruitflies and other insect pets and to extend the shelf of produce by restricting growth of pathogenic microorganisms and slowing down the rate of metabolism of the produce. The advantages of gamma-radiations over other procedures include its use as a continuous and more effective process without residual radioactivity or chemical residue. The preservation and extention of the refrigerated life of fresh fruits by irradiation is of considerable intrest to post-harvest technologists. However, it is dependent on a number of factors such as cultivar maturity, permeability of packaging materials, preirridiation chemical treatment, and storage temperature employed. Among the vitamins, ascorbic acid is the most radiosensitive vitamin. Use of radiation (up to 1 KGY) in disinfestation of Cepatitis capitata prevented its eggs from hatching. Based on triangle tests, no difference in sensory qualities in plum with or without irradiation could be detected except for the change in the color. However, before the progress can be utilized, it should meet the criteria laid down: There must be higher tolerance of the host to radiation than spoiling organisms or metabolities, it must be more economical than other treatments and the wholesomeness of the treated fruits must be confirmed.

Processing

Most plums are dehydrated to plums. Some are canned and few are used for preparation of jams, jellies, beverages, wines and brandies.

Jam

Like all other fruits, jam or jelly can be prepared from plums, but most of the times it is used in the preparation of mixed-fruit jam. The procedure is similar to that for to other fruit. Analysis of plum jam detected the presence of sugars such as α-and β-maltose and glucose besides late-ripening varieties are generally more suited than early ripening ones, mainly due to their acid sugar balance and high dry matter content.

Juice

Prune juice is not a fruit juice in the usual sense but rather a water extract of dried prunes. It is rich in minerals and acts as a mild laxative. Prune juice has becomes a popular breakfast drink in the United States. Prune juice is produced either by diffusion or a disintegration process. In the diffusion process, soluble solids of the prunes are leached out in a succession of hot water extractions, each lasting for 2-4 h. to produce juice by the disintegration process; the prunes are cooked for 2-3 h, then put through a disintegrator: a filter aid is added to the slurry, which is filtered to get the juice. To extract the juice, the fruits could be steamed treatment for 6-12 h. the juice is filtered, adjusted to 22.5-23° Brix, flash-pasteurized to 88°C, and filled hot into cans at 82-85°C. to sterilize the top, the containers are inverted and then cooled. The juice can be bottled and sterilized alternatively. Addition of pectinolytic and acidity and a drastic decrease in apparent viscosity and improving the color and clarity without affecting the flavor.

Processed prune juice posses unique characteristics, such as predominance of Y-amino butyric acid, citrulline, *O*-phosphoethanolamine and quinic acid and absence of anthocyanins, (-) epicatechin, phloridzin and acids such as citric and tartaric acid. The presence of anthocyanin in prune juices would be a clear,mark of adulteration, while its degradation was found to be enhanced by hear, oxygen and hydroxymethyl furfural (HMF).

Beverages

Plum fruits are utilized to a limited extent in the preparation of beverages of various types, such as ready-to-serve beverages, nectar, squashes and appetizers. Because it is acidic, not more than 40 per cent juice can be used in sweetened plum juice. To prepare nectar, 20 per cent juice extracted enzymatically, with 15 per cent Brix, was found to be optimum. Another product from such a juice is plum appetizer. The product has in it a spices extract which includes mint extract (0.4 per cent), ginger (0.5 per cent), slat (1.3 per cent), black salt (0.7 per cent), cumin (0.25 per cent), cardamom (0.25 per cent), citric acid (0.5 per cent) and sodium benzoate (0.065 per cent).

Concentrates

Prune juice concentrate is made from prune juice or extract and most of it is preserved by heat processing. However, frozen concentrate is superior to heat-processed product, bottled or canned. The extract is depectinized using pectinol enzyme and the filtered juice is then evaporated to 60 per cent Brix. The product is preserved by freezing and marked as frozen prune juice concentrate. Preparation of plum juice concentrate is in of the best methods utilizing fruit during glut seasons. When subjected to concentration at 50-60°C under vacuum of 69-76cm Hg, plum juice can be concentrated to 73° Brix. The increased concentration enhances total soluble solids, reducing sugars, browning and viscosity, while it decreases acidity and pectin content slightly.

Canned Plums

Canning of plums is an important industry in several countries such as Great Britain Plum canning consists of washing and destemming fruits simultaneously, followed by grading. The graded plums are inspected for imperfections and filled into cans or glass jars, either by hand packing or using automatic filling machines. Either hot syrup or water at 88-93°C is then added. The concentration of syrup may range from 0 to 55 per cent. The filled cans, which are double seamed, are exhausted at 88-93°C for 8-10 min. the average time for processing the cans in boiling water is 10-35 min depending upon the can size. The heat-processed cans are cooled in a rotary cooler with a water spray to 38°C or air cooled.

Prunes are a less important crop than peaches and pears with respect to canning. Italian cultivars of prunes are canned in the fresh state and labeled as purple plums. The process of canning is similar to that for plums with the exception that the prunes are placed in 40 per cent Brix syrup in enameled cans. For canning, the prunes are harvested at firm ripe stage. Large-size dried prunes of the French cultivar prune de Agon are used for canning, before canning, prunes with 18 per cent moisture are blanched in boiling water for 4-5 min, rinsed and packed into cans after sorting and making allowances for the dried prunes to take up part of the syrup during processing and storage. Usually, the syrup used for canning is 20 per cent Brix. The rest of the procedure is the same as for plums.

Prunes can also be canned as "dry pack prunes" in a similar manner except that syrup is not used and the cans after roller operation are exhausted for 20 min in live steam, sealed and allowed to cool in air. Before serving, these are cooked in water and re used as a between-meals snack.

Mature plums are usually picked before optimal ripeness and maximum color developments based on texture rather than color as then more resistant to bruising mechanical harvesting and soften less during pasteurization.

In sensory tests, firmer plums are generally preferred to softer fruits. Sensory intensity ratings of the color of canned plum halves show good agreement with total anthocyanin content. To avoid artificial coloring, ripe fruits of anthocyanin-rich plum cultivars should be used. The anthocyanin concentration of canned plums increased considerably when plum cultivars with high initial anthocyanin content were canned, but successive softening and sloughting of ripe plums could be prevented by the addition of 500 ppm $CaCl_2$. The increase in firmness is due to the reaction of divalent ions with the free carboxylic groups of pectates. However, the use of various pure can liquids (distilled water, sucrose, glucose compound syrup in distilled water or combined with other additives, *e.g.* ascorbic acid, citric acid, EDTA, SO_2 and Ca^{2+} added separately) did not stabilize the anthocyanin content. Anthocyanins diffuse from the skin into the can liquid during storage, as revealed by the analysis of color by absorption spectroscopy. Overripe plums, which are usually considered unsuitable for industrial canning, produced canned plums with higher anthocyanin content than the conventional product, necessitating reconsideration of present industrial quality criteria.

Upon comparison of traditionally canned with skin in pouches (with or without added sugar), it was observed that stability of total anthocyanin increases with decreasing amount of fluid surrounding the fruit. Preservation of plums in this type of packing requires very little liquid and therefore results in better retention of color and sensory quality of product.during processing and storage of plums the decrease in pigmentation has been attributed to polymerization and the condensed anthocyanins have better stability than monomeric molecular diffusion. The overall loss of pigment from skin by diffusion can be reduced considerably by processing skins in pouches without added liquid.

Dried Plums

About 75 per cent of the world's supply of dried prunes is produced in California and in the Pacific Northwest. Although sun drying was very common earlier, today they are dehydrated, to prevent spoilage during the rains. The French prune cultivar is most commonly selected for dehydration, as well as imperial sugar varieties. Prunes are prepared for drying by cleaning with air blast and water sprays, dipping in cold or hot water, followed by spreading on a single layer in trays and dehydration to 18 per cent moisture in a forced-air tunnel dehydrator.

Conventional dried prunes are sometimes dried to a very low moisture content (4.0 per cent) in a vacuum shelf drier. enabling free flowering and hygroscopic, dehydrated prunes require very careful handling and packaging. Pitted prunes are dehydrated after perforation to achieve better dehydration and reconstitution character. Dried prunes, prune flakes, nuggets and granules are also available commercially for use in fillings and other bakery spreads, while low-moisture prune powder is utilized as a sweetening and flavoring agent in whole wheat or rye bread.

As innovative process called osmotic dehydration has been described for drying fruit pieces/slices or chunks when a exposed to concentrated sugar syrups or salt to remove about 50 per cent water from the fruit by osmosis. The partially dried fruit can be further dried by conventional dehydration technology including vacuum; in the latter case it is called Osmovac. The process has several advantages, including reduction of time of exposure to high temperature, mineral heat damage to color and flavor, acid reduction in some fruit and incorporation of more sugar, thereby in creasing the acceptability of the final product.

Pretreatment of Santa Rosa plums with a boiling solution of Na_2CO_3 (3 per cent) + ethloleate (2 per cent) for 45 s followed by osmotic dehydration produced an acceptable product. Similarly, osmotically dried prunes were found to be excellent in physicochemical characteristics and sensory qualities. Thermal treatment of the fruit reduced the overall quantity and quality of the o=isolated wax, converted the cuticular layer of the plum from hydrophobic to hydrophilic, increased the osmotic rate and shortened the drying time.

During dehydration, degradation of dihydroxycinnamic acid was found to be related directly to the evolution of polyphenol oxidase activity and the degradation was worse when the drying temperature was low; the reverse is true with flavanoids. However, anthocyanin changed only during the initial hours of drying.

A recent approach in drying is the thin-layer dehydration of plums at a temperature of 60-120°C, an air moisture content of 0.008-1.8889 kg/water and an air of velocity of 0.5-2 m/s. Drying of prunes in multishelf dryers using solar-heated air in turbulent flux over the material has also received some attention, although in variable climates some conventional energy may be required to give additional air heating.

Purees

Puree is an intermediate product which can be used for processing into different products. It is packaged aseptically. An aseptic bulk storage system for acid fluid products was developed which includes pulping, concentration, preheating, heating to achieve microbiological and enxymatic stabilization and cooling prior to storage of the sterilized product. With 30.2 per cent moisture 47.7 per cent fat and 13.6 per cent sugars (including lactose) a product of excellent organoleptic quality was obtained by mixing unsalted sweet cream butter (57 per cent) with plum puree (25 per cent), skin (10 per cent) and granulated sugar (8 per cent). The prune pulp is also used in ice cream mix, confectionery products and meat sauces.

Steep-Preserved Plum

The preservation of fruits by steeping does not require costly equipment or packaging materials. It is of practical interest to the fruit industry in terms of extending the seasonal availability of the fruits. However, plums cannot be preserved in steeping solution for long, as they lose color, texture and flavor during storage.

Wine

Plum fruit gives with an appealing color and acceptable quality. The pulp is fermented to produce the wine. Compared to grapes, the sugar content originally present in the fruits is not attractive, but the color and flavor make plums acceptable. Plum wines are quite popular in many countries, particularly in Germany and Pacific coastal states.

To prepare plum wine, for every pound of plum, 1 liter of water is added followed by addition of starter culture. The mixture is allowed to ferment for 8-10 days before pressing. It is practically impossible to press the fruit before fermentation. Addition of pectolytic enzyme before fermentation facilitates pressing, increasing the yield of juice and hastening wine clarification. Additional sugar may be added to the partially fermented juice, depending on the type of wine required *i.e.* table or desert. Aging, filtration, bottling and processing is similar to that for grape wine. For preparation of wine, the plum fruits should be fully ripe, diluted with water in a 1:1 ratio (whole-fruit basis) and 0.3 per cent pectinol, 150 ppm SO2, and enough sugar to raise TSS of the must ot 24° Brix should be added. Ac considerable improvement in the quality of the plum wine takes place upon raising the sugar content of the wine 12° Bix. Potassium metabisulfite (KMS) is commonly usedin most wine fermentations, but it reduces the color of plum wine. The use of sodium benzoate intead of KMS produices wine with better color and sensory qualities without affecting the physicochemical characteristics. Some of the physico-chemical characteristics of a plum wine are given in table.

Physico-chemical Characteristics of Plum Wine

Characteristics	*Range*
Ethanol (per cent v/v)	8.5-11.0
Total soluble solids (°Brix)	8.0-12.0 (sweets)
Titratable acidity (per cent malic acid)	0.62-0.88
Volatile acidity (per cent acetic acid)	0.028-0.040
Esters (mg/liter)	104-109
Color (tintometer color units)	
Red	6.10
Yellow	10
Total coloring matter and tannins (mg/100 ml)	119

Mineral Content of Plum Wine of Different Varieties

Variety	*Mineral Content (ppm)*							
	Na	*K*	*Mg*	*Ca*	*Fe*	*Cu*	*Zn*	*Mn*
Methey	26	1450	20	93	4.16	0.50	18	0.98
Green gage	15	1258	20	134	6.97	0.22	1.05	1.00
Santa Rosa	20	1008	18	82	12.73	0.20	1.04	0.95

Among the cultivars evaluated, Santa Rosa plums made the wine of best quality. The mineral content of wine from Santa Rosa, methley and green gage plums was desirable from enological and nutritional points of view. Plum fruit is acidic in nature and o make palatable wine, besides dilution of the pulp with water, the alternative available is use of a yeast, Schizosaccharomyces pombe, as a deacidiying agent in plum must besides Saccharamyces by this yeast was independent of the sugar content of the must but was adversely affected by 2.5 pH and SO_2 content of more than 150 ppm, while addition of ammonium sulfate enhanced the malic acid degradation by this year. However, the activity of the yeast was susceptible to higher concentration of ethanol and the color of the wine depends on the quantity of the acid metabolized. More information on these aspects is needed before making use of this approach to prepare plum wine.

To produce quality wine acids, a proper balance of tannin and good fruity are prerequisite. Application of osmotic techniques after water blanching the fruit increased TSS and decreased acids though a little loss in anthocyanin and mineral content occurred. The sensory quality of wine prepared from water blanched and somatically treated fruit was the best. The process could become economical if it is coupled with other fruit and vegetable processing.

Vermouth

Vermouth of commercial an be prepared from plum. Vermouth with increased alcohol contents has increased aldehyde, ester and phenol contents, but acidity and vitamin C declined. The spices extracted contributed to the increase of total phenols, esters and aldehyde content. The best product and about 15 per cent alcohol and was sweet in nature.

Plum Brandy

Brandies derived from fruits other than grapes are referred to as fruit brandies. several countries produce true fruit brandies, often highly regarded for their distinctive flavor.

The generic name of plum brandy is Silvovitz and is a national drink of middle European cpoiuntries such as Romania, Hungary and Yugoslavia. Imported Yugoslavian Silvovitz is often aged in oak, mulberry or acacia cooperage.

Most of the samples are light yellow and have little evidence of wood-aged character. Plum branbdies produced in the Alsace region of France are mostly named after the variety of the fruit used. The varieties used are similar to those used for dried prunes. Besides overproduction, a significant amount of cullage fruit, usually having defects of size or side splits made the growers think to utilize the fruit for making Slivovitz from plum. The method of making plum brandy is similar to that of grape brandy in large quantities.

The aroma components of an experiment brandy include aliphatic fatty acids, free acids, free fatty acids, aromatic acid, esters, alcohols, ethyl acetate lactones and unsaturated aliphatic aldehydes, Fruit traits affected the HCN content in fruit brandies and the prune fruit had the lowest HCN content. Among the volatiles were the carboxylic acid, alcohol and aldehydes found in plum brandy.

REFERENCES

Amerine, M. A, H.W berg, R.K Kunkee, C.S. Ough, V.L. Singhleton and D. Webb. 1980. Technology of wine making, 4th ed., AVI, Westport, CT.

Bhutani, V.P., V.K. Joshi and S.K Chopra. 1989. Mineral composition of experimental fruit wines, *J Food Sci. Techonol.* 26; 332.

Biondi, G, C. Partell, A. Marchi and R. Bassi. 1990. Maturity indices for harvesting plums as a function of quality, *Rivista di fruitticultura e di Arbofloricoltura 52: 65.*

Borecka, H, H. Zdyband and B. Wajtas.1984. Preliminary experiments in plum storage, *A. Prace Doswiaadjalne-""-Zakresu-Sadownictwa* 25: 197.

Cambrink, J.C. 1993. Plums and related fruits, Encyclopedia of food science, food technology and nutrition (R. Macre, R, K Robinson, and M.J Sadler, Eds.) Academic Press, London, 1993, p.3630.

Chopra, S.K, S.S.Mishra, V.P.Bhutani, and A.S. Kashyap. 1986.Advances in research on temperature fruits, Parmer University of Horticulture.p.319.

Costa G. F. Succi. R. Baisi, M. Bolognesi and M. Taghivini, 1990. Further studies on mechanical harvesting of European plum. *Rivista di Fruttiicoltura e de Ortofloricoltura 52: 55.*

Dev, S.S., P.S. Khalon and C.S. Sidhu.1989. Effect of various treatments on the shelf life of Plum. National Seminar on Recent Advances on post-harvest Management of Temperate Fruits, Vegetables and Ornamental Plants, Dr. Y.S. Parmer University of Horticulture and Forestry, Solan, India, p. 22(Abstr.).

Dinamarea, E.A, F.G. Mitchel and A.A. Kader. 1989. Use of sucrose esters as retardants of ripening of pears and plum, *Rev. Fructicola* 10(3): 116.

Gorsel, H, L.C. Van, E.L. Kerbal, M. Smiths and A.A.Kader, 1992. Compositional characterization of prune juice. *J. Agr. Food Chem.* 40: 784.

Hartmann, P.E.O, V.A. Dekock and M.A. Taylor.1988. Picking maturities and cold storage requirements of Songold plums, Deciduous fruit growers 38(5): 161.

Hohn, E.1987. Maintaining Plum quality. Schweiz. Z. Obst. Weinbau. 123 (19): 520.

Hulme, A. C. 1978. Biochemistry of fruits and their products, Vol. 1, Academic Press, London.

Jindal, K.K. and M.P.Dwivedi, 1986. *Advances in research on temperate fruits*, Parmer University of Horticulture and Forestry, Nauni, Solan India, p.287.

Jindal, K.Kand N. Sharma. 1986, *Advances in research on temperate fruits*, Parmer University of Horticulture and Forestry, Solan India, p. 281.

Joshi, V. K, P.C. Sharma and B.L. Attri. 1991. A note of deacidifaction activity of scizosaccharomyces pombe in plum musts of variation composition of J. APPL. Bacteriol.70: 385

Joshi, V. K, S, K, Chauhan and B.B Lal. 1993. Evaluation of enzymatically extracted plum juice from preparation of beverages. *J. Food Sci. Technol.* 30: 208(1993)

Joshi, V.K and V.P, Bhutani. 1990. Effect of preservation on the fermentation behavior and quality of plum wine *XXIII Int. Hort. Congress*, Italy, Absrt 2585.

Joshi, V.K, S.K. Chauhan, and B.B. Lal.1991. Extraction of juice from peaches, plums and apricots by pectinolytic treatment. *J. Food Sci. Techonol.* 28: 6.

Joshi.V.K, B.L. Attri and B.V.C. Mahajan.1991. Studies on preparation and evaluation of vermouth from plum. *J. Food Sci. Technol.*28/: 138.

Kaul, J.L., and R.L. Munjal, 1982, Fruit wrappers and skin coating for control of post-harvest decay of apple, *Indian J. Mycol. Plant Pathol.* 12: 179.

Kozlov, V.N and A.S. Destik. 1985. Butter with apple and other fruit fillers. *Tovarovedenie* 18: 29

Krishniah, J, C. Satyaprasad, T. Girdhar Singh, V. Thirupthaiah and B. Dave. 1985. Decco food grade fruit coating for control of post-harvest decay of Alubukhara fruits, *Indian J. Mycol. Plant Pathol.* 15: 274.

Landenhoven. M.L, M. kruger, F. Gouws and M. Faber.1991. *MPC* food composition tables, South African Medicial Research Council, Parow, South Africa.

Mehta, K and K.K. Jindal. 1986. Effect of some nutrient sprays on fruit maturity and quality of Japaneseplum (*Prunus salicina* L.) cv. Santa Rosa. Advances in Rsearch on Temperate Fruits (T.R. Chadha, ed). DR. Y.S. Parmar University of Horticulture and Forestry, Solan. India, p.203.

Moy, J.H, K.Y. Keneshiro, A.T. Ohta and N. Nagai. 1983. Radiation disinfestation of California stone fruits infested by medfly effectiveness and fruit quality. *J. Food Sci.* 48: 928.

Mudhar, G.S and G.S. Bhatia.1983. Steeping preservation of fruits. *J. Food Sci. Technol.* 20(2): 77.

Pangelova, S.I and I. Vittanova. 1987. Seasonal change in the weight and in dry matter, organic acid and tannin contents of Plum fruits, *Fiziogiya na Rasteniyata* 13: 61.

Raynal, J and M. Moutounet. 1989. Intervention of phenolic compound in plum technology I, *Mechanism of Anthocyanin Degradation. J. Agri, Food Chem.* 37: 1051.

Rein, S.J, K.Otto and H. Jacob. 1988. On the baking quality of important plum and damson varieties. *Erwerbsobstbau* 30(7): 180.

Salunkhe, D.K and B.B. Desai.1984. *Post-harvest Biotechnology of fruits*, Vol. 1, CRC Press, Boca Raton, FL, p.157.

Salunkhe, D.K, H.R. Bolin and N.R Reddy. 1991. *Storage, processing and Nutritional Quality of Fruits and Vegetables*, 2nd ed., CRC press, Boca Raton, FL, p. 243.

Sharma, K.R and N.S. Thakur. 1992. Evaluation of wooden boxes fabricated from lesser valued farm tree species for packaging and transportation of plum, *J. Food Sci. Technol. 29: 235.*

Stosic, D, M. Gorunovic and B. Popovic. 1987. Preliminary toxicological of the kernel and oil of several species of the genus *Prunes, Plantees medicinales et Phytotherapie* 21(1): 8.

Techasena, O, A. M. Lembort and J. J. Bimbenet. 1991. Simulation of plum drying in deep bed. *Drying technol.*9 (4): 947.

Tinker, L.F, B.O. Schneeman, P.A.Lavis, D.D. Gallacher and C.R. Wagganer.1991. Consumption of prunes as a source of dietary fiber in men with mild super cholesterolemia. *Am.J. Clin. Nutr.*53: 1259.

Tonini, G, S. Brigati and D. Caccioni. 1992. Air precooling of Plums: *Technological and Biological Aspects*, 52(6): 71,

Tormann, H and H.J.V. Zyl. 1982. Maturity stands for export plums. *Deciduous of fruit grower* 32(1): 22.

Truter, A.B and J. C. Combrink. 1992. Controlled atmosphere storage of peaches nectarines and plums. *J.S. Afr. Soc. Hort. Sci.* 2(1): 10.

Voldirch, M and V. Kyzlink. 1992. Cyanogenesis canned stone fruit. *J. Food Sci.*57: 161.

Vyas, K.K and V. K. Joshi. 1988. Deacidifaction activity of scizosaccharomyces pome in plum musts. *J. Food Sci. Technol.* 25: 306.

Vyas, K.K and V. K.Joshi. 1982. Plum wine making: Standardization of a methodology. *Indian Food Packers*.36 (5): 80(1982).

Vyas, K.K, R.C Sharma and V.K. Joshi. 1989. Application of osmotic technique in plum wine fermentation –Effect on physo chemical and sensory quality. J. Food Sci. Tech, 26: 126.

Wani, M. A and S.P.S. Saini. 1990. Processing of plum. *J. Food Sci. Technol*: 27; 304.

Weinert, I. A. G, J. Solms and F. Escher.1990. Quality of canned plums with varying degree of ripeness chemicals characterization of color and changes. L*ebensum. Wisse-u—Techno 23: 445.*

Weinert. I. A. G, J. Solms and F. Escher. 1990. Diffusion of anthocyanins during processing and storage of canned plums. *Lebensum.wisse-u-Techno*.23: 396.

Weitz, D.A, E. A. Laque and R.D. Piacentini.1991. Solar drying simulation of prunes arrangements in thin layers, *Drying Techol.* 9(4): 947

Wills, R.B.H, F.M. Scriven and H. Greenfield. 1983. Nutrient composition of stone fruit cultivars. *J. Sci. Food Agr*, 34: 1383.

Wills, R.B.M, T.H. Lee, D. Grahm, W.B., McGlasson, and E.G. Hall. 1982. *Post-harvest,* AVI Westport, CT, p.24.

Woodroof, J.G and B.S Luh. 1986.Commercial fruit processing. AVI West port, ct.

Chapter 12

Walnut

Introduction

Persian (English) walnut (*Juglans regia*) is a member of the family Juglandaceae, and is by far the most important commercial walnut. Etymologically, the word walnut derives from the Germanic *wal-* and Old English *wealhhnutu*, literally "foreign nut", *wealth* meaning "foreign" (*wealh* is akin to the terms Welsh and Vlach), in Hindi it is known as Akhrot. The Romans called walnuts *Juglans regia*, "Jupiter's royal acorn." The genus *Juglans* contains three other groups (the Butternut, East Asian species and Black Walnut species) that produce edible kernels. English walnut has its origins in Eastern Europe, Asia Minor, and points eastward to the Himalayan Mountains. However, there are native *Juglans* in North, Central, and South America, Europe and Asia. Although native to Europe, it probably was not utilized there until improved forms were imported from Persia. The species came to the new world with English settlers and to California via missionaries. Walnuts were first

introduced to California in the 1700's by the Spanish padres. It wasn't until the 1860's, however, that the industry was launched with the pioneering work of nurseryman Joseph Sexton, known as the father of the commercial walnut industry. The wood of black, English, and other walnuts are close-grained, dark-brown colored, and very strong. Walnut wood is used to manufacture lumber and veneers for fine furniture and cabinets, and it is sometimes carved into components for artisanal furniture.

The best-known edible fruits harvested from species in the walnut family are those of the European or English walnut. Nuts are borne singly or in clusters of 2-3 on shoot tips. A green, fleshy shuck surrounds the nut, which splits irregularly at maturity. The shell is rough, wrinkled or furrowed, and thin. Nuts are ovoid to round, ½ -2" in diameter, containing two kernels separated by a thin, papery central plate extending from the inner layer of the shell. The most important use of the fruits of the walnut is directly for eating. However, fresh walnut kernel contain about 50 per cent of their weight as oil, which can be expressed from these fruits and used as edible oil or to manufacture soap, perfume, cosmetics, or paint. Walnuts have sometimes been used as minor folk medicines. The inner bark of the walnut can be used as a laxative, while the fruit rind has been used to treat intestinal parasites, ulcers and syphilis. An infusion of boiled leaves has been used to get rid of bedbugs. The shells are used for a variety of cleaning and polishing operations. The green hulls or shucks are used to make tinctures helpful with intestinal parasites and fungus conditions.

Walnut is a high energy food, rich in oil including omega-3 fatty acids, vitamins and minerals, and valued as healthy snack food and bakery ingredients. Its alpha linolenic acid has substantial cardio protective effects. It contains 'melatonin' an antioxidant produced by pineal gland and responsible for inducting and regulating sleeps. It also reduces the incidence of cancer and, delays aging. Western Himalayan region of India produces high quality walnuts. Walnuts are rich in proteins, fats and minerals and are a concentrated source of energy. These contain a good amount of vitamin B group and are the richest in vitamin B_6 among all the nuts. Immature fruits of walnut can be utilised for preparing various products like pickles, chutneys, fresh juices and syrups. The fruit has excellent flavour and is mainly consumed as a dry fruit. Commercially, it is used for preparation of bakery products, chocolates, ice-creams, ornaments, oils, and confectionary and salad products.

The major walnut growing countries are China, USA, Iran, Ukrain, Turkey and Mexico. World production of walnut is 3.28 Million MT (FAO 2012). Walnuts are produced commercially in 48 countries on 834.9 thousand hectares. Production has increased 48 per cent since 1992, largely in response to a 40 per cent increase in acreage. Yields have increased slightly in the last decade, and average about 2000 lbs/acre. Chinese production has increased over 3-fold since 1980, and China overtook the USA as the world leader in walnut production in 1994.

In India, walnuts are grown in Jammu and Kashmir, Arunachal Pradesh, Himachal Pradesh and Uttarakhand. Total area under walnut cultivation in India is 149.5 thousand hectares with an annual production of 284.4 thousand M T in 2011-12 (APEDA 2012). The country exported 7249.00 MT to world worth Rs.

199.80 crores in 2012-13. Major exporting countries are UK, Egypt, Arab Republic, US, Germany, USA, Australia and Taiwan. Walnut is an important crop grown in Jammu and Kashmir and contributes around 98 per cent of the country's output. The state produces about 86,263 tonnes from an area of 61,723 hectares.

Persian walnuts are best adapted to Mediterranean climates, with dry, hot summers and mild winters. Cold hardiness is a major limiting factor Persian walnut. Walnut is highly sensitive to the extremes of winter and summer temperatures as well as to its duration. It requires a climate which is free from frost in spring and from extreme heat in summer. A temperature of even 2 or 3 degree below freezing point (0°C) kills leaves, shoots and flowers and thus resulting into a crop failure. High temperature of about 38°C occurring in the early season causes sun-burning of hulls and shrivelling of kernels resulting sometimes in empty nuts and if it occurs later in the season, the kernels may be partially shrivelled, dark in colour or they may stuck to the shell. In warmer areas, walnuts do not receive sufficient chilling causing them to leaf out and bloom late. An annual rainfall of about 80 cm is considered sufficient for the cultivation of walnut which can be supplemented drier regions with irrigations, particularly for young plants. Chilling requirement is 400 to 1600 hr. With the need for late leafing characteristics to avoid walnut blight, cultivars with higher chill requirements have been produced. A well drained, deep sill loam soil containing an abundance of organic matter is the most suitable for walnut cultivation. Soil analysis of top soil and sub-soil is also essential as the walnut requires a fertile and well drained top soil and the sub-soil should be free from solid rock, impervious clay or gravel layers which restrict root growth. A soil depth of 2-3 metres gives the best results, because walnut roots penetrate up to a depth of about three metres. It requires a soil pH of neutral range *i.e.* 6 to 7.

Status of Cultivation

The worldwide production of walnut seeds has been increasing rapidly in recent years, with the largest increase coming from Asia. The world produced a total of 3.28 million metric tonnes of walnut seeds in 2012; China was the world's largest producer of walnut seeds, with a total harvest of 1.70 million metric tonnes. The other major producers of walnut seeds are Iran, United States, Turkey, Ukraine, Mexico, Romania, India, France and Chile.

The average worldwide walnut seed yield was about 3 metric tonnes per hectare, in 2010. Among the major producers, eastern European countries have the highest yield. According to the FAO, the most productive walnut seed farms in 2010 were in Romania, with yields above 23 metric tonnes per hectare.

The United States is the world's largest exporter of walnut seeds. The Sacramento and San Joaquin valleys of California produce 99 percent of the nation's commercial English walnut seeds.

Production of Walnut in World

Walnuts are produced commercially in 48 countries on 1.6 million acres. Yields average about 2000 lbs/acre. (FAO Stat. 2004.).

Table 12.1: Top 10 Walnut Producing Countries – 2012

Rank	*Country*	*Production (Tonnes)*
1	China	1,700,000
2	Iran	450,000
3	United States	425,820
4	Turkey	194,298
5	Mexico	110,605
6	Ukraine	96,900
7	India	40,000
8	Chile	38,000
9	France	36,425
10	Romania	30,546
—	World	3,282,398

Source: FAO, 2012.

Indian Scenario

In India, walnuts are grown in Jammu and Kashmir, Arunachal Pradesh, Himachal Pradesh and Uttarakhand. Jammu and Kashmir contributes around 98 per cent of the country's output. The total area under walnut in India is 149.5 thousand hectares with an annual production of 40,000 tonnes during 2011-12. The country has exported 7249 MT of walnuts mostly to France, Germany, Spain, Portugal, Austria, United Kingdom, Kuwait, Bahrain, Dubai and Saudi Arabia for the worth of Rs.199.80 crores during the year 2012-13. The Jammu and Kashmir State of India alone accounts for >98 per cent of India's total production with an average productivity of 2.69 metric tonnes/ha from an area of 83613.80 ha and production of 224595.85 metric tonnes. Besides, domestic market of worth of US $ 140-200 million for kernel and in shelled walnuts is also fulfilled by the state (Per. Com. with J&K Walnut Exporters Association in 2012). The state produces about 86,263 tonnes from an area of 61,723 hectares. Walnut cultivation in Jammu and Kashmir is mostly in Badrawah, Poonch,Kupwara, Baramulla, Bandipora,Gandarbal Budgam, Srinagar, Anantnag and other hilly areas.

Walnut is also grown in Himachal Pradesh and Uttarakhand to a limited extent. In Himachal Pradesh, it is grown in a couple of districts. Around 50 hectares are under its cultivation in the Mandi district. Chamba district has just over 55 hectares under it. Also, there is some area in Shimla district under this crop. To promote walnut cultivation in Himachal Pradesh, a walnut cultivation development station has been established at Nohra in Sirmour district.

Nutritional Composition, Uses and Medicinal Value

Walnuts are rich in proteins, fats and minerals and are a concentrated source of energy. These contain a good amount of vitamin B group and are the richest in vitamin B6 among all the nuts. Immature fruits of walnut can be utilised for preparing various products like pickles, chutneys, fresh juices and syrups. The fruit has excellent flavour and is mainly consumed as a dry fruit. Commercially, it is used for preparation of bakery products, chocolates, ice-creams, ornaments, oils, and confectionary and salad products. Shells are used in glue and plastics and for making solutions for cleaning and polishing surfaces.

Walnuts have a number of medicinal and non-food uses, as well as some toxic properties. Juglone is excreted by the roots of walnuts, and is toxic to many other plants (*i.e.*, it is allelopathic). Even dead roots can release juglone for years after the tree is gone. Shells are ground and used as anti-skid agents for tires, blasting grit, activated carbon, and sometimes as an adulterant of spices. The husk yields valuable oil and a yellow dye when pressed; the oil is used in soaps, paints, and dyes. The oil from walnut kernels is high in unsaturated fats and can be used in cooking. The wood is heavy and fine-grained, used mostly for furniture and gun stocks.

Juglone from walnut fruit and bark acts against dermatomycosis (skin fungi), being first used for this purpose by Greeks and Romans. Walnut tincture, an extract made with grain alcohol, is derived from fresh green hulls of the black walnut tree. It is said to kill adult and developmental stages of at least 100 parasites. It is touted as a great antiseptic that is high in iodine, and excellent for treatment of any kind of fungus condition. Also a good vermifuge for pin worm, ringworm, and other parasites, and even removes warts and treats psoriasis. The tincture is generally used in conjunction with wormwood and cloves as part of a complete parasite program.

Ellagic acid is found in leaves and fruits; it is being studied for use as a cancer therapy drug, in addition to having many other biological effects.

Nutritional Composition

Walnuts are a nutrient-dense food: 100 grams of walnuts contain 15.2 grams of protein, 65.2 grams of fat, and 6.7 grams of dietary fibre. The protein in walnuts provides many essential amino acids.

Table 12.2: Nutritional Composition of English Walnut per 100g

Nutrient	
Calories (kcal)	654
Calories from Fat (kcal)	586.89
Calories from Saturated Fat (kcal)	55.13
Protein (g)	15.23
Carbohydrates (g)	13.71
Dietary Fiber (g)	6.7
Soluble Fiber (g)	2.21
Total Sugars (g)	2.61
Monosaccharides (g)	0.18
Disaccharides (g)	2.43
Other Carbs (g)	4.4
Fat (g)	65.21

Nutrient	
Saturated Fat (g)	6.13
Monounsaturated Fat (g)	8.93
Polyunsaturated Fat (g)	47.17
Vitamins	
Vitamin A - (IU)	20
Vitamin A - (RE)	2
Vitamin A - (RAE)	1
Carotenoid (RE)	2
Beta-Carotene (mcg)	12
Vitamin B1 (mg)	0.34
Vitamin B2 (mg)	0.15
Vitamin B3 (mg)	0.12
Vitamin B3 - Niacin Equiv (mg)	3.86
Vitamin B6 (mg)	0.54
Choline (mg)	39.2
Biotin (mcg)	19
Vitamin C (mg)	1.3
Vitamin E - Alpha-Toco (mg)	0.7
Folate (mcg)	98
Folate, DFE (mcg)	98
Vitamin K (mcg)	2.7
Pantothenic Acid (mg)	0.57
Minerals	
Calcium (mg)	98
Copper (mg)	1.59
Iron (mg)	2.91
Magnesium (mg)	158
Manganese (mg)	3.41
Molybdenum (mcg)	29.5
Phosphorus (mg)	346
Potassium (mg)	441
Selenium (mcg)	4.9
Sodium (mg)	2
Zinc (mg)	3.09
Other fats	
Omega 3 Fatty Acid (g) - Alpha Linolenic Acid (ALA)	9.08
Omega 6 Fatty Acid (g)	38.09

Source: USDA. United States Standards for Grades of Shelled Walnuts (*Juglans regia*). (Source: USDA National Nutrient database) 25, Nov. 2012

While English walnut is the predominant commercially distributed nut because of the ease of its processing and its nutrient density. Unlike most nuts that are high in monounsaturated fatty acids, walnut oil is composed largely of polyunsaturated fatty acids (47.2 grams), particularly alpha-linolenic acid (18:3n - 3; 9.1 gram) and linoleic acid (18:2n - 6; 38.1 gram). Seven phenolic compounds (ferulic acid, vanillic acid, coumaric acid, syringic acid, myricetin, juglone and regiolone) have been identified in walnut husks. Walnuts also contain the ellagitannin pedunculagin.

Medicinal Uses/Health Benefits

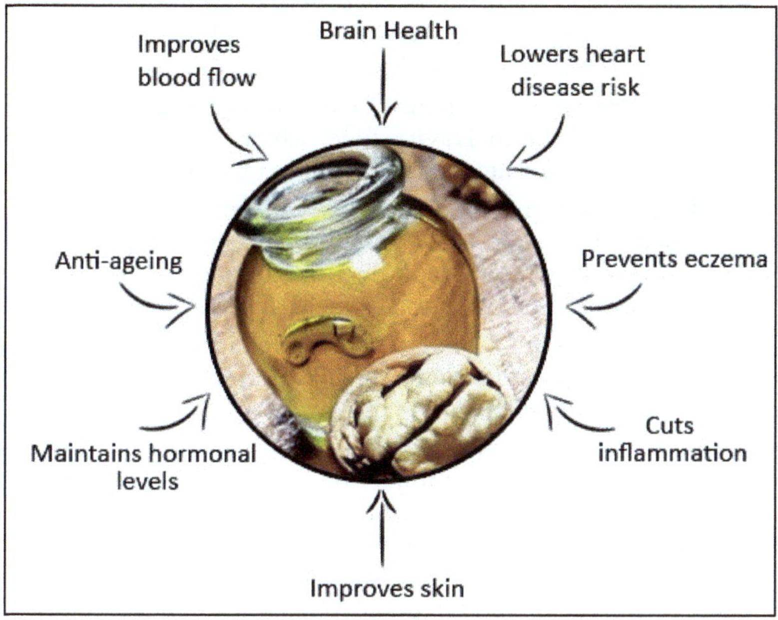

Walnut and Health

a) ***Smart For the Heart:*** High values of omega 3 fatty acidsin walnut are known to reduce the bad cholesterol in the body and encourage the production of Good cholesterol. Eating a handful of walnuts tastes great, and is a heart-healthy addition to your diet.

b) ***Walnuts and Diabetes:*** Higher walnut consumption is associated with a significantly lower risk of type 2 diabetes in women which may be attributed to the favorable unique fatty acid and nutritional profile of walnuts providing the plant based omega-3 fatty acid alpha-linolenic acid (ALA), along with numerous bioactive compounds with beneficial effects on Coronary Artery Diseases (CAD) risk, including dietary fiber, folate, and antioxidants.

c) ***Walnuts and Antioxidants:*** According to an evidence-based review, antioxidants help to protect from certain chronic diseases of aging, including cardiovascular, neurological and anticarcinogenic ailments due to their ability to control free radicals — known to negatively influence healthy aging. Walnuts contain several antioxidants including selenium, melatonin, gamma-tocopherol (a form of vitamin E) and several polyphenols. Walnuts ranks second only to blackberries in terms of antioxidant content.

d) ***Beneficial Nutrients:*** In addition to antioxidants and essential ALA/ omega-3 fatty acid, an ounce of walnuts provides a convenient source of protein and fiber.

e) ***For Immunity:*** Walnuts are really beneficial for immunity. Walnuts have high amounts of anti-oxidants in them that help in keeping your immune system healthy and prevent the onset of diseases.

f) ***For Brain Health:*** The Omega 3 Fatty acids present in the Walnuts are also good for the brain. Having food rich in Omega 3 fatty acids keep the nervous system working smoothly and improves your memory.

g) ***For Breast Cancer:*** American Association for Cancer Research has published in its 2009 research volume that consuming a few walnuts each day can help keep the risk of the most common breast cancer away. So having a few of them will reduce the chances of Breast Cancer.

h) ***For Inflammatory Diseases:*** Those with inflammatory diseases like Asthma, Arthritis and Eczema can benefit from the Walnuts as they have high amounts of Fatty Acid in them.

i) ***For Bone Health:*** Walnuts contain an essential fatty acid in them called alpha linolenic acid. This alpha linolenic acid and its compounds have been associated with bones that have grown stronger and healthier. Also as you intake omega 3 fatty acids through walnuts, inflammation decreases and this is related to keeping bones stronger for longer.

j) ***For Better Sleep and Stress:*** Walnuts have melatonin and this compound is said to be associated with better sleep patterns. Also the omega 3 fatty acids keep blood pressure low helping to relive stress.

k) ***For Pregnancy:*** Those who are expecting can gain a lot from having walnut daily. Walnut contains healthy Vitamin B Complex groups present in them like folates, riboflavin, hiamine and more. These are necessary for a pregnant woman and the fetus.

l) ***For Constipation and Digestive System:*** Walnuts are rich in fiber and they are a great way to keep your digestive system functioning right. All humans require fiber on a daily basis to keep their bowels functioning correctly. Having Walnuts daily can help with the digestive problems and keep your bowels functioning properly.

m) ***Walnuts for Healthy and Beautiful Skin***

i. Delay skin ageing with these nuts: Walnuts for skin are good as it is packed with B-vitamins that make them a great treat for your skin. Vitamin B is an excellent stress and mood manager. Lower stress levels mean better skin. Increased stress levels can result in earlier onset of wrinkles, thus inducing faster ageing. The presence of the B-vitamins, together with Vitamin E, a natural antioxidant, helps in fighting the free radicals induced due to stress. This further delays the ageing process.

ii. Walnuts for a moisturized skin: Those with dry skin should try applying warm walnut oil regularly. Walnut oil helps to keep the skin moisturized. It nourishes the skin from within, enhancing the growth of healthier and radiant cells.

iii. Reduces chances of dark circles: Regular application of warm walnut oil is known to lighten the dark circles. It is a wonderful soothing agent. The oil extracted from walnuts is known to ease puffiness and relax your eyes, retaining its glow and color.

n) Eat walnuts for healthier hair: These days, with factors like pollution, hectic lifestyle and poor eating habits, our hair is more dull and damaged than ever before. One can now get healthy, glowing hair by incorporating walnuts in your daily hair care regimen. Walnuts, whether consumed or applied externally as oil, are known to keep your hair healthy and glowing.

i. Longer, stronger hair: Walnuts are good sources of potassium, Omega 3, Omega 6, and Omega 9 fatty acids. All these ingredients strengthen the hair follicles.

ii. Natural anti-dandruff agent: Walnut oil is widely used in producing hair oils due to its rich moisturizing properties. Hence, it is widely suggested as a natural anti-dandruff agent.

iii. Healthier scalp: Regular application of walnut oil keeps the scalp moisturized and hydrated, thus keeping medical conditions such as dermatitis at bay. The anti-fungal properties of walnut oil are beneficial in keeping infections triggered by ringworm at bay. This further ensures a healthier and cleaner scalp. A healthy scalp will lead to healthy hair.

iv. Highlight your hair color naturally with walnuts: The husk of walnuts is a natural coloring agent that could accentuate the natural highlights of your hair. Walnut oil contains a rich amount of assorted proteins. These help in improving the color of the hair, while giving it a healthier glow.

Major Commercial Varieties

Most of the production in India comes from the seedling trees scattered in the temperate region of the northwestern Himalayas from 1200 m to about 3500 m above sea level. Many of these trees are as good as any improved cultivars. Therefore, known cultivars grown in Himalayan region of India have resulted from local selections by the farmers based on their preferences. However, a good amount of exotic germplasm has been maintained in the field gene bank at National Bureau of Plant Genetic Resources, Regional Station, Shimla in Himachal Pradesh. The promising selections from this exotic collection were 'Lake English', 'Blackmore',

'Tutle-16' and 'Izvor'. Recently, few indigenous selections were also made as 'Hamdan' and 'Sulaiman' from Jammu and Kashmir; 'Govind' and 'Partap' from Himachal Pradesh and 'Chakrata seedling 13' from Uttaranchal.

Table 12.3: Varieties Grown in Different States

State	*Varieties*
J&K	Lake English, Drainovsky and Opex Caulchry
Himachal Pradesh	Gobind, Eureka, Placentia, Wilson, Kashmir Budded
Uttarakhand	Chakrata Selections

However, a large number of cultivars have been evolved in the USA which is the main walnut exporting country of the world. Description of some important varieties is given here.

1. **Hartley:** It is one of the most popular commercial cultivars of California. It is a selection from a seedling. The nuts are large with broad flat base and pointed tip. The shell is light coloured, thin and seals well. The variety is tolerant to codling moth and blight disease.
2. **Payne:** It is tile second leading cultivar of California which originated as a seedling. Nut is medium to small in size with a good seal. Trees are moderately vigorous, round in shape and require heavy pruning to maintain vigour.
3. **Franquette:** It is an old and leading cultivar of France. Nut is small, good shell seal and kernel is light. in colour. Tree is large and upright in nature and is known for its late bud break thus escaping injury from frost during late spring.
4. **Serr:** It was evolved from a cross of Payne X PI 159568. It is heavy yielding and well adapted to warm conditions. The tree is very vigorous and gives poor yield on very fertile soils. The kernel is light in colour and good in quality. It is susceptible to codling moth and blight disease.
5. **Ashley:** It is a high yielding, early bearing cultivar which requires heavy pruning to keep the tree vigorous. Kernel is of high quality, good in flavour and light tan in colour. This variety is unsuitable for high rainfall areas due to blight problem.
6. **Sunland:** It is a mid-season variety obtained from a cross of Lompoc X PI 159568 in California. The nut is very large, long, oval with good seal and kernel colour is good. It is susceptible to blight-and late spring fros
7. **Chico:** It is evolved from a cross of Sharkey and Marchetti. The variety is fruitful on lateral buds and the nut is small but excellent in quality with light coloured kernel. Heavy pruning is required to maintain nut size.
8. **Vina:** It was evolved from a cross of Franquette and Payne. The kernels are light in colour, high in quality and shelling percentage is 48. It requires good pruning to maintain its vigour and nut size.

9. **Howard:** Howard is a lateral fruiting and high yielding cultivar. Nuts are large, round and smooth with a good seal. Kernel is light in colour and shelling percentage is 50. Tree is small, semi upright and tolerant to blight.
10. **Pedro:** It is an early bearing variety due to which it has become popular cultivar in many walnut growing regions. The nuts are large shell heavy and seal fair. It sheds its pollen over a long period becoming a good pollinizer.
11. **Chandler:** It is a very promising cultivar which has been evolved from a cross of Pedro X VC 56 -224. The nut is large, oval, smooth with a good seal and excellent colour. Kernel percentage is 49. Tree is moderately vigorous and semi-upright.
12. **Tehama:** It is a heavy yielding midseason and lateral fruiting cultivar evolved from a cross of Waterloo and Payne. Nut size is large but shell seal is not so good because sometimes it cracks at the suture exposing the kernel to damage. Kernel colour is good witl1 50 per cent shelling. Tree is large and vigorous and mainly serving as a pollinizer.

Ten varieties of walnut are being developed at Central Institute of Temperate Horticulture, Srinagar, J&K, India and released recently (Anon. 2010). The description of these is as:

CITH Walnut-1: High yielding variety with largest nuts (27.98g) kernel size (14.79g), good kernel recovery (52.6 per cent), light shell kernel color. The nuts are long trapezoidal in shape with easy to remove kernel halves suitable for export market.

CITH Walnut-1

CITH Walnut-2: Nuts are medium large, ovate, medium shell texture, medium shell color, strong shell seal, intermediate shell strength, complete shell integrity, plumpy, easy to remove kernel halves and light kernel colour. It gives 16.51 g nut weight and 6.61 g kernel weight.

CITH Walnut-3: Nuts are large, round, medium shell texture, medium shell colour, strong shell seal, strong shell strength, complete shell integrity, well filled kernel, plumpy, difficult to remove kernel halves and light kernel colour. It gives nut weight 16.75 g and kernel weight of 7.69 g.

CITH Walnut-2

CITH Walnut-3

CITH Walnut-4: Nuts are medium large, long trapezoide, rough shell texture, light shell colour, strong shell seal, intermediate shell strength, complete shell integrity, thin, well filled kernel, moderately plumy, very easy to remove kernel halves and light kernel colour. It gives nut weight of 14.24 and kernel weight of 6.92 g.

CITH Walnut-5: High yielder, having extra light kernel color, suitable for export market, bigger nut (19g) and kernel (9.5g) size, good kernel recovery (48.9 per cent), light shell color, ovate in shape, moderately easy remove the full kernel halves

CITH Walnut-6: Nuts are large broad ovate in shape, shell colour light, intermediate shell seal, Intermediate shell strength, well filled kernel, moderate

CITH Walnut-4

CITH Walnut-5

plumy and easy to remove kernel halves. It gives in shell nut weight (25.6 g), kernel weight (12.4 g) with kernel recovery (50.8 per cent).

CITH Walnut-7: Nuts are large in size, Elliptic in shape, rough shell texture, medium coloured shell, strong shell seal, intermediate shell strength and well filled kernel, plumpy, moderately easy for removal of kernel halves. It gives inshell nut weight of 24.7 g, kernel weight of 12.26 g with kernel recovery of 49.60 per cent.

CITH Walnut-8: Nut are having light kernel colour, nut weight (21.96g), and kernel weight (11.11g), good kernel recovery (54 per cent), light shell colour, long

CITH Walnut-6

CITH Walnut-7

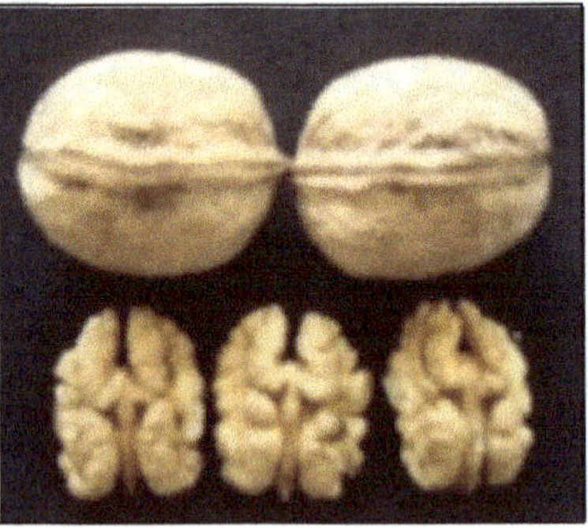

CITH Walnut-8

trapezoidal in shape, very easy to remove kernel halves, rough shell texture, strong shell seal and strong shell strength

CITH Walnut-9: Nuts are large in size, round in shape, light in colour, strong shell seal, intermediate shell strength and well filled kernel, plumy, moderate to remove the kernel halves. It gives in shell nut weight 21.23 g and kernel weight (10.65 g) with kernel recovery of 50.9 per cent.

CITH Walnut-10: Heavy bearing, Nuts are medium, long trapezoidal in shape, smooth shell texture, medium coloured shell, intermediate shell seal, intermediate

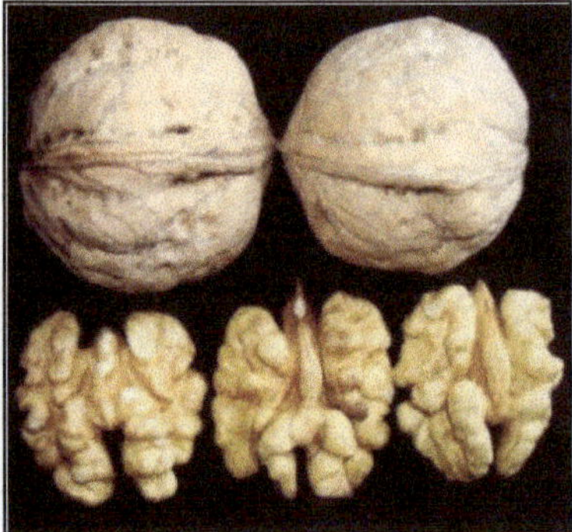

CITH Walnut-9

CITH Walnut-10

shell strength, well filled kernel, plumy and easy to remove kernel halves. It gives in shell nut weight (19.95 g) and kernel weight (11.08 g) with kernel recovery (55.5 per cent).

Major Constraints in Cultivation

Compulsions that account for present low productivity and inferior quality have been observed to be of production, protection and post-harvest nature including processing. The following bottlenecks have been observed as limiting factor for production of quality produce and improper marketing.

Lack of Suitable Methods of Propagation and Availability of Adequate Vegetatively Propagated Plant Material

The availability of grafted plant material is a major limiting factor in the cultivation of walnut. This is because of low and variable success obtained with different methods of vegetative propagation in various studies as under: It is reported 50 per cent success with side tongue grafting. However, the success as high as 85.50 per cent was obtained with veneer grafting when attempted in the last week of July. In dormant grafting under controlled conditions, it was 71.7 and 66.7 per cent with cleft and tongue grafting, respectively. In a study on the effect of plant growth regulators on stooling in walnuts reported that better response was obtained with 5000 ppm IAA + 10000 ppm IBA + 5000 ppm NAA and 10000 ppm NAA. It is reported that stool shoots treated with IBA+IAA+NAA each at 7500 ppm in Lanoline at incisions above the debarked portion, gave 80 per cent rooting compared with 13.3 per cent in the control. However, Samuel and Sharma (1991) obtained no success with hardwood cuttings, air layering and mound layering even with the application of growth regulators.

Lack of Clonal Rootstocks

The clonal rootstock are not available and generally seedlings of the same or related species are used. These seedlings exhibit lot of variation among the plants. Due to hard shell, the seed germination is also a problem in these nuts. The studies have indicated that nuts having 0.17 cm thick shell gave 74.50 per cent germination and with 0.35 cm thick shell gave only 33.33 per cent germination. In the earlier studies it is also indicated that good germination largely depends upon good cracking quality of the shell of walnut. It is reported that walnut seeds of medium size (3-4cm long) gave a germination of 90 per cent as compared with only 71 per cent for smaller seeds and 85 per cent for larger ones. We obtained better germination and growth of seedlings with 125 to 250 ppm GA. A large number of factors such as nut type, size, position of seed, planting depth, soil moisture and temperature and seed treatment influence the germination of seeds. The planting of large nuts at 7 cm depth in soil in vertical position gives better germination and seedling growth.

Lack of Standard Cultivars

There is lack of suitable cultivars of walnut in India. The existing plantation is of seedling origin. The produce obtained from these non-descript trees is not uniform and as such production cannot be geared to various market outlets where uniform crop of better quality is in demand. It has been seen that in general, nuts produced from present plantation belong predominately to thick shell group where shelling percentage is low. Hence, there is a need to identify the suitable trees from native seedling populations or introduce cultivars from other countries which are suitable for different climatic conditions prevailing in India.

Problem of Re-Establishment of Nursery Plant in the Orchard

There is high rate of mortality when nursery plants are planted in the orchard. This may be due to the tap root system which is disturbed while uprooting the plant from the nursery. This needs extra care after planting to get better success in the field. Another method of raising seedlings in polythene bags have also been suggested. These seedlings can be grafted while in the bags and later these can be planted at a desired place without any disturbance to the root system.

Specific Climatic Requirements

The climate is a limiting factor in the expansion of walnut cultivation. The high temperature of about 38°C causes sun burning of hulls and shrivelling of kernels resulting sometimes into blank nuts. Severe damage is further aggravated if the humidity is low and temperature exceeds 40°C. If low temperature persists in summer, the filling of nuts is not proper. The cool or short growing season is a limiting factor in walnut production.

Pollination Behaviour of the Species and Lack of Suitable Pollinizers

The walnut tree is monoecious and the problems in pollination and fruit set are due to dichogamy. Dichogamy varies with the age and is more pronounced in young trees. The mature trees have more catkins which are produced over a longer period and thus may better overlap the female bloom. The climatic conditions also play an important role. The catkins respond more quickly than the female blossoms during warm period. This makes dichogamy more complete in protandrous cultivars. The non-coincidence between the flowering periods of male and female flowers is the major cause for low yields in walnut and to overcome this problem top working of branches that bear flowers producing pollen at the required time is also recommended by some researchers. The pollen can be used for pollinating the same cultivar or other cultivars at the same location or different locations. Improvement in productivity can be achieved with the planting of suitable pollinizer cultivars or by hanging the catkins of male flowers on trees where the male flowers are not yet mature or have shed pollen but stigmas of female flowers are ready to receive pollen or by dusting the pollen or instead planting of homogamous cultivars.

Unfavourable Climate

Pollination in walnut is carried out by wind. It is self fertile but pollination is not satisfactory in certain varieties mostly because these varieties fail to mature their pollen at the time when the female flowers are receptive. Hot spring weather hastens the development of catkins and makes them shed their pollen quickly. It does not have great effect on the development of pistillate flowers. The pollination difficulty in the established orchards can be overcome temporarily by bringing catkins from the neighbouring plantations and hanging them in the trees. It can be permanently overcome by top working certain limbs with the desirable varieties. The new plantations of walnuts should be located near the existing bearing trees. Snowfall or very cold weather at the time of flowering causes poor setting of the crop.

Lack of Proper Package of Practices for Cultivation

There is no proper technology available with the growers for taking up walnut cultivation on scientific lines under Indian conditions. This is due to lack of well planned orchards and no systematic trials have been laid out to study the various aspects of walnut cultivation. The trees are planted mostly on marginal and neglected land and do not receive any fertilizer and irrigation. The trees are not trained and pruned which results in giant sized trees often difficult to manage and create problem for harvesting. Thus, there is a need to standardize techniques to develop low headed trees of desired size giving optimum production.

Long Juvenile Period

The walnut has long juvenility period and takes about 8 to 10 years to come into fruiting on seedling rootstocks. Due to long juvenility period, the growers are not tempted to plant walnut trees. However, use of less vigorous rootstock may help in reducing the juvenile period.

Inadequate Post-Harvest Management Practices

Improper storage, drying, grading practices at farm level are the most serious factors effecting post-harvest management of walnut. There is lack of awareness about sanitary condition to handle nuts and well planned integrated handling system to manage the nuts.

International standards on grades are not adopted. There is lack of processing, storage and orchard management facilities.

International Competition

There is increased competition from overseas suppliers. External competition from California, Mexico, China and other countries which are competing with India especially in the EU market.

Horticultural Maturity Indices

Walnuts are mature as soon as the husk cuts free from the nut, but they usually are not harvested until rains crack the husk, causing the nut to drop, usually in October. Kernels are considered mature when oil accumulation is complete. This is generally indicated by browning of the internal packing tissue. However, harvest should not begin until the husk is well split and separated from the shell. In the hottest growing regions of California, kernels may be mature 3 weeks prior to husk "maturation" (*i.e.*, dehiscence). Low temperatures and high RH, as in some growing regions or at night, advance dehiscence. Ethephon applications are used to advance the harvest and make nut maturation more uniform throughout the tree/orchard.

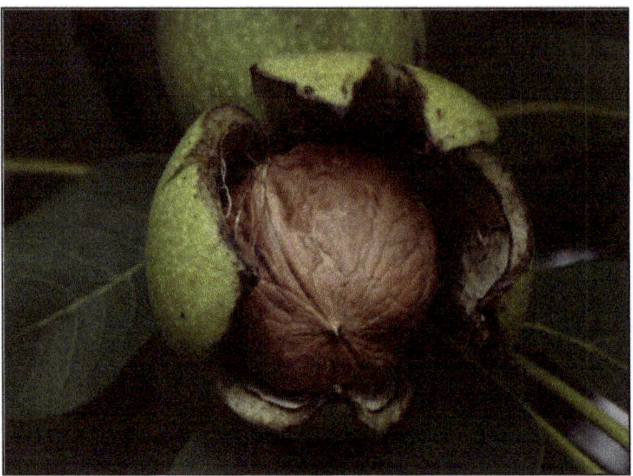

Mature Walnut

Harvesting

Walnut farmers wait until late August when the green husks of the walnut begin to peel back due to their drying in the summer heat. "Mature walnuts lose quality rapidly after they have fallen, so gather fallen nuts frequently to prevent mold, discoloration and decay.

In order to prepare for picking, the ground must be cleared of debris around the trees. Then, as with other nuts, mechanical shakers are attached to the tree and shake clear all of the walnuts which can number in the thousands per tree. Nuts are collected from the ground between the months of September and October. After collecting nuts, these are cleaned, washed and dried by spreading them on sheets or floor. Sometimes in order to improve the appearance of nuts, these are bleached with either alkali oracid solution. Nuts which fall down with their husks intact are generally second-grade. After removal of the husks, cleaning and drying, they should be stored and marketed separately to fetch a higher price. Put gloves on to prevent the hulls from staining. Step on the walnuts before picking them up to loosen their hulls. Use a pocket knife to pull the hulls off of English walnuts if you prefer to not step on the nuts. Rinse off the nuts to remove any remaining bits of hull. Lay the walnuts on a tarp in your backyard in a single layer and cover with plastic netting or a screen. Screening the nuts keeps the wildlife from pilfering your harvest. Avoid placing them in direct sunlight. Cover with a second tarp in rainy conditions. Leave the walnuts in place for three to four days or until the kernels inside become brittle. Open one or two walnuts with a nutcracker after three days to check for adequate drying. Once dry, place the English walnuts into an airtight container.

Drying of Walnuts

Delay in drying causes rapid loss in nut quality and makes walnuts susceptible to the mould. Drying of nuts stabilises the product's weight and prolongs storage life. Once the hulls have been removed from either type of walnut, the nuts must be dried. There is a certain amount of natural moisture that is within the nut as it grows. This moisture must be removed before the nutmeats are stored. Leaving the moisture invites mould to develop; mould that contains toxins that could prove to be dangerous if consumed.

The best method of drying the nuts is to place them in a single layer on a screen. An old screen door or window would be perfect for this task. Allow the nuts to remain in dry, cool area that does not receive direct sunlight for a minimum of two weeks. After the drying time has been completed, crack the shell of one of the nuts open and remove the meat. Grasp both sides of the nutmeat and press backwards; if the nutmeat cracks crisply in half, the nuts are fully dried. If the break seems too soft, allow the weeks to air dry for one more week before trying again.

Drying of Walnuts

Sorting and Grading

Walnuts are usually sorted by size and colour, which can be manually or through a conveyor belt as too dark and too green by use of an electronic eye. A blower will remove any worm infested or empty walnuts as their lack of weight lets them be blown away. Then the walnuts are sorted by hand and then stored in large cargo containers. These containers have large grates in the bottom that allow air to flow through. A hot air blower is attached to the container and the walnuts are then dried. This is to reduce the acidity that naturally occurs in a too-green walnut.

Walnuts are transported to a packing plant where they are graded based on usage, in-shell or shelled. Shelled walnuts are further graded by color. Walnut kernels are screened and separated into different sizes.

IN-SHELL WALNUTS – Following drying, sizing of the in-shell nut occurs. In-shell walnuts are sized as jumbo, large, medium, or baby according to USDA standards.

SHELLED WALNUTS – Walnuts are mechanically cracked as needed. The shelled material is air-separated and screened for size and sent to electronic laser-sorting units for kernel color and shell removal. The walnuts are certified to meet USDA grade standards and customer specifications. Product is then packed for shipment to the market place.

The Indian walnuts are categorized in to 4 categories *viz.*, paper shelled, thin shelled, medium shelled, and hard shelled. Walnuts produced in India have different sizes and shapes and are categorised into paper-shelled, thin shelled, medium-shelled and hard shelled. These are classified on the basis of their size and shell cracking rate also. The main grades are:

1. Indian special light half, 2. Indian light broken, 3. Indian light pieces, 4. Indian light crumbs, 5. Indian light my-fire, 6. Indian light amber halves, 7. Indian light amber broken, 8. Indian light amber pieces.

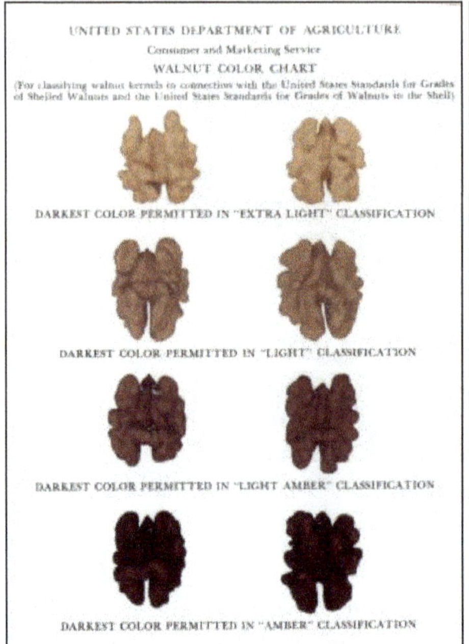

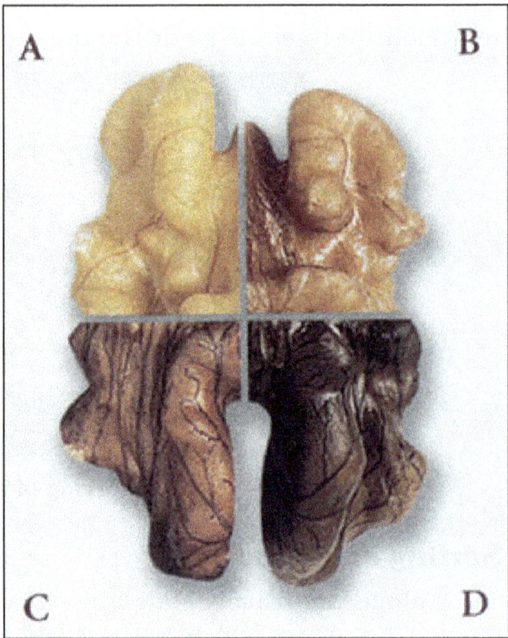

Different Grades of Walnut
(*Source*: USDA. United Standards dor grades of Shelled Walnuts).

Packaging

Walnuts are the packaged by weight in giant sacks for shipping. Generally walnuts are packaged in, net bags (1.5 - 2.5 kg capacity), poly sacks (5 and 10 k capacity g), cartons (10 kg capacity) and flat jute fabric bags (55 kg capacity). For export purpose, these are packed in double gunny bags. Walnuts are consumed in the winter season, so the problem of their shelf life is seldom felt. The quality of nut meat, however, deteriorates due to darkening and rancidity which are affected by air, moisture, heat and light.

Shelled nuts are packaged in, among other things, boxes lined with parchment paper, each containing 12.5 kg (net).Walnut kernels are ideally stored at a water content of 2 - 3 per cent in packaging which is impermeable to water vapor. Vacuum packaging excludes atmospheric oxygen, which promotes rancidity.

In-shell walnut kernels derive protection against oxidative changes from the intact shell and kernel skin. Packaging should be moisture-proof. Shelled products should be packaged in airtight, moisture-proof, opaque or foil packages to maximize shelf-life. Unroasted kernels are less likely to take up moisture than roasted kernels. FDA-approved antioxidants such as butylated hydroxyanisole (BHA) and butylated hydroxytoluene (BHT) can be applied to the kernels in vegetable oil to enhance stability. Edible coatings may be used as O_2 barriers to retard kernel rancidity.

Correct selection, design and construction of the packaging is just as important for loss-free transport as are the requirements placed on the product itself. Savings are often made on packaging in order to reduce total costs. These packages often do not fulfil the requirements placed upon them, such that they cannot withstand the mechanical, climatic, biotic and chemical stresses to which they are exposed during transport, storage and cargo handling. Simply because packaging is incapable of withstanding such stresses does not, of course, mean that a loss will inevitably occur, but the risk will in some cases be greatly increased.

Packaging of Walnuts

Products that contain walnuts, especially roasted walnuts, are subject to oxidation which can reduce shelf life and cause rancidity. To prevent rancidity and extend finished product shelf life, the product should be packaged in materials that do not allow light or air to come into contact with the product. Packages that can be resealed are ideal for snack mixes which may be opened and closed several times by the consumer. Nitrogen flushing (replacing the oxygen with nitrogen) can also extend the shelf life of unopened sealed product. Walnuts that are formulated with coatings like chocolate, starches, gums, egg whites and sugar will have a longer shelf life than plain or roasted walnuts because they are protected by the coated exterior.

Fresh taste of walnuts can be maintained by keeping them cold. Heat causes the fat in walnuts to change structure, which creates off odors and flavours. Fresh walnuts smell mildly nutty and taste sweet.

Storage

Optimum Storage Conditions

Low water content and high fat content of the kernel make it relatively metabolically stable and able to tolerate low temperatures. The primary objectives of storage are to maintain the low water content attained after preliminary drying (for enzyme activity suppression, retention of texture and reduction of microbial activity) and limit exposure to O_2 to minimize rancidity. The optimum temperature range for storage is 0 to 10°C (32 to 50°F), with the lower temperature being better. Within this temperature range, a 50 to 65 per cent RH will maintain walnuts at 4 per cent moisture. The best place to store walnuts is in refrigerator or freezer, depending on when to use them. If walnuts are going to use right away, place them in your refrigerator. If walnuts to be store for a month or longer, store in freezer.

Storage Tips

Follow these tips to maintain the high quality, freshness, flavor and nutritional value of California walnuts:

- ☆ Store in air-tight packaging.
- ☆ Store away from foods with strong odors, like cabbage and onions.
- ☆ Rotate inventory. Practice FIFO – First In, First Out.
- ☆ For storage up to six months: refrigerate at 0°C (32°F) to 5°C (41°F) at 65 percent relative humidity.
- ☆ For storage longer than six months: freeze at 18°C (0°F).
- ☆ Shelf-life can be extended by storage in < 1 per cent O_2. O_2 < 0.5 per cent (balance N_2) or CO_2 levels above 80 per cent in air can be effective in insect control.

Transportation

Walnuts and their products are transported by means of ship, railroad, truck and aircraft.

If the lower limits are set for the water content of goods, packaging and flooring and the oil content of the goods are complied with and if protection against solar radiation is ensured (risk of self-heating) ventilated containers are essentially required.

In damp weather (rain, snow), the cargo must be protected from moisture, since it may lead to mold, spoilage and self-heating as a result of increased respiratory activity.

No hooks should be used with bagged cargo, so as to prevent damage to the bags and loss of volume.

Stowage factor required for bagging of walnut is usually 2.10 m^3/t (flat jute fabric bags, 55 kg), essentially kept in cool, dry, good ventilation and segregated by fiber rope, thin fiber nets

In order to ensure safe transport, the bags/cartons must be stowed and secured in the means of transport in such a manner that they cannot slip or shift during transport. Attention must also be paid to stowage patterns which may be required as a result of special considerations, such as ventilation measures.

In the event of loading as general cargo, dunnage should be used to protect against damage:

- ☆ Floor dunnage: criss-cross dunnage and packing paper
- ☆ Side dunnage: lining with wooden dunnage and mats or jute coverings: protection from metal parts of the ship, since traces of metal promote cargo rancidity due to autoxidation.
- ☆ Top dunnage: important for voyages to cold regions (winter), since sweat may drip onto the cargo.

Risk Factors and Loss Prevention during Transportation

Temperature: Walnuts require particular temperature, humidity/moisture and ventilation conditions.

For this reason, precise details should always be obtained from the consignor as to the travel temperature to be maintained.

Humidity/Moisture: Moisture damage to both unshelled and shelled walnuts occurs under particularly unfavorable conditions, such as direct contact with rain or by dripping cargo sweat, if the product has not been protected by wooden dunnage, mats or jute coverings. Such protection is also important in order to minimize the risk of rancidity due to contact between the nuts and metal parts of the ship's hull or of the container.

Ventilation: Recommended ventilation conditions: air exchange rate: at least 10 changes/hour (airing)

Biotic activity: Walnuts are living organs in which respiration processes predominate, because their supply of new nutrients has been cut off by separation from the parent plant.

Care of the cargo during the voyage must be aimed at keeping decomposition processes at the lowest possible level, so as to keep within limits any losses in quality caused by the emission of CO_2, heat and water vapor.

Gases: In walnuts (particularly when fresh), metabolic processes continue even after harvesting. They absorb oxygen and excrete carbon dioxide (CO_2). If ventilation has been inadequate (frost) or has failed owing to a defect, life-threatening CO_2concentrations or O_2 shortages may arise. Therefore, before anybody enters the hold, it must be ventilated and a gas measurement carried out. The TLV for CO_2 concentration is 0.49 vol. per cent.

Self-heating/Spontaneous combustion: Heating damage occurs in particular in shipments from India, Chile and other tropical regions due to incorrect stacking before and during the voyage and results in rancidity of the nuts, which will subsequently at best be suitable for industrial processing.

Walnuts should not be stowed together with fibers/fibrous materials as oil-soaked fibers may promote self-heating/spontaneous combustion of the cargo. Fat decomposition in walnuts leads to the risk of self-heating and, ultimately, to a cargo fire.

Odor: Walnuts have a very slight, pleasant odor. Walnuts are sensitive to unpleasant and/or pungent odors. They should be stored and transported in clean air away from foodstuffs with an intense odor.

Contamination: Risk of contamination of other goods by fats and oils. Walnuts are sensitive to dust, dirt, fats and oils.

Mechanical influences: Walnuts are impact- and pressure-sensitive. If stack pressure is too high, the walnuts are crushed, so promoting self-heating. In addition, if the nuts are transported in jute bags, the oils accumulate in the fabric and accelerate the process.

Toxicity/Hazards to health: Walnuts may contain aflatoxin. The molds Aspergillus flavus and Aspergillus parasiticus produce the toxin aflatoxin, which may be present in the cargo. In general, this is "country damage", *i.e.* the toxin is already present in the walnuts at the time of harvesting. As a rule, aflatoxin is only found in individual walnuts. If batches intended as a human foodstuff are affected by this toxin, the product can no longer be approved for human consumption. Walnuts affected by aflatoxin cannot readily be distinguished from the other nuts in a batch. The toxin may be detected using UV light.

Shrinkage/Shortage: Weight loss of up to 2 per cent due to moisture loss is tolerated.

Loss of volume may be caused by tears in bags and by theft. The risk is increased by repeated cargo handling and extended storage.

Insect infestation/Diseases: On acceptance, attention must be paid to inspecting the cargo for insect infestation. The time of infestation may be determined with some certainty by the stage of development of the pests, with infestation generally occurring in the country of origin.

The proportion of empty, bad or pest-infested walnuts should not exceed 15 per cent.

Mites, cockroaches, sawtoothed grain beetles, flour beetles, meal moths, dried fruit moths and rats and mice may attack walnuts. Walnuts from the previous year's harvest have a particular tendency to beetle infestation. The quarantine regulations of the country of destination must be complied with and a phytosanitary certificate and fumigation certificate may have to be enclosed with the shipping documents. Information may be obtained from the phytosanitary authorities of the countries concerned.

Quality Standards

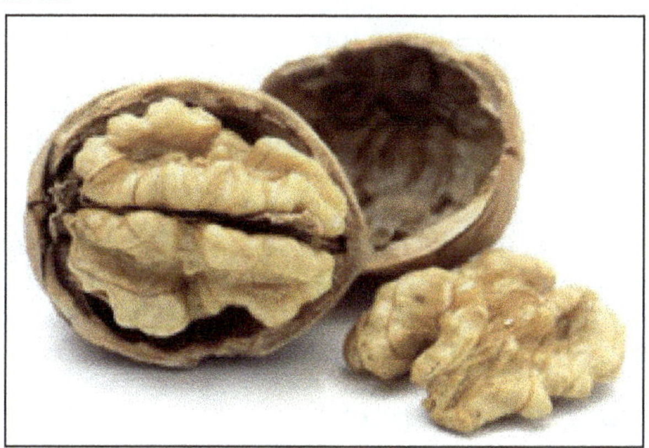

Quality Characteristics and Criteria

The primary quality criterion is a high oil content (55 to 65 per cent dry weight) that is free of off-flavors caused by oxidation of polyunsaturated fatty acids. Thus, an important criterion is maintaining kernel water content below 4 per cent. Not only does this retard the progression of events that lead to rancidity, it also prevents mold growth and maintains the kernel's crispness. If water content drops too low, however, damage to the kernel's covering can enhance O_2 penetration and rancidity.

The skin covering the kernel contains chemicals that protect fatty acids in the kernel from becoming rancid. Light-colored kernels earn a higher price because the light color indicates that the kernel still has a relatively long shelf-life. An important criterion for evaluating new varieties is yield of light-colored kernels. Retention of light-color is influenced by the integrity of the seal between the two halves of the shell, because the shell is an important barrier to O_2 entry. Shell strength and seal integrity are also important in protecting kernels from insect damage and infections from fungi that often follow insect damage. Additional characteristics of "processing quality" are ease of shelling and yield of intact kernel halves.

Quarantine Issues

Insect infestation is a potentially important problem as are the fungal infections that often accompany insect damage. Fumigation with methyl bromide or phosphine has been used for disinfestation but the use of the former is being restricted and insect resistance to phosphine has been reported. An initial disinfestations treatment of 0.4 per cent O_2 for 6 days followed by a combination of "protective" treatments that includes 10°C (50°F) storage in 5 per cent CO_2 and application of the Indian meal moth granulosis virus for control of the navel orangeworm and the Indian meal moth (*Plodia interpunctella* Hübner). These walnut storage pests are susceptible to the *Bacillus thuringiensis* insecticidal crystal protein and researchers have successfully introduced the gene for the protein into walnut embryos and demonstrated its efficacy in suppressing larval development

Special Considerations

This presentation of quality maintenance guidelines for walnuts has introduced some discussion of changes of the fundamental character of the nut that may be brought about by the use of genetic engineering. The anticipated changes in nut quality and stability could have important impacts on the potential for use of walnuts in a variety of food products. However, it is not certain that the work will be successful or that engineered products will be acceptable, particularly since an important part of the crop is destined for export. Whether or not new walnuts are introduced, it is certain that the most important aspects of post-harvest quality will be based on rapid harvest with minimal exposure to field heat, forced-air drying at relatively low temperature, and cold storage at a RH designed to maintain low nut moisture in a reduced O_2 atmosphere.

Codex Standards for Export

Marketing and Commercial Quality Control Standards

This standard applies to walnut kernels from varieties (cultivars) grown from *Juglans regia* L.

Quality

The purpose of the standard is to define the quality requirements for walnut kernels at the export control stage, after preparation and packaging.

A. Minimum Requirements 1

(i) In all classes, subject to the special provisions for each class and the tolerances allowed, walnut kernels must be:

- ☆ sufficiently dry to ensure keeping quality;
- ☆ sound;
- ☆ produce affected by rotting or deterioration such as to make it unfit for consumption is excluded;
- ☆ firm;
- ☆ sufficiently developed; shriveled kernels are to be excluded;
- ☆ clean, practically free from any visible foreign matter and from shell;
- ☆ free from insects or mites whatever their stage of development;
- ☆ free from damage caused by pests;
- ☆ free of any rancidity or oily appearance;
- ☆ free from mould;

☆ free of abnormal external moisture;

☆ free of foreign smell and/or taste.

The condition of the walnut kernels must be such as to enable them:

☆ to withstand transport and handling, and

☆ to arrive in satisfactory condition at the place of destination.

(ii) Moisture content

The walnut kernels shall have a moisture content of not greater than 5 per cent.

B. Classification

Walnut kernels are classified in the three classes as defined below according to their quality and colour.

(i) "Extra" Class

Walnut kernels in this class must be of superior quality, uniformly light-coloured with practically no dark straw and/or lemon-yellow colour and with no dark brown. They must be characteristic of the variety and/or commercial type5. They must be practically free from defects with the exception of very slight superficial defects provided that these do not affect the general appearance of the produce, the quality, the keeping quality or its presentation in the package. Scuffing is allowed on:

☆ quarters and all pieces,

☆ halves, provided it covers no more than 10 per cent of the surface area of the skin.

(ii) Class I

Walnut kernels in this class must be of good quality, of a colour not darker than light brown and/or lemon-yellow.

They must be characteristic of the variety and/or commercial type.5 Slight defects may be allowed provided that these do not affect the general appearance of the produce, the quality, the keeping quality or its presentation in the package. Scuffing is allowed on:

☆ quarters and all pieces,

☆ halves, provided it covers no more than 20 per cent of the surface area of the skin.

(iii) Class II

This class includes kernels which do not qualify for inclusion in the higher classes, but satisfy the minimum requirements specified above. Walnut kernels in this class must be of a colour not darker than dark brown. Defects may be allowed, provided that the walnut kernels retain their essential characteristics as regards general appearance, quality, keeping quality and presentation. This class also includes mixtures of kernels of different colours and designated in the marking by the words 'mixed colours.

Scuffing is not considered as a defect.

Sizing (Styles)

Walnut kernels are classified by style as follows:

- ☆ Halves: kernels separated into two more or less equal and intact parts;
- ☆ Quarters: Kernels separated lengthways into four more or less equal pieces;
- ☆ Large pieces: portions smaller than a "chipped kernel" but larger than a "broken piece";
- ☆ Broken pieces: Portions of kernels which can pass through a 8mm sizing screen but not through a 3mm sizing screen;
- ☆ Large pieces and halves: a mixture of kernels corresponding to the styles large pieces halves and of which the proportion of halves may be specified in the marking.

Tolerances

Tolerances in respect of quality, colour and type shall be allowed in each package for produce not satisfying the requirements of the class indicated.

Table 12.4: Quality and Colour Tolerances

Defects allowed	*Tolerances Allowed (Per cent by weight of kernels)*		
	Extra	*Class I*	*Class II*
(1) Kernels not satisfying the minimum requirements, which include not more than:	4	6	**8**
Rotten kernels 0.5 1 b 2 b	0.5	1[b]	2[b]
Mouldy kernels 0.5 1 b 2 b	0.5	1[b]	2[b]
Shell fragments or foreign matter	0.1	0.1	**0.1**
(2) Kernels darker in colour	8	9	**10**
(3) Scuffing (halves only)	10	10	-

B. Mineral Impurities

Not greater than 1g/kg acid insoluble ash.

C. Size Tolerances (Styles)

For all styles, a minimum percentage of kernels corresponding to the style indicated in the marking is required and a maximum percentage by weight of kernels different from the style indicated is tolerated:

Table 12.5: Size Tolerance

Style	Minimum Percentage and Tolerances Allowed (Per cent by weight of kernels)					
	Halves	Chipped Kernels	Quarters	Large Pieces	Broken Pieces	Frag-ments
Halves	85a	15b	5c		1c	1c
Quarters			85a	15b	5c	1c
Large pieces				85a	15b	1c
Broken pieces				10b	90a	1d
Large pieces and halves	20 b			65a	15b	1c

a: Minimum percentage; b: Tolerances allowed; c: Included in 15 per cent tolerance; d: Included in 10 per cent tolerance.

Presentation

A. Uniformity

The contents of each package may be uniform and contain only kernels of the same origin, crop year, quality, style and when applicable of the same variety and commercial style. Uniformity of colour is compulsory for Extra Class and Class I. However, with regard to shape, "halves" which pass through a 15 mm mesh and "chipped kernels" may be included without limitation in consignments of "large pieces". The visible part of the contents of the package must be representative of the entire contents.

B. Packaging

Walnut kernels must be packed in such a way as to protect the produce properly.

If wooden packaging is used, the produce must be separated from the bottom, sides and lid by paper or suitable protective material.

The materials used inside the package must be new, clean and of a quality such as to avoid causing any external or internal damage to the produce. The use of materials and particularly of paper or stamps bearing trade specifications is allowed provided that the printing or labelling has been done with non-toxic ink or glue.

Kernels may be packed in airtight sealed containers, in a vacuum or in an inert gas.

C. Presentation

Kernels must be presented:

In small unit packages of uniform weight intended for sale directly to the consumer.

Packaged in bulk.

Marking

Each package must bear the following particulars in letters grouped on the same side, legibly and indelibly marked and visible from the outside:

A. Identification
B. Nature of produce
C. Origin of produce
D. Commercial specifications

Definition of Defects for Walnut Kernels

Any defect adversely affecting the appearance or edibility of the kernel including:

- Staining or discolouration: abnormal colouration which covers more than one eighth of the surface of the kernel and which is of a colour in pronounced contrast with the colour of the rest of the kernel (dark blemishes or areas of discolouration);
- Embedded dirt: kernels or portions of kernels with dirt or other foreign material embedded into the flesh of the kernel.
- Crushing of more than 5 per cent of the volume of the kernel;
- Drying defect: the kernel is moist, soft or leathery.

Source: www.nutfruit.org/stand

Marketing and Export

Marketing

Walnuts are marketed as nuts or kernels. These arrive into the market from September onwards and the kernels follow two to three weeks afterwards; the peak arrival season being from November to January. Walnuts produced in Himachal Pradesh and Uttarakhand are consumed almost locally, whereas in Jammu and Kashmir the produce is brought to the assembly market in Jammu, which is the biggest market for walnuts in India. Efforts have been made to assemble quality nuts in Shahia market of Chakrata hill (Dehradun) and send to Delhi market. The nuts in the market are roughly sorted out and empty, stony, highly blighted, shrivelled, moldy and darkened nuts removed. Thin shelled nuts are packed in wooden boxes, while medium-shelled nuts are packed in gunny bags. The product then moves either to the commission agents or the exporters go downs.

In international market walnuts are typically sold as a snack item or for use as an ingredient in candies, cereals and baked goods. More than 70 percent of walnuts are sold as shelled (NASS 2013). Diamond Foods, a former cooperative that went public in 2005, is one of the largest U.S. processors of walnuts (ERS 2005).

The California walnut industry is made up of more than 4,000 walnut growers and about 80 walnut processors. Two main organizations oversee industry advertising efforts and regulation—the Walnut Marketing Board, established by a Federal Marketing Order for walnuts in 1948, and the California Walnut Commission, established through the California State Legislature in 1987. The Walnut Marketing Board is responsible for the U.S. quality control regulation, which mandates that all walnuts be inspected and certified as meeting strict USDA specifications. The Walnut Marketing Board also provides industry analysis and general domestic marketing services. The California Walnut Commission is primarily responsible for international market development.

Prior to 1993, per capita consumption of walnuts remained relatively stable at about 0.5 pounds per capita. In 1993, consumption fell to under 0.4 pound per capita, and after increasing in 1994, fell again in the two succeeding years. However, since the low in 1996 of 0.3 pound, per capita consumption of walnuts has been on the rise. In 2009 per capita consumption of all walnuts was 0.55 pound.

Industry-supported research found that walnut consumption provides health benefits because it is a good source of omega-3 fatty acids, vitamin E and other antioxidants associated with a healthy heart and a potential reduction of cancer cell growth (ERS 2005). The publicity surrounding these results has helped stimulate walnut demand. In addition, a 2005 report by the USDA's Economic Research Service (ERS) suggests that the increased consumption in 2004 was due in part to the introduction of McDonald's fruit and walnut salad, which had an impact on consumption both directly, by increasing sales, and indirectly, by reminding consumers of alternative uses of walnuts.

Export

As far as the world trade in walnuts is concerned, it is traded in two forms, *viz.*, shelled and in-shell. USA is a major exporter of both shelled and in-shell walnuts. The United States is the world's largest exporter of walnuts. In 2012, the nation exported walnuts valued at more than $1.1 billion. Of that amount, $645.7 million, or X percent, was for shelled walnuts. Top destinations were South Korea, followed by Germany, Japan and Canada. The United States also exported $466.4 million of in-shell walnuts. Top buyers were China and Turkey. (FAS 2012)

In shelled walnut exports, USA is followed by Chile, Mexico, China, Republic of Moldova, India and others. Whereas in in-shell walnut exports, USA is followed by France, Mexico, Chile, China, Ukrain, *etc.* In shelled walnut imports, Germany ranks first followed by Japan, Spain, Turkey, France and the UK. While for in-shell walnuts, Italy is the major buyer followed by Spain, Mexico, Germany, Turkey and the Netherlands. Table shows that India exported 7,249 metric tonnes of walnut valued at Rs 1924 million (37 million US$) mostly to UK (911.37 MT), Egypt (682.22 M T), Netherland (405.22 M T), USA 291.91 MT), China (470.50 MT), Australia (238.63 MT), Taiwan (197.00 MT), France (283.17 MT), Spain (184.81 MT) during the period 2012 (APEDA 2012),

As a whole, walnut exports from India have increased over the years. However, this is insufficient as normally 45-50 per cent of the production is consumed domestically. Even though the domestic and external demand for walnut has increased over the years, in India walnut cultivation could not develop rapidly due to a number of constraints including insufficient scientific research, improper and random classification, long gestation period and low free density. Apart from these, there are several pre-harvest and post-harvest problems in this sector. Constraints in walnut trade include awareness of maturity indices, method of harvesting and non-scientific dehulling.

Table 12.6: Major Exporting Countries of Walnut 2012

Sl.No	Country	Qty	Value
1	Usa	2,01,460.00	1,138.00
2	Turkey	1,42,665.00	1,012.00
3	Chile	26,873.00	196.00
4	France	35,170.00	175.00
5	Ukraine	34,796.00	118.00
6	Italy	15,075.00	112.00
7	Rep. of Moldova	12,824.00	85.00
8	Georgia	15,393.00	83.00
9	Areas, Nes	9,149.00	71.00
10	Germany	7,241.00	68.00
11	Azerbaijan	10,408.00	49.00
12	Netherlands	5,035.00	40.00
13	**India**	**7,249.00**	**37.00**
14	Romania	6,147.00	36.00
15	China	4,300.00	30.00

Table 12.7. Exports from India (Value in Rs. Lacs Quantity in MT)

Sl.No	Country	Qty	Value
1	U K	911.37	3,529.34
2	Egypt	682.22	3,319.41
3	Netherland	405.22	1,618.78
4	Germany	415.49	1,450.07
5	U S A	291.91	1,272.20
6	China	470.50	1,255.84
7	Australia	238.63	1,080.75
8	Taiwan	197.00	884.33
9	France	283.17	862.43
10	Spain	184.81	762.37

Post-harvest Disorders, Diseases and Insects/Pests

Physiological Disorders

The most serious post-harvest physiological disorder that affects walnut quality is oil rancidity. The problem appears to be caused by poor seed storage conditions; elevated temperature and RH, failure to use CA with reduced O_2 concentration.

Insect Problems

Insect damage can contribute to walnut handling and quality problems. The major insect pest is the codling moth (*Cydia pomonella*). While nuts attacked in mid-season are likely to be rejected, minor early codling moth damage can lead to more serious infestation with navel orange worm (*Amyelosis transitella* [Walker]), and this is more difficult to detect if the nut is not shelled. The walnut husk fly (*Rhagoletis completa*) is a serious pest in mid to late season. Larval feeding damages the husk tissues, leading to staining of the shell and failure of the husk to split. This reduces the harvest yield and nuts that remain in the orchard serve as a reservoir of insects to threaten the next crop cycle. In all cases, insect damage will tend to increase problems with pathogen infection.

Post-harvest Pathology

Most infections with pathogens are initiated in the orchard and transferred to the post-harvest environment. In-shell product is protected unless the shell has been broken or penetrated by insects. The most serious pathogens are fungi such as *Aspergillus flavus* and *A. parasiticus* which can produce aflatoxins that are both toxic and carcinogenic. It is important that damaged kernels be discarded prior to storage and that the low temperature and RH conditions discussed earlier be maintained in order to reduce the chance for mold growth. Toxin-producing Penicillium sp. have also been found on walnuts.

Processing

Hulling

Once the walnuts are harvested, pre-cleaners are used to clean the walnuts so they are ready for hulling. A huller removes the outer green hull and the nut is mechanically dehydrated (air dried) to the optimum 8.0 percent moisture level, preventing deterioration of the nut and protecting its quality during storage. Hulled walnuts are transported to nearby packing plants and are stored until needed for cracking.

Final Preparations

The walnuts are then returned to the processing center and put into a sorter that again sorts them by size. After they are sorted, they drop down into a giant cracking machine. This machine is conical in shape with a solid rotating center. As the walnuts fall into a more narrow space, they are cracked against the side of the rotator's bin and fall below to a secondary rotator which captures the walnuts that have not quite been cracked. The products shown in Table 12.8 are prepared and marketed.

Novel Value Added Products

Now a day's process technology is available for making excellent novel value added products of walnut.

Spiced Snack Mix

Walnut Coated Cheese Truffles

Samosas with Walnuts

Spiced Walnut

Table 12.8: Different Walnut Products

Walnut Halves	Walnut halves in products demonstrate its natural shape and visual appearance. Applications include trail mixes, confectionary, garnishing and decorating baked goods, and whole nut snack mixes.	
Large Pieces	Typical applications include trail mixes, energy bars, frozen dairy inclusions and baked good toppings.	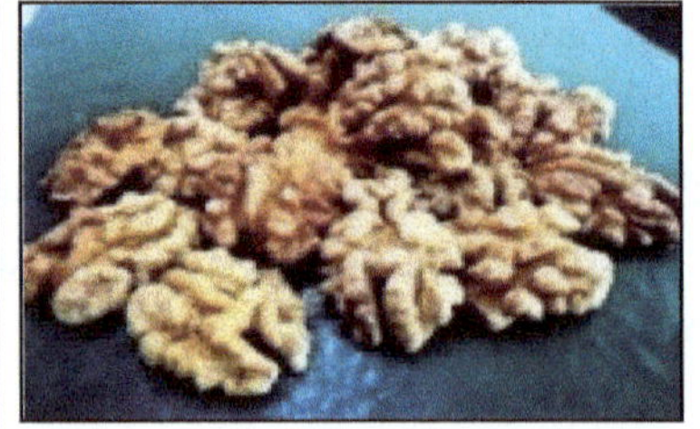

Medium Pieces	Use medium walnut pieces in products where equal flavor and texture is needed throughout the product. Baking mixes, pizza toppings, chocolate based candy bars.	
Small Pieces	Use small pieces in seasoning blends, baking mixes, breading mixes and frozen dairy toppings. Small pieces work especially well in premixed bakery blends because the small size allows for even distribution throughout the entire mix allowing the nut flavor and texture to come through with every bite.	
Roasted Walnuts	Roasting walnuts heightens the flavor and aroma while increasing the crunchy mouth-feel. Roasted walnuts work well in applications that will not be further baked such as no-bake energy bars and ready-to-eat snack mixes. For items that will be subjected to further heating (like cakes and muffins) roasted nuts can be used but should be limited to the interior of the baked item and not part of the surface area that will be directly exposed to the heat.	
Walnut Meal and Gluten Formation	The fat in walnut meal shortens or "interrupts" the gluten strands preventing them from getting longer and will yield tender products. A combination of gluten free flour and walnut meal may result in a product without enough structure, thus R and D work is always necessary to achieve final desired texture.	
Blanched Walnuts	Walnuts cannot be easily blanched because their general shape and contour makes it difficult to remove the skin. The best way to blanch walnuts is to either first warm them in the oven or boil them in water and then rub the skins off. Blanched walnuts can be used to help developers create refined baked good products that will not have any discoloration from the walnut skins. Sometimes lighter colored walnuts are available and those can be used as an alternative to blanching.	

Future Opportunities and Researchable Issues

Measures Needed

The domestic and external demand in India has been increasing over the years. It is projected to reach around 75,000 tonnes by 2020. So the country needs to bring in additional area to meet the projected demand. As in Jammu and Kashmir suitable land for its cultivation is scarce, we need to expand walnut production to other states like Uttarakhand and Himachal Pradesh which have suitable agro-climatic conditions for growing walnut.

Future Outlook

The walnut industry in India is definitely going to grow rapidly. Because the fresh domestic market is limited due to the lack of proper marketing facilities as well as great demand for inshell walnuts of high quality with cream colored kernels in the international market. Keeping in view defined constraints, efforts are required to place the walnut industry on sound footing. A strategic research and development program and ensured implementation are needed to popularize walnut cultivation in India in order to compete within the World Trade Organization regime. If the following measures are not implemented exports will be negatively affected, resulting in loss of foreign exchange.

- Identification and testing of superior genotypes based on inshell nut weight, creamish kernel color and identification of superior trees from naturalized populations from the walnut growing areas need to be initiated to extend the new plantations. Many select trees are exceptionally good for nut size and kernel quality compared with exotic material.
- The important and key element in expanding nut cultivation among the large number of potential growers in India is to make available good nursery stock at a reasonable cost. Therefore, there is an urgent need to establish a few good nurseries in the hill region to supply the planting demands of farmers.
- Sound research programs on cultural techniques, propagation, pollination, nutrition and post-harvest handling and storage of walnuts need to be strengthened and initiated for commercialization.
- Clonal rootstocks and dwarf cultivars of walnut need to be introduced for adaptive research and high-density plantation culture.
- Expansion of walnut growing in areas of northeastern hill region of India especially in the states of Meghalaya, Arunachal Pradesh, Sikkim and Darjeeling district of West Bengal need to be commercialized as scattered plantings of walnuts in farmers' backyards were already established.
- The existing walnut marketing system needs to be strengthened.
- Walnut growing may be diversified in other similar agro-climatic regions like the northeastern hill region states.

REFERENCES

Anon. 2010. Hort. Horizon: New Walnut Varieties released. *Ind. J. Hort.* **67**(1): i-vii.

APEDA, 2012. http: //agriexchange.apeda.gov.in/product_profile/prd_profile.aspx?categorycode=0203.

Elaine, B and Feldman, M. D. 2002. The Scientific evidence for a beneficial health relationship between walnuts and coronary heart disease. *J. Nutr.* **132**: 1062S–1101S.

Beuchat, L.R. 1978. Relationship of water activity to moisture content in tree nuts. *J. Food Sci.* **43**: 754-755, 758.

Bhat, A.R. and Ahmad, F. 2002. Hamdan and Sulaiman–new walnut varieties for North-Western Himalayas. *Indian Hort.* 47(**4**): 33-34.

Bhat, A.R., Ahanger, A.U., Sofi, A.A. and Mir, N.A. 1992. Evaluation of some walnut selections for quality parameters in Jammu and Kashmir. Proc. Natl. Symp. Emerging Trends in Temperate Hort. *NHB Tech.* Commun. 1. pp. 27-29.

Chauhan, J.S. and Sharma, S.D. 1980. Veneer grafting, A new technique for walnut propagation. *Fruit Science Reports*, 7(**2**): 45-48.

Chauhan, J.S. and Sharma, S.D. 1982. Influence of different dates of veneer grafting on the success of walnut propagation. *Punjab Hort. J.* 22(**3-4**): 177-80.

Chauhan, J.S. and Shrama, S.D. 1979. Phenotypic variability in walnut from Kinnaur district of Himachal Pradesh. I*ndian J. Agr. Sci.* 49(**6**): 420-22.

Dandekar, A.M., McGranahan, G.H., Vail, P.V. Uratsu, S.L., Leslie, C.A. and Tebbets, J.S. 1998. High levels of expression of full-length cryIA(c) gene from Bacillus thuringiensis in transegnic somatic embryos. *Plant Sci.* **131**: 181-193.

Dhuria, H.S., Bhutani, V.P. and Dhar, R.P. 1977. Standardization of propagation techniques in walnut. Preliminary studies on grafting. *Indian J. Hort.* 34(**1**): 20-23.

Diousse, L., Pankow, J.S., Eckfeldt, J.H., Folsom, A.R., Hopkins, P.N, Province, M.A., Hong. and Ellision, R.C. (2001) Relation between dietary linoleic acid and coronary artery disease in the National Heart, Lung and Blood Institute Family Heart Study. *Am J Nutr.* **74**: 612-619.

Elaine Hardman, 2011. PhD. Breast cancer risk drops when diet includes walnuts, Marshall researchers find. Marshall University. 2011 September 1.

ERS, 2005. http: //www.agmrc.org/commodities__products/nuts/english-walnuts-profile/

FAO, 2012. http: //www.fao.org/statistics/en/.

FAO, 2004. http: //www.fao.org/statistics/en/.

FAS, 2012. http: //www.agmrc.org/commodities__products/nuts/english-walnuts-profile/

FAO Stat. 2004. http: //faostat3.fao.org/faostat-gateway/go/to/download/Q/QC/E

Gangwar, R.P. and Kumar H. 1975. Studies on germination capacity of walnut (*J. Regia* L) in relataion to their size and position in the soil. *Haryana J. Hort. Sci.* **4**: 142-144.

Gaur N.V.S. 1980. Effect of gibberellin, sulphuric acid and cracking on the germination and growth of seedlings of walnut. *Indian J. Hort.* **37**(1): 16-19.

Gautam, D.R. 2000. Selection of Persian walnut from Shimla district of Himachal Pradesh. *Indian J. Plant Genet. Resources* **13**(3): 294-297.

Hendricks, L.C., Coates, W.C., Elkins, R.B., McGranahan, G.H., Phillips, H.A., Ramos, D.E., Reil, W.O. and Snyder, R.G. 1998. Selection of varieties. In: Walnut Production Manual, D.E. Ramos (ed) Univ. of Calif., Div. *Agric. Nat.* Res., Pub. No. **3373**, pp. 84-89.

Hu, F.B., Stamfer, M.J., Manson, J.E., Rimm, E.B., Wolk, A., Colditz, G.A., Hennekens, C.H. and Willett, W.C. (1999) Dietary intake of α-linolenic acid and risk of fatal ischemic heart disease among women. *Am J Clin Nutr.* **69**: 890-897.

Johnson, J.A., Vail, P.V., Soderstrom, E.L., Curtis, C.E., Brandl, D.G., Tebbets, J.S. and Valero, K.A. 1998. Integration of nonchemical, post-harvest treatments for control of Navel orange worm (*Lepidoptera: Pyralidae*) and Indian meal moth (*Lepidoptera: Pyralidae*) in walnuts. *J. Econ. Ent.* **91**: 1437-1444.

Chandy, K.T. 2012. Agricultural and Environmental Education I. Introduction Walnut (*Juglans regia*). Walnut 189. Walnut Booklet No. 189 Nuts Production: NPS- 6.

Lal H., Nautiyal M.C. and Sharma R.M., 1984. Walnut seed germination - Effect of planting depth and seed position in soil. *Prog. Hort.* **16**(1): 6-8.

Lal, H. and Singh, R.D. 1978. Some promising walnut strains in Chakarata hills of Uttar Pradseh. *Prog. Hort.* **1:** 61-65.

Lal, H., Ratoore, D.S. and Prasad, Y. 1993. Evaluation of some promising walnut (*Juglans regia* L.) cultivars grown at Pithoragarh. *Prog. Hort.* **25**(1 and 2): 5-8.

Lopez, A., Pique, M.T., Romero, A. and Aleta, N. 1998. Modelling of walnut sorption isotherms. Ital. *J. Food Sci.* **10**: 67-74.

Maté, J.I., Saltveit, M.E. and Krochta, J.M. 1996. Peanut and walnut rancidity: Effects of oxygen concentration and RH. *J. Food Sci.* **61**: 465-468, 472.

McGranahan G.H., Leslie, C. 2012. Walnut. In: Badenes ML, Byrne DH (eds) Fruit Breeding. Springer New York Dordrecht Heidelberg London.

NASS, 2013. http: //www.agmrc.org/commodities__products/nuts/english-walnuts-profile/

Noncitrus Fruits and Nuts, National Agricultural Statistical Service (NASS), USDA.

Olson, W.H., Labavitch, J.M., Martin, G.C. and Beede, R.H. 1998. Maturation, harvesting and nut quality. In: Walnut Production Manual, D.E. Ramos (ed) Univ. Calif., Div. *Agric. Nat.* Res., Pub. No. **3373,** pp. 273-276.

Pan An., Qi Sun., Jo Ann, E., Manson., Walter, C., Willett. and Frank, B. Hu. 2013. Walnut consumption is associated with lower risk of type 2 diabetes in women. *J. Nutr.* **143**: 512–518.

Pandey, D., Tripathi, S.P. and Sinha, M.M. 1982. Walnut propagation through stooling in U.P. hills. *Punjab Hort. J.* **22**(3-4): 169-72.

Patel, G. 2005. Essential fats in walnuts are good for the heart and diabetes. *J Am Diet Assoc.* **105**: 1096–1097.

Puttoo, B.L., Razdan, V.K. and Baschadhary, K.C. 1990. Constraints in the production of walnuts in India. *Acta Horticulture*. 284.pp 327.

Qureshi, A.S. and Dalal, M.A. 1990. Status of nut crops in Jammu and Kashmir state. *Prog. Hort.***17**: 195-205.

Rana, J.C., Singh, D., Yadav, S.K., Verma, M.K., Kumar, K., Predheep, K. 2007. Genetic diversity collected and observed in Persian walnut (*Juglas regia* L.) in the western himalayan region of India. *Plant Genet Resour News Letter*. **151**: 68-73.

Rashid, A. 1978. Effect of plant growth regulators on stooling of walnut. *Indian J. Hort.* v. **35** (3).

Rathore, D.S. and Mithal, S.K. 1986. Conservation and preliminary evaluation of the genetic resources of temperate fruits. Silver jubilee seminar on agri-horticulture plant genetic resources for the development of hill agriculture, *NBPGR, Shimla*, pp. 181-90.

Reiter, R.J., Manchester, L.C., Tan, D.X. 2005. Melatonin in walnuts: influences on levels of Melatonin and total antioxidant capacity of blood. *Int J Appl Basic Nutr Sci.* 21(**9**): 920-924.

Samuel, R. and Sharma, S.D. 1991. Clonal propagation of walnut. *Agric. Sci.* Digest 11(**4**): 185-89.

Sharma, R. 2012. Area and production database of fruit crops. Directorate of Hortic State Department of Horticulture, Jammu and Kashmir Government, India.

Sharma, S.D. and Chauhan, J.S. 1980. A note on variability in nut characters of walnut seedling. *Punjab Hort. J.* 20 (**1-2**): 100-101.

Sharma, S.D. and Chauhan, J.S. 1981. Effect of shell thickness on seed germination and growth of seedlings obtained from the nuts of seedling walnuts. *South Indian Hort.* 29(**2**): 87-89.

Thaper, A.R. 1961. Nut tree culture in India, Farm Bull. 62.

Vail, P.V., Tebbets, J.S., Hoffmann, D.F. and Dandekar, A.M. 1991. Responses of production and storage walnut pests to Bacillus thuringiensis insecticidal crystal protein fragments. Biol. Control **1**: 329-333.

van Steenwyk, R.A. and Barnett, W.W. 1998. Insect and mite pests. In: Walnut Production Manual, D.E. Ramos (ed) Univ. Calif., Div. Agric. Nat. Res., Pub. No. **3373,** pp. 247-253.

Vigneshwara Varmudy January 2011. Market Survey. FACTS FOR YOU.

http: //vc.vivekanandaedu.org/wp-content/uploads/pdf/Walnut_Jan11.pdf

Walnuts: Second Biggest Nut Crop Produced in the United States, Fruit and Tree Nuts Outlook, ERS.

Woodroof, J.G. 1979. In: Tree Nuts Production Processing Products, 2nd ed. pp. 604-55. AVI Pub. Co. INC Westport, CT.

Zettler, J.L., Halliday, W.R. and Arthur, F.H. 1990. Phosphine resistance in insects infesting stored peanuts in the southeastern United States. *J. Econ. Ent.* **82**: 1508-1511.

Index

D

E

F

G

H

I

Figure 3.1a: Apricot Varieties being Grown at CITH Srinagar, J&K. (p. 69)

Figure 3.1a: Apricot Varieties being Grown at CITH Srinagar, J&K. (p. 70)

Figure 3.2: Apricot Oil from Chuli Apricots in Leh.(p. 72)

Figure 3.3: Apricot Cultivation in Leh.(p. 73)

Figure 3.4: Maturity and Ripeness Stages of Apricot. (p. 78)

Figure 3.5: (p. 83)

Figure 3.5 (p. 83)

Figure 3.7: Osmo Dehydrated Product of Apricot. (p. 85)

Apricot Nectar (p. 85)

Apricot Jelly (p.87)

Figure 3.9: Quality Apricot Fruit Bar. (p. 89)

Figure 3.9: Quality Apricot Fruit Bar. (p. 89)

(p. 241)

CITH Walnut-1 (p.151)

CITH Walnut-2 (p. 252)

CITH Walnut-3 (p. 252)

CITH Walnut-4 (p. 252)

CITH Walnut-5 (p. 253)

CITH Walnut-6 (p. 253)

CITH Walnut-7 (p. 253)

CITH Walnut-8 (p. 254)

CITH Walnut-9 (p. 254)

CITH Walnut-10 (p. 254)

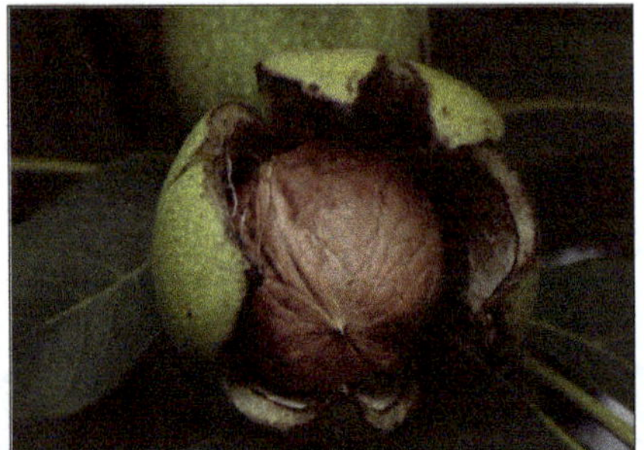

Mature Walnut (p. 258)

Drying of Walnuts (p. 259)

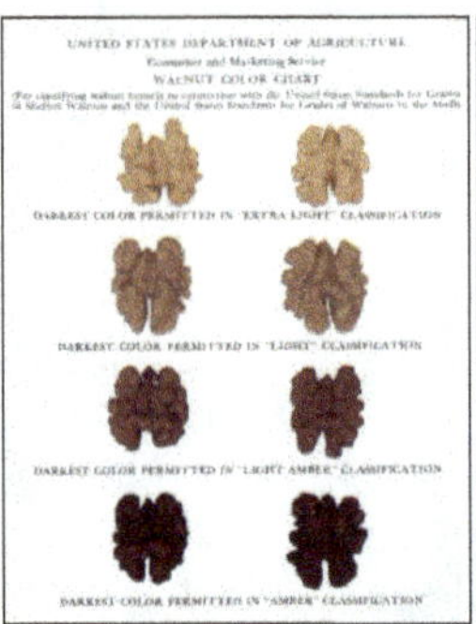

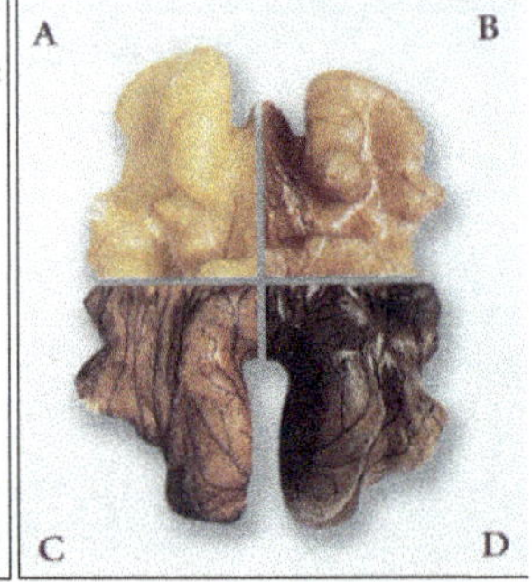

Different Grades of Walnut
(*Source*: USDA. United Standards dor grades of Shelled Walnuts).(p. 260)

Packaging of Walnuts (p. 260)

(p. 265)

www.ingramcontent.com/pod-product-compliance
Ingram Content Group UK Ltd.
Pitfield, Milton Keynes, MK11 3LW, UK
UKHW021009290726
14059UKWH00001BA/47